International Conference on Dynamics of Systems on the Nanoscale

(DySoN 2012)

Journal of Physics: Conference Series Volume 438

St. Petersburg, Russia
30 September – 4 October 2012

ISBN: 978-1-62748-670-5
ISSN: 1742-6588

Printed from e-media with permission by:

Curran Associates, Inc.
57 Morehouse Lane
Red Hook, NY 12571

Some format issues inherent in the e-media version may also appear in this print version.

Printed by Curran Associates, Inc. (2013)

For permission requests, please contact the Institute of Physics at the address below.

Institute of Physics
Dirac House, Temple Back
Bristol BS1 6BE UK

Phone: 44 1 17 929 7481
Fax: 44 1 17 920 0979

techtracking@iop.org

Additional copies of this publication are available from:

Curran Associates, Inc.
57 Morehouse Lane
Red Hook, NY 12571 USA
Phone: 845-758-0400
Fax: 845-758-2634
Email: curran@proceedings.com
Web: www.proceedings.com

Papers

Journal of Physics: Conference Series **438** (2013) 011001 doi:10.1088/1742-6596/438/1/011001

Preface

AV Solov'yov[1,*]

[1]Frankfurt Institute for Advanced Studies, Ruth-Moufang Str. 1, 60438, Frankfurt am Main, Germany
[*]On leave from A.F. Ioffe Physical-Technical Institute, Polytechnicheskaya 26, 194021, St. Petersburg, Russia

E-mail: solovyov@fias.uni-frankfurt.de

The Second International Conference "Dynamics of Systems on the Nanoscale" (DySoN 2012) took place in Saint Petersburg, Russia during September 30 – October 4, 2012. The venue was the Courtyard by Marriott St. Petersburg Vasilievsky Hotel, 2nd line of Vasilievsky Island 61/30A, 199178. The conference was organized by the Frankfurt Institute for Advanced Studies – Goethe University, A.F. Ioffe Physical-Technical Institute and Saint Petersburg State Polytechnic University.

This DySoN conference has been built upon a series of International Symposia "Atomic Cluster Collisions: structure and dynamics from the nuclear to the biological scale" (ISACC 2003, ISACC 2007, ISACC 2008, ISACC 2009 and ISACC 2011). During these meetings it has become clear that there is a need for an interdisciplinary conference covering a broader range of topics than just atomic cluster collisions, related to the Dynamics of Systems on a Nanoscale. Therefore, in 2010 it was decided to launch a new conference series under the title "Dynamics of Systems on the Nanoscale". The first DySoN conference took place at the National Research Council, Rome, Italy in 2010. The DySoN 2012 is the second conference in this series.

The DySoN 2012 Conference promoted the growth and exchange of interdisciplinary scientific information on the structure formation and dynamics of animate and inanimate matter on the nanometer scale. There are many examples of complex many-body systems of micro- and nanometer scale size exhibiting unique features, properties and functions. These systems may have very different nature and origin, e.g. atomic and molecular clusters, nanoobjects, ensembles of nanoparticles, nanostructures, biomolecules, biomolecular and mesoscopic systems. A detailed understanding of the structure and dynamics of these systems on the nanometer scale is an important fundamental task, the solution of which is necessary in numerous applications of nano- and biotechnology, material science and medicine.

Although mesoscopic, nano- and biomolecular systems differ in their nature and origin, a number of fundamental problems are common to all of them: What are the underlying principles of self-organization and self-assembly of matter on the micro- and nanoscale? Are these principles classical or quantum? How does function emerge on the nano- and the mesoscale in systems of different origin? What criteria govern the stability of these systems? How do their properties change as a function of size and composition? How are their properties altered by their environment? Seeking answers to these questions is at the core of a new interdisciplinary field that lies at the intersection of physics, chemistry and biology, a field called Meso-Bio-Nano (MBN) Science.

Both experimental and theoretical aspects of the mentioned problems were discussed at the DySoN 2012 Conference. Particular attention was devoted to dynamical phenomena and many-body effects taking place in various MBN systems, which include problems of structure formation, fusion and fission, collision and fragmentation, collective electron excitations, reactivity, nanoscale phase transitions, nanoscale insights into biodamage, channeling phenomena and many more.

This volume is a collection of the contributions received from the participants of the DySoN 2012 Conference. It provides an overview of the topics, new results and ideas that have been discussed at the conference. I would like to thank all the authors of these proceedings, as well as all the participants of the conference for making it so successful.

The third DySoN Conference will be held in Edinburgh in May 2014.

1. DySoN 2012Conference photo

2. Topical Areas of DySoN Conference

- Structure and dynamics of clusters, nanoparticles, biomolecules and bionanosystems
- Clustering and self-organization on the nanoscale
- Cluster and biomolecular ensembles, complexes, nanostructured materials
- Nanoscale phase transitions
- Reactivity and nanocatalysis
- Thermal, optical and magnetic properties of nanosystems
- Electron/spin transport in molecular systems
- Collision processes, fusion, fission, fragmentation
- Propagation of particles through a medium

3. Conference Program

Sunday, 30 September 2012

$16^{00} - 18^{00}$	Participants registration
$18^{00} - 21^{00}$	**Conference reception**

Monday, 1 October 2012

$9^{00} - 9^{30}$	**DySoN 2012 Opening** **Andrey V. Solov'yov**, Frankfurt Institute for Advanced Studies, Germany *Dynamics of systems on the nanoscale: challenges for theory and experiment*
$9^{30} - 11^{00}$	*Morning session I: Structure and dynamics of clusters, nanoparticles and biomolecules* Chair: **Jean-Patrick Connerade**, Imperial College, UK **Julius Jellinek**, Argonne National Laboratory, USA *Novel analysis of dynamics at the nanoscale* **Philippe Bréchignac**, Universite Paris Sud, Orsay, France *New investigations of coronene clusters* **Eric Suraud**, Universite Paul Sabatier, France *On electronic emission from irradiated clusters and nano structures*
$11^{00} - 11^{30}$	**Coffee break**
$11^{30} - 13^{00}$	*Morning session II: Nanoscale dynamical transitions and transformations* Chair: **Julius Jellinek**, Argonne National Laboratory, USA **Shiv N. Khanna**, Virginia Commonwealth University, USA *Spin accommodation and geometric effects in catalysis: breaking polar covalent bonds via complementary active sites* **Jerry Bernholc**, North Carolina State University, USA *Structural transformations in ferroelectric polymers and metal-protein complexes* **Andrey Lyalin,** Hokkaido University, Japan *Role of the active and inert supports in enhanced catalytic activity of gold clusters*
$13^{00} - 14^{30}$	**Lunch**
$14^{30} - 16^{00}$	*Afternoon Session I: Nanoscale insights in biodamage* Chair: **Bernd Huber**, CEA-CIMAP, France **Ilia I. Fabrikant**, University of Nebraska-Lincoln, USA *Electron attachment to molecules in a cluster environment* **Pablo de Vera**, Universitat d'Alacant, Spain *Ionization of biomolecular targets by ion impact: Input data for radiobiological applications* **Oral presentation of posters**

Journal of Physics: Conference Series **438** (2013) 011001 doi:10.1088/1742-6596/438/1/011001

16^{00} – 16^{30}	**Coffee break**
16^{30} – 18^{00}	*Afternoon Session II: Clustering and self-organization on the nanoscale* Chair: **Catherine Bréchignac**, CNRS, France **Kit Bowen**, Johns Hopkins University, Baltimore, USA *$(PbS)_{32}$: a baby crystal* **Michael Wilkinson**, The Open University, Milton Keynes, UK *Ostwald ripening and a test-tube model for rainfall* **Sergiy Perepelytsya**, Bogolyubov Institut for Theoretical Physics,Ukraine *Dynamic ordering of ions around DNA helix*

Tuesday, 2 October 2012

9^{30} – 11^{00}	*Morning session I: Collision processes involving nanosystems* Chair: **Eric Suraud**, Universite Paul Sabatier, France **Bernd Huber**, CEA-CIMAP, France *Ion-induced chemical reactions in finite size systems* **Rafael Garcia Molina**, Universidad de Murcia, Spain *Interaction of proton beams with carbon nanotubes* **Bing Zhang**, Wuhan Institute of Physics and Mathematics, China *Non-adiabatic dynamics in polyatomic molecules by femtosecond photoelectron imaging*
11^{00} – 11^{30}	**Coffee break**
11^{30} – 13^{00}	*Morning session II: Thermo-mechanical, optical and magnetic properties of nanosystems* Chair: **Shiv N. Khanna**, Virginia Commonwealth University, USA **Florent Calvo**, Université de Lyon 1, France *Assessing quantum thermal baths for finite systems* **Alexander Yakubovich**, Frankfurt Institute for Advanced Studies, Germany *Diffusion and nanoindentation processes in titanium and nickel materials* **Mikhail Petrzhik**, National University of Science and Technology "MISIS", Moscow, Russia *Dynamics of martensitic structure at TiNb-based quenched alloys under heating and loading*
13^{00} –13^{15}	**Conference photo**
13^{15} – 14^{30}	**Lunch**
14^{30} – 16^{00}	*Afternoon session I: Clustering and self-organization on the nanoscale* Chair: **Kit Bowen**, Johns Hopkins University, Baltimore, USA **Catherine Bréchignac**, CNRS, France *Atmospheric degradation of silver nanofractal by trace amount of chlorine* **Artur Böttcher**, Karlsruhe Institute of Technology, Germany

	Supported mass-selected carbon clusters: thermally and chemically driven surface modifications **Ilia A. Solov'yov**, Beckman Institute for Advanced Science and Technology, University of Illinois at Urbana-Champaign, USA *Dynamics of nanofractal formation and fragmentation*
$16^{00}-16^{30}$	**Coffee break**
$16^{30}-18^{10}$	*Afternoon session II: Oral presentation of posters* Chair: **Gustavo Garcia**, Instituto de Fisica Fundamental, Spain **Open Discussion, 30 min** Interdisciplinary DySoN conference: current status and perspectives

Wednesday, 3 October 2012

$9^{30}-11^{00}$	*Morning session I: Nanoscale insights in biodamage* Chair: **Rafael Garcia Molina**, Universidad de Murcia, Spain **Eugene Surdutovich**, Oakland University, USA *Multiscale physics of ion induced radiation damage* **Sandrine Lacombe**, Universite Paris Sud, Orsay, France *Lanthanide nanoparticles: toward theranostic in hadrontherapy* **Gustavo Garcia**, Instituto de Fisica Fundamental, Spain *Electron and positron scattering with nanostructures (atomic clusters and macromolecules)*
$11^{00}-11^{30}$	**Coffee break**
$11^{30}-13^{00}$	*Morning session II: Collision processes involving nanosystems* Chair: **Andrei Korol**, Frankfurt Institute for Advanced Studies, Germany **Vadim Ivanov**, St. Petersburg State Polytechnic University, Russia *Photoionization of clusters and fullerenes* **Alexey Verkhovtsev**, Frankfurt Institute for Advanced Studies, Germany *Study of collective electron excitations in fullerenes* **Andrej Grubisic**, JILA, University of Colorado, Boulder, USA *Coherent multiphoton photoelectron emission from individual, supported metallic nanoparticles*
$13^{00}-14^{30}$	**Lunch**
$14^{30}-16^{00}$	*Afternoon session I: Propagation of particles through medium* Chair: **Vadim Ivanov**, St. Petersburg State Polytechnic University, Russia **Hartmut Backe**, Institute of Nuclear Physics, Germany *X-ray production with a μm-scale crystal undulator at the Mainz microtron MAMI* **Andrei Korol**, Frankfurt Institute for Advanced Studies, Germany *Crystalline undulator as a source of coherent radiation* **Vincenzo Guidi**, Universita di Ferrara, Italy *New trends for manipulation of charged particle beam via coherent interactions in crystals*

16^{00} –16^{30}	**Coffee break**
16^{30} –17^{30}	*Afternoon session II: Structure and dynamics of clusters, nanoparticles and biomolecules* Chair: **Andrey V. Solov'yov**, Frankfurt Institute for Advanced Studies, Germany **Yurii Lozovik**, Institute of Spectroscopy, Russia *Mesoscopic supersolid of dipole clusters* **Jean-Patrick Connerade**, Imperial College, UK *From Pauli's birthday to confinement resonances - a potted history of quantum confinement*
17^{30} – 18^{00}	**Open Discussion** Interdisciplinary DySoN conference: current status and perspectives **Conference Closing**
19^{00} - 22^{30}	**Conference Dinner**

Thursday, 4 October 2012

9^{30} – 15^{30}	**Conference Tour**

4. International Advisory Committee

- C. Bréchignac (Laboratoire Aime Cotton, CNRS, Orsay, France)
- M. Broyer (University of Lyon, France)
- J.-P. Connerade (Imperial College London, London, UK)
- F. Gianturco (University of Rome "La Sapienza", Rome, Italy)
- J. Jellinek (Argonne National Laboratory, Argonne, Illinois, USA)
- S. Khanna (Virginia Commonwealth University, USA)
- Yu. Lozovik (Institute of Spectroscopy, Russian Academy of Sciences, Russia)
- N. Mason (The Open University, Milton Keynes, UK)
- A.V. Solov'yov (Frankfurt Institute for Advanced Studies, Germany)
- E. Surdutovich (Oakland University, USA)

5. Organizing Committee

- V.K. Ivanov (Saint Petersburg State Polytechnic University, Russia)
- A.V. Korol (Frankfurt Institute for Advanced Studies, Germany)
- D. Radulescu (Frankfurt Institute for Advanced Studies, Germany)
- A.V. Solov'yov (Frankfurt Institute for Advanced Studies, Germany)
- A.V. Verkhovtsev (Saint Petersburg State Polytechnic University, Russia)
- A.V. Yakubovich (Frankfurt Institute for Advanced Studies, Germany)

6. Sponsors

The conference was held under the auspices of the following sponsors:

- Frankfurt Institute for Advanced Studies, Goethe Universität
- Saint Petersburg State Polytechnic University
- A.F. Ioffe Physical-Technical Institute
- EU Commission

7. Conference Chair

Prof. Dr. Andrey V. Solov'yov
Frankfurt Institute for Advanced Studies
Goethe University
Ruth-Moufang-Str. 1
D-60438 Frankfurt am Main, Germany

8. Conference website

http://fias.uni-frankfurt.de/dyson

9. Conference Venue

The Conference was hosted by Courtyard by Marriott St. Petersburg Vasilievsky Hotel, 2nd line of Vasilievsky Island 61/30A, St. Petersburg, Russia 199178.

The photo (right) shows the main building of the hotel.

10. Conference tour

On October, 4 there was the conference tour to the famous suburb of Saint Petersburg, the summer residence of Russian tzars, the town "Pushkin" (also known as Tsarskoe selo). There the conference participants visited the Catherine the Great Palace - a splendid example of the Russian Baroque, and saw the well-known Amber Room, enjoyed the parks during the beautiful time of golden autumn.

Journal of Physics: Conference Series **438** (2013) 012001 doi:10.1088/1742-6596/438/1/012001

From Pauli's birthday to '*Confinement Resonances*' – a potted history of Quantum Confinement.

JP Connerade[1]
Quantum Optics and Laser Science Group, Physics Department, Imperial College London SW7 2BW, UK and European Academy of Sciences Arts and Letters, Résidence le Monte Carlo, 3 Bd Albert Ier, 06600 Antibes, France.

E-mail: jean-patrick@connerade.com

Abstract. Quantum Confinement is in some sense a new subject. International meetings dedicated to Quantum Confinement have occurred only recently in Mexico City (the first in 2010 and the second, in September 2011). However, at least in principle, the subject has existed since a very long time. Surprisingly perhaps, it lay dormant for many years, for want of suitable experimental examples. However, when one looks carefully at its origin, it turns out to have a long and distinguished history. In fact, the problem of quantum confinement raises a number of very interesting issues concerning boundary conditions in elementary quantum mechanics and how they should be applied to real problems. Some of these issues were missed in the earliest papers, but are implicit in the structure of quantum mechanics, and lead to the notion of Confinement Resonances, the existence of which was predicted theoretically more than ten years ago. Although, for several reasons, these resonances remained elusive for a very long time, they have now been observed experimentally, which puts the whole subject in much better shape and, together with the advent of metallofullerenes, has contributed to its revival.

1. Introduction

Until the development of quantum confinement and the discovery of metallofullerenes, the best ways of investigating atomic structure were via the excitation and/or ionisation of one or more electrons or else via electron attachment.

However, excitation and/or ionisation, by reducing electronic screening, actually increase the influence of the nuclear charge as compared with electron-electron interactions.

Most of the unsolved problems of atomic physics concern electronic correlations (the many-body problem). Increasing the influence of the nuclear charge means moving towards the part of the problem which is mostly considered as solved. In this respect, excitation and/or ionisation act against correlation forces. Electronic attachment, from this perspective, is better, but results in a more fragile system with very few quantum states available for study. Until fairly recently, there seemed to be no other viable approach

[1] To whom any correspondence should be addressed.

Quantum confinement provides a much richer alternative. It consists in trapping the atom within a cavity (endohedrally, from the greek 'endo' meaning inside' and 'hedron' which denotes the geometrical face of the cavity) whose dimensions are small enough to alter its fundamental properties quite significantly. Ideally, one should also be able to control the geometry of the cavity and to 'tune' the strength of the confining force, which of course is difficult experimentally.

As regards probing the system experimentally, the aim is still to use the traditional methods of optical spectroscopy in the first place to probe the resulting system, but the observer now has an extra tool in hand: compressing the atom under study and watching how its properties are changed.

2. Defining the 'Ideal Problem'

As just explained, the aim is to trap a many-electron atom inside and preferably at the centre of a spherical cavity, so we can easily calculate its properties just by altering either the boundary conditions or the radial potential in the Schroedinger equation of the test atom. In principle, one aims for a many-electron atom if the purpose is to probe correlations, although it is advisable (as all atomic physicists know) to start by understanding hydrogen. Of course, such an ideal situation does not exist (as usual in physics). Even if the 'confining cavity' is C_{60}, one must remember that it is not a sphere (a perfect sphere, in quantum mechanics, would not even be able to rotate). Also, in 'real' experiments, one can only tune the confining force rather clumsily (by ionising the fullerene, or adding other shells around it) and one can only alter the size of the confining cavity in rather large steps (by using another fullerene, for example). Then, we have no guarantee that the confined atom will simply sit at the centre of the confining cavity. Usually, it does no such thing, but will rattle around off-centre. Last but not least, the cavity also has quantum properties, i.e. its own intrinsic excitation spectrum.

Nonetheless, there are well-established virtues in Physics of studying ideal problems, whose basic property is that the equations describing them can be solved exactly even if the systems cannot actually be made or observed. As examples, one need only think of the perfect black-body or of Einstein's two-level atom. Similarly, one cannot actually observe a hydrogen atom in the absence of any radiation field. The exact solution for hydrogen is nonetheless a fundamental example in Quantum Mechanics. Indeed, idealised systems with exact solutions usually turn out to be of extreme importance. The solutions themselves can often be adapted to match 'real' problems, and their properties define what regions of parameter space should be explored. Thus, the whole basis of quantum defect theory in Atomic Physics comes from phase shifting regular and irregular solutions of the hydrogen atom problem with respect to each other to match other spherical boundary conditions.

3. The Founding Paper of the Physics of Confined Atoms

It happens that the problem of Hydrogen confined in a sphere with infinite non-penetrable walls can be solved exactly. This property was first noticed by Sommerfeld and Welker [1]. Of course, the whole problem was treated in absence of any radiation field, and indeed the sphere, being impenetrable, renders the whole system unobservable. There were no fullerenes in the literature at that time, and quantum confinement did not even exist. The only reason for considering this problem in the first instance was that it is tractable, i.e. it can be solved. That was reason enough to work it out theoretically, and it was fortunate for the authors that Wolfgang Pauli's sixtieth birthday occurred around this time, so they were able to bring the solution along as a birthday present and have it published in a special issue of the *Annals of Physics*. Unfortunately, the date of publication was not a very favourable one, and there are comparatively few copies remaining of this original paper, written in German. It is, however, the seminal paper for the whole subject of Quantum Confinement.

By using this simple example, Sommerfeld and Welker [1] established a few important facts. For instance, they showed that the optical electron, when a confining cavity is small enough, becomes ***delocalised***, i.e. no longer bound to the atom, but only confined by the cavity. They compared this mechanism with the formation of the conduction band in a solid, in which an atom is confined within a cage defined by its neighbours and they also related their study to the properties of atoms under very high pressure, reported by Michels *et al* [2] at an earlier birthday party, in honour of van der Waals.

Interestingly, what prevents electrons from penetrating a real conducting shell is the Pauli force, and the van der Waals correction to the ideal gas law involves the volume occupied by atoms. Sommerfeld and Welker's [1] paper implied that atomic volume itself would be dependent on the confining pressure, a quantum effect which would rather spoils the van der Waals correction at very high pressures, but nobody (not even the authors) noticed that, so all was well. Much more detailed and realistic parametric models for the conduction band in solids soon became available. While Sommerfeld and Welker [1] had uncovered a good picture of the mechanism involved in delocalising to the conduction band without any adjustable parameters, their model could not readily be adapted to the representation of an infinite solid, so this application was discarded..

Eventually, however, the subject of quantum confinement was to surface again, but many years after this somewhat neglected (and soon forgotten) first step. Indeed, it seems likely that the problem was reinvented independently many years later by the other researchers, since the work of Sommerfeld and Welker [1] was rarely cited.

4. The Rebirth of Quantum Confinement

The main difficulty with the new subject invented by Sommerfeld and Welker [1] was the lack of real applications. Because the impenetrable sphere is rather like a piston compressing the atom from outside, they had discussed atoms under extreme pressure, which was as close as they could come to a real application. A first step was naturally to extend such ideas to molecules as done by Ley Koo and Cruz in Mexico in 1981 [3], but this still did not provide a real experimental system to probe.

There is also a situation where applying pressure is relevant: when one considers the *reversible* insertion of an ion into a solid. In a lithium ion battery, Li insertion is necessary to make useable electrodes. In fact, this is a very topical problem today, connected with the need for better high power batteries to power electric motor cars, and the study of reversible insertion is a key technological issue.

So perhaps, the first application of confined atoms to a problem of practical importance was the suggestion of modelling reversible insertion via a chemical pressure, using quantum confinement as the driving 'piston' [4,5]. This is an important problem, which raises many interesting questions about the confinement of many-electron atoms (as opposed to hydrogen). I will return later in the present paper to considering the special properties of confined many-electron atoms

Exploring the theory of confined atoms also led to a revival, in which the problem studied was the same as originally set up, except that more values of angular momentum were considered (Sommerfeld and Welker [1] had only included *s*-states), and the possibility that the confined atom might lie away from the centre of the confining sphere was also raised [6,7].

The reason for the renewed interest so long after the work of Sommerfeld and Welker [1] had been put aside as a mere curiosity was of course the discovery of the fullerene molecule and the realisation that, because it is hollow, it would be possible to trap an atom (A) inside it. A distinction was drawn between a molecule consisting of a fullerene with an atom stuck to it on the outside and a fullerene with the atom encapsulated within it. The latter (endohedral trapping) was denoted as $A@C_{60}$. The geometry of the fullerene was of course close enough to that of a confining sphere, although it soon turned out that the nature of the confinement in an endohedral fullerene is rather different from the problem considered by Sommerfeld and Welker [1].

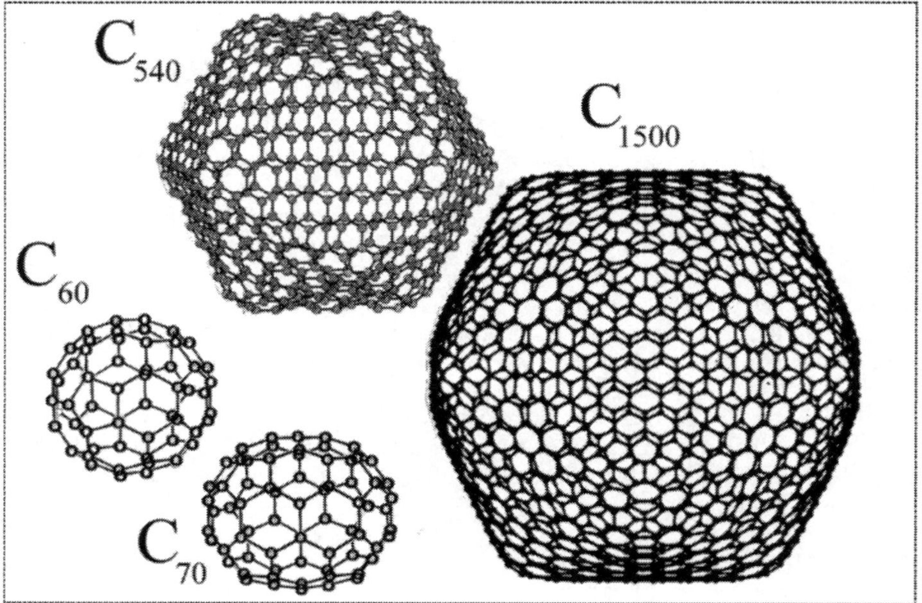

Figure 1. Structure of some of the closed shell fullerenes, which are hollow and can contain endohedral atoms or molecules (from ref [18]).

Experimentally, making endohedral systems and in particular the metallofullerenes in quantity proved difficult. Initially, they could readily be produced for detection of their presence only by very sensitive techniques such as mass spectroscopy, using methods of synthesis which range from collection in discharges to ionic bombardment [8], but producing them in large enough quantity for probing their physical properties, for example by using spectroscopy, was a more challenging task. The development of new chemical pathways holds much promise. Another option for the future will be trapping them. Then, even a single one will become accessible by spectroscopy.

The delay involved in resolving this experimental issue has provided an opportunity for theoretical work on endohedral systems with the challenge of predicting their properties ahead of observation

5. The nature of the confining sphere

Although endohedral systems were elusive, much progress was made in undertanding the properties of C_{60} itself, and experiments led to a better knowledge of the properties of the confining sphere.

In photoionisation studies, interference effects were observed, resulting in oscillations of the cross section [9]. From the frequency of these oscillations, it proved possible to deduce a lot of useful information about the geometry of the average potential experienced by electrons escaping from the C_{60} sphere. Of course, such information is relevant for metallofullerenes only, insofar as one assumes that this sphere remains relatively unchanged by the inclusion of any particular endohedral atom. One may anticipate that such an approximation must work best for smaller sizes of endohedral atoms.

Furthermore, in scattering experiments (collisions of C_{60} with Rydberg atoms), the formation of the C_{60}^- negative ion was observed [10]. This observation implies that the potential due to the sphere is not repulsive (as in the work of Sommerfeld and Welker [1]) but attractive. Endohedral confinement by an attractive shell may seem surprising at first sight if one thinks purely classically. It is in fact a quantum

Journal of Physics: Conference Series **438** (2013) 012001 doi:10.1088/1742-6596/438/1/012001

effect, which is best understood by reference to the well known problem of reflection at an attractive potential barrier in textbook scattering problems.

An exact result immediately followed,, but only for repulsive confinement, within penetrable or impenetrable walls It is in some sense quite an obvious result for confinement by a repulsive cube or sphere that the endohedral atom must live at the geometrical centre in its ground state [11]

6. Quantum states of the confining sphere

One aspect of the problem does not usually intrude much in confinement by a repulsive well and was actually missed by Sommerfeld and Welker in their treatment of the problem. It is the influence of quantum states of the confining well. These began to emerge as 'glitches' in numerical work by Baltenkov and also by Connerade and Dolmatov, performed in Tashkent, but their physical origin remained mysterious until systematic studies by Connerade, Dolmatov, Lakshmi and Manson [12].

Originally, these 'glitches' were suspected to be artifacts of the numerical method used, but Connerade, *et al* [12] varied the strength of the confining potential (which of course one is free to do theoretically), thereby tuning the quantum states of the confining shell and controlling the energy degeneracy between them and the excited states of the confined atom. This revealed Wigner repulsion or avoided crossings, and thus demonstrated that they are, in fact, genuine quantum states of the combined system, originating from those of the confining spherical attractive shell. In Fig.2 some of these avoided crossings are shown.

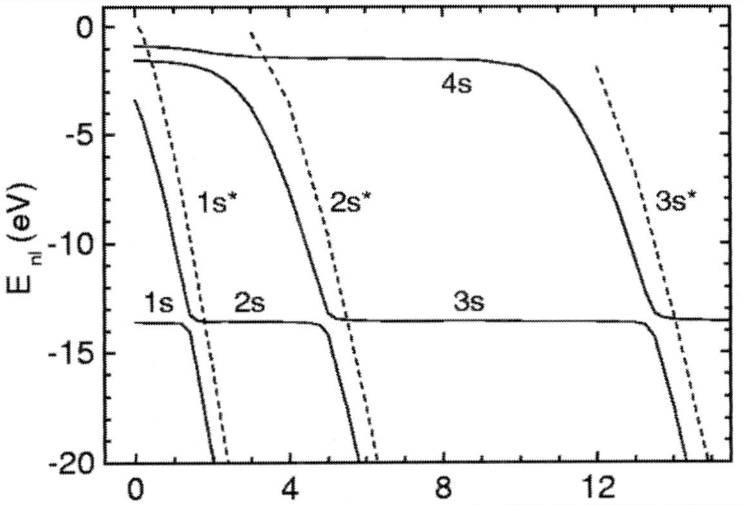

Figure 2. Avoided crossings in H@C_{60} from reference [12]. Note that labelling of the states by principal quantum number is somewhat arbitrary, depending on whether one follows a curve according to node structure or according to the binding energy.

A further aspect was confirmed in calculations by Kengkan Seehamart and Connerade [13] performed in Khon Kaen: these authors used one dimensional modelling of the same system and demonstrated that the avoided crossings are actually independent of geometry. They originate solely from the energy spectrum due to the double-well potential of the combined system. Figure 3 shows how this was established: the points are from *ab initio* one-dimensional calculations and the curves are from the tree-dimensional calculations by Connerade, Dolmatov and Manson [12]. Quite clearly, there is no discernible difference between the two. This is precisely what one expects for avoided crossings because their properties are controlled by energy degeneracy and the range of the interactions.

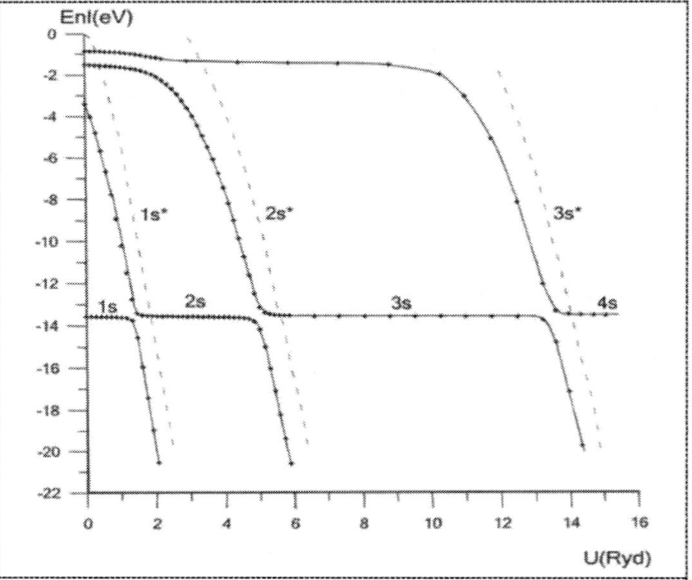

Figure 5. Comparison between one- and three-dimensional calculations of the avoided crossings in H@C_{60} after ref [13].

7. The classification of resonances

It thus became apparent that several kinds of resonance were present simultaneously in the spectrum of the confined atom system (which of course must include the cavity). Fullerene is not an ideal sphere (as the approximation made in the theory omits), but is a proper molecule, which can rotate and which possesses excited states of its own. Taking this into account, the issue of how to classify all the resonances was addressed by Connerade, Dolmatov and Manson [14] on the basis of their origin, which gives rise to the following distinct types for the A@C_{60} system.

(a) atomic resonances modified by confinement ;
(b) confinement resonances due to the cavity, modified by the inclusion of the atom and
(c) molecular resonances due to the fullerene, modified by the presence of the endohedral atom.

8. The Quest for Confinement Resonances

Theoreticians at large accepted these ideas, and calculations of endohedrally confined atoms thenceforth allowed for the occurrence of confinement resonances. There was, however, a problem which gradually became a worry. Although calculations often revealed many such resonances, they never seemed to be observable experimentally, and this remained a mystery for some time.

Eventually, in a paper entitled *Confinement resonances in photoionization or endohedral atoms: a myth or reality?* Korol and Solov'yov [15] found that the structure of confinement resonances in the photoionisation cross section is extremely sensitive to the mean displacement of the atom from the cage centre. This at least provided some hint as to why confinement resonances might, in practice, never be observed. If one likens the confined atom system to a 'quantum mechanical rattle', the endohedral atom might be always on the move, and confinement resonances would be washed out by phase cancellation effects. This hypothesis provided a nice way out, but had the unfortunate consequence of leaving in the literature the theoretical prediction of unobservable quantum effects, which seemingly could never be verified. Quite apart from the need to check theory against experiment, this would have left the question open whether or not their presence should be included in calculations and then averaged out or whether confinement resonances should be left out completely.

9. The First Observation of Confinement Resonances

So matters remained until 2010, when Kilkoyne *et al* [16] announced the first observation of Confinement Resonances in the photoionization spectrum of $Xe@Cl_{60}^{+}$. The experiment was performed by using a mass and charge selected ion beam of the endohedral atomic ion and merging it with a beam of synchrotron radiation. The phenomenon was detected in the $Xe@Cl_{58}^{3+}$ product ion channel. Figure 6 shows the resulting spectrum.

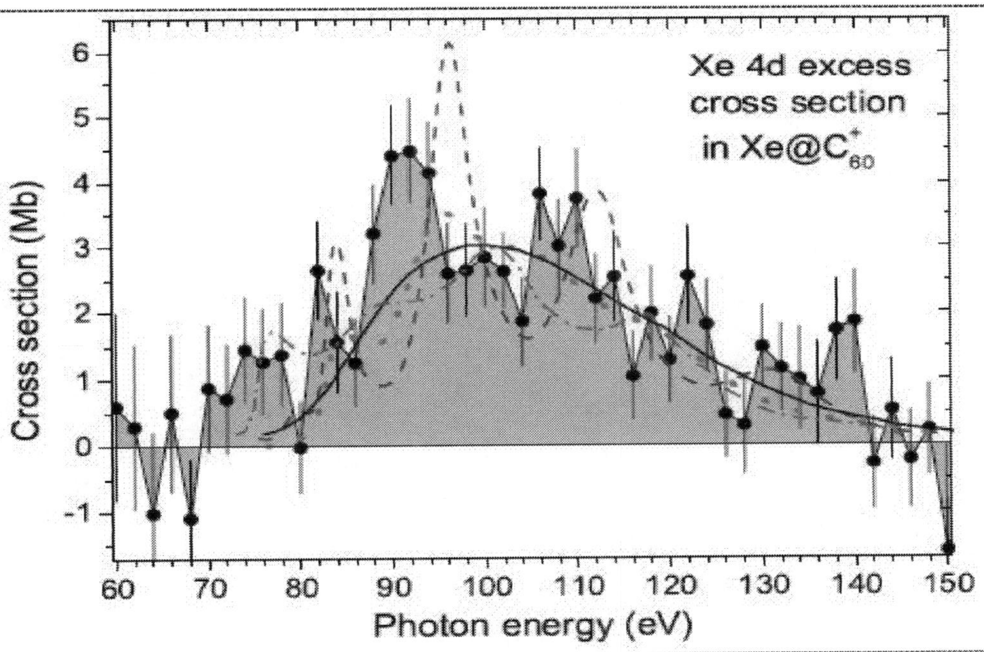

Figure 6. Confinement resonances in the spectrum of endohedral Xe in C_{60} after ref [16]. The experimental points (large dots) carry statistical error bars of one standard deviation. A many-body calculation for the giant resonance of the free atom is shown for comparison (black curve with a single broad maximum) and the dashed curve shows many-body calculations

The Confinement Resonances are observed in the giant 4*d* resonance of Xe, a very broad and strong feature as its name indicates, which facilitates the observation, as it is predicted by theory to fragment into four components. The structure is not exactly as predicted by theory, but the fragmentation is indeed convincingly revealed. It is interesting that giant resonances were initially believed by some to be a property of condensed matter, because of their resemblance to band structure and their persistence in the spectra of solids. The Russian physicist Zimkina from St Petersburg, who studied them in the soft X-Ray spectra of solids, argued strongly that they are of atomic origin [17] and the experiment which established this fact[18] in 1974 was also performed by using synchrotron radiation. When first reported, they were described in [18] as an '*atomic plasmon*' because they are dominated by many-body effects. Giant resonances, because they persist in many different atomic environments ranging from the free atom to condensed matter, provide an excellent tool for comparative studies of atoms in different environments (see [19]).

Returning to the issue of why Confinement Resonances become conspicuous in this case, it is clearly relevant that the size of a Xe Atom is large enough that it only just fits inside the confining fullerene sphere. Thus, it has much less freedom to rattle around, to use the descriptive analogy made above.

The fact that Xe@C_{60} is more favourable for the observation of Confinement Resonances than other comparable systems has led to many new calculations being performed. Thus, in Figure 7, we show some recent theoretical data for generalised oscillator strengths [20] by both independent particle (Hartree-Fock) and many body theories (RPAE). It is interesting that the structure due to Confinement Resonances is much more prominent in many-body calculations. Again, as mentioned in the introduction to the present paper, endohedral trapping provides a new tool to probe correlations in atoms and other small quantum systems.

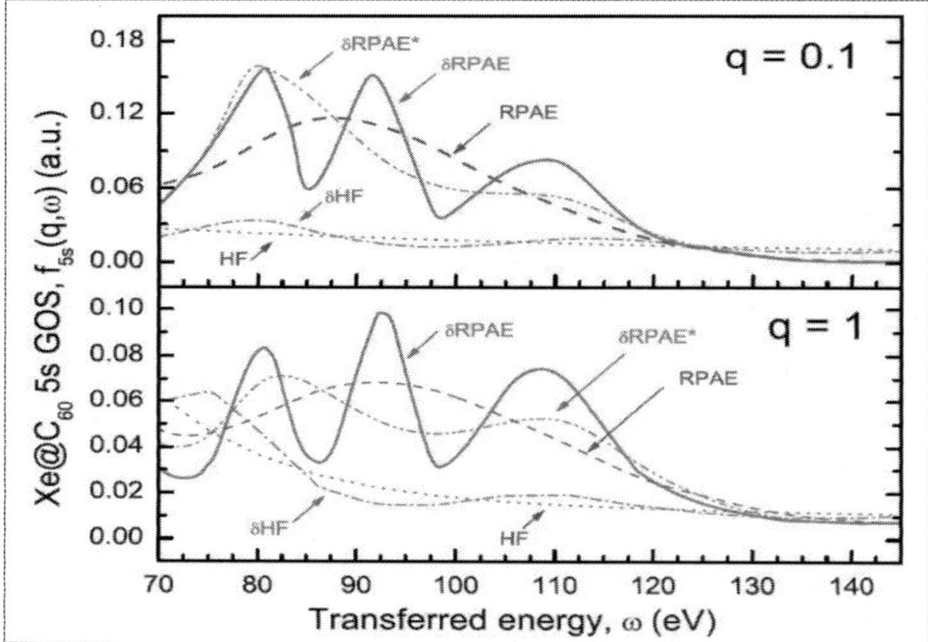

Figure 7. Comparisons between many-body calculations and independent particle model calculations for structure due to confinement Resonances in the generalised oscillator strength of XeC_{60} (see text and also reference [20])

The prediction of 'confinement resonances' by Connerade, Dolmatov and Manson [14] was published in the year 2000. For 10 years, the reality of such resonances remained in doubt and their observability was called into question. Because they were so elusive, they became a real test of Quantum Confinement as a subject: the issue was whether the approximations made in describing confinement by using spherical shells and neglecting many molecular and inter-molecular forces were too crude to represent reality, or whether the model was indeed well-adapted to the situation and could capture the basic physics involved. The observation of Confinement Resonances puts paid to such doubts by validating the approximations involved.

10. The plasmon of the Fullerene

Plasmons, when they appear, are not just a property of a specific system, but a general property of any system when its dynamics are dominated by a collective oscillation. The atomic giant resonance is simply an overdamped version of the *Mie resonance* in classical physics, which is a collective motion of all the electrons of a shell. This is not only a property of the atom, but also of *any* quantum sphere or shell. The fullerene shell has 240 electrons available, and therefore can be considered as a highly conducting shell of electrons. The fullerene plasmon is observed experimentally to occur around 20eV excitation energy..

The ***dynamical polarisability*** of the charge shell is responsible for the optical giant resonances or plasmons, which arise simply from the collective motion of delocalised electron. So, when one observes a Mie resonance, it is a signature of a good conductor, for example a metallic sphere or shell. Of course, a highly conducting sphere should must obey the laws of electrodynamics, and so one would expect it to act like a Faraday cage, screening the atom placed inside it from any external fields.

The fact that the fullerene can be considered as a highly conducting cage led Harneit [21] to suggest a very neat solution to a problem which occurs in quantum computing, namely how to make a compact and simple register without using atomic traps, by exploiting the screening due to this conducting cage to isolate the spins of individual atoms and thus to enable both the construction and the preservation of q-bits.

Figure 8. After ref [21] The scheme proposed for the fabrication of a register for a quantum computer involving a chain of endohedral fullerenes deposited on a surface.

11. The quantum Faraday cage

For full screening, it is considered important to use endohedrals with the atom close to the centre of the confining fullerene. Nitrogen has been proposed as the best case. One problem of course is to avoid distortions of the sphere when placed close to a surface. Other questions which arise are: how good is the screening provided by the Faraday cage? How would such a quantum Faraday cage really work? Is it true that full screening occurs at the quantum level?

One senses that that, if full screening were really to occur at the quantum level, electromagnetic radiation could never penetrate the confining shell, and endohedral atoms would become unobservable. In fact, in atomic physics more generally, there would be a strong attenuation which could inhibit the excitation of inner shell spectra in atoms completely. Clearly, this is not the case. But why not?

This issue was addressed by Connerade and Solov'yov [22], who investigated the dynamical screening of a confined atom by a fullerene, and who included in their model the effects of the fullerene Plasmon on the dynamical polarisability of the system.

They found that the photoabsorption of the endohedral atom screened by the cage can be expressed as the product of two factors, namely the response of the atom itself and a dynamical enhancement factor $\boldsymbol{F}(\omega)$ which appears as a pure number, due entirely to the polarizability of the metallic sphere. This enhancement factor is frequency dependent and, not surprisingly, maps out the profile of the fullerene giant resonance, as shown in Figure 9.

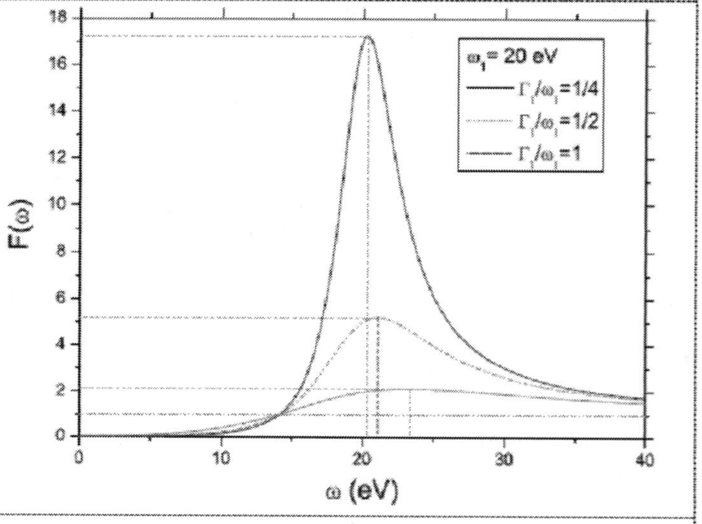

Figure 9. General form of the dynamical enhancement factor [22].

From this result, we see that:

(i) at very low frequencies, we have full screening (classical Faraday situation);

(ii) at the plasmon frequency, the confining cage acts like an antenna, *amplifying* the electromagnetic field around the confined atom;

(iii) at very high frequencies, the shell has no effect, i.e. the atom sees undisturbed radiation, which is how atomic inner shells can be excited.

This behaviour can broadly be described as a quantum 'skin effect' because of its frequency dependence. Clearly, should the shell actually amplifies the effect of external radiation rather than screen the atom inside, this would become a concern for making stable q-bits, which depend on screening for their survival. One can show that these results apply for the whole space within the confining sphere, which is in some sense a generalisation of the usual theorems applying to the Faraday cage in electromagnetism.

It is of course not obvious that the spherical cage will not be distorted by the presence of a surface, and this may also spoil the quality of screening. Creating a viable register for a quantum computer is thus no easy task.

12. Confinement by Oblate and Prolate Fullerenes

The fullerenes are not all spherical, of course, and even the near-spherical ones may become distorted by (for example) interactions with a nearby surface, so it is interesting to consider how the system behaves when this happens. The case of an elongated or squashed surrounding cage can be handled on the same principles as for a sphere by applying a geometrical distortion to the problem [23] which turns the sphere into a prolate or oblate spheroid. When one applies this method to hydrogen confined at the centre, states which were energy-degenerate in spherical symmetry of course become split and the degeneracy is resolved. In consequence, the spectrum begins to resemble that of a diatomic molecule. One can deduce that the spectrum of such a system carries all the necessary information to measure the extent of the deformation.

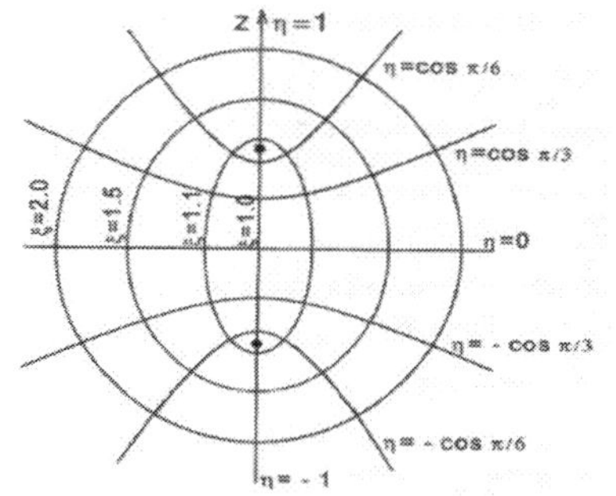

Figure 10. The spheroidal coordinate system used in ref. [21] to represent distortion of the confining cage. In the situation illustrated here, the distortion produces a prolate spheroid, bounded by two ellipsoids of revolution. The oblate case is handled in the same way.

13. Atomic Size Effects

As the reader will have noticed, Confinement Resonances were discovered experimentally for Xe, a rather large atom, whereas most of the description of effects due to quantum confinement above has been based on the example of confined hydrogen.

In fact, size effects must clearly play an important role once the confining cavity becomes small enough to be comparable to the endohedral atom itself. As pointed out in ref. [15], it is very unlikely that small atoms will remain at the centre of an attractive confining sphere (even less so if it is distorted, of course) and the quest for the elusive Confinement Resonances indeed illustrates this point. If one approaches quantum confinement from the angle of probing correlations, the heavier atoms are the most interesting. Furthermore, from the angle of applications, it turns out that heavier atoms are more relevant.

Once the size of the atom approaches that of the confining cage, there is the question: what effect does confinement have on the size of the atom, as well as the related question: what effect does confining an atom have on the size of the cage? The second question is difficult to answer theoretically but, for endohedral fullerenes, it can probably be addressed by analysing the properties of Confinement Resonances, since they carry information about the geometry of the confining shell.

As regards the first question (What happens to many-electron atoms under confinement?) Hydrogen is a very uninteresting atom in its response: all one can do is to reduce the number of quantum states or to detach the electron by confinement. It is much more worthwhile is to study atoms whose properties (chemistry) can be significantly altered under compression.

Such atoms belong to the sequences known as the transition elements, rare-earths, lanthanides and actinides, which were arranged into a quasi-periodic table describing their localisation-delocalisation properties (the first order Mott transitions in solids) by Smith and Kmetko [24]. For fundamental reasons [18], these also turn out to be atoms which exhibit giant resonances, which frequently possess valence instabilities and which violate the simple *aufbau* principle of Bohr and Stoner.

An atom in a key position from this point of view is of course La, and it is interesting to consider theoretically what happens to La under confinement by a fullerene cage. This problem becomes rather more complex than the cases treated above. The behaviour of the lanthanides is extremely sensitive, because of the development of a double-well radial potential, due to centrifugal barrier effects which have been discussed extensively in the literature (see ref [19] for details). In this situation, the behaviour of the atom becomes strongly influenced by relativistic effects, and a proper treatment is correspondingly more difficult.

This case has been addressed by using fully relativistic Dirac-Fock theory [25], and the outcome indeed demonstrates that the atom is 'soft' under confinement, i.e. that the outermost electron depends very sensitively for its mean radius on the properties of the confining sphere. The treatment is rendered even more difficult by the fact that boundary conditions must be applied with some caution to the fully relativistic problem. The reason one must be cautious is best understood by reference to the confinement problem first studied by Sommerfeld and Welker [1], in a cavity with infinitely high repulsive walls. If one applies the same Dirichlet boundary conditions for this case to the Dirac equation, they turn out to violate relativity by forcing both the large and the small component of the relativistic wavefnction to zero at a point other than the origin (which would artificially lead to spontaneous electron-positron pair creation as in Klein's paradox). To avoid this unphysical situation, one has to apply the boundary condition in a different way. Fortunately, one may either (a) apply one of the other boundary conditions suggested by ref [26] or (b) deal only with problems in which the step height at the boundary remains very small compared to the rest mass of the electron, and make the approximation that the boundary condition applies only to the large component, which turns out to be sufficient for a umber of situations.

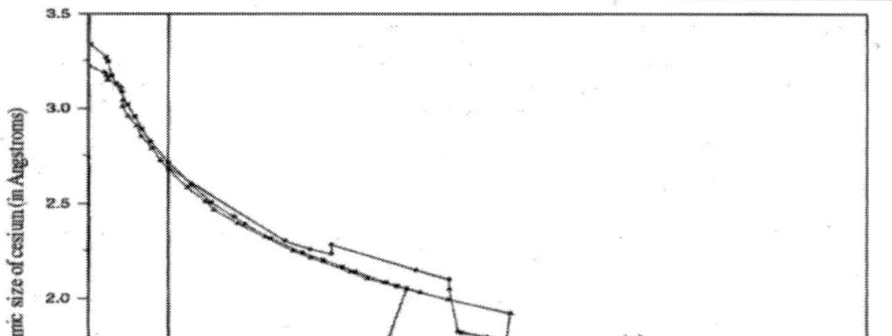

Figure 11. (after ref. [25]) Variation of atomic size with applied pressure (a) experimental (b) relativistic and (c) non-relativistic limit of the relativistic calculation. The large discontinuity is due to 6s-5d orbital rearrangement. The small discontinuity in the experimental curve is due to recrystallisation..

By using the latter approach, Connerade and Semaoune [27] have studied Cs in a cavity with repulsive walls, and shown that the *ab initio* Dirac-Fock method correctly reproduces the switch between 6*s* and 5*d* orbitals under pressure which is observed experimentally as a discontinuity in the atomic size of Cs compressed by a diamond anvil..

Connerade and Semaoune [25] applied the same method to study the case of La in an attractive spherical shell with the geometrical parameters of C_{60}, and found that the 5*d* wavefunction is indeed extremely sensitive to confinement. Figure 12 shows how the mean radius of the 5*d* wavefunction varies between the free atom and the atom confined in an attractive shell. Note the very large change of radius of the main maximum.

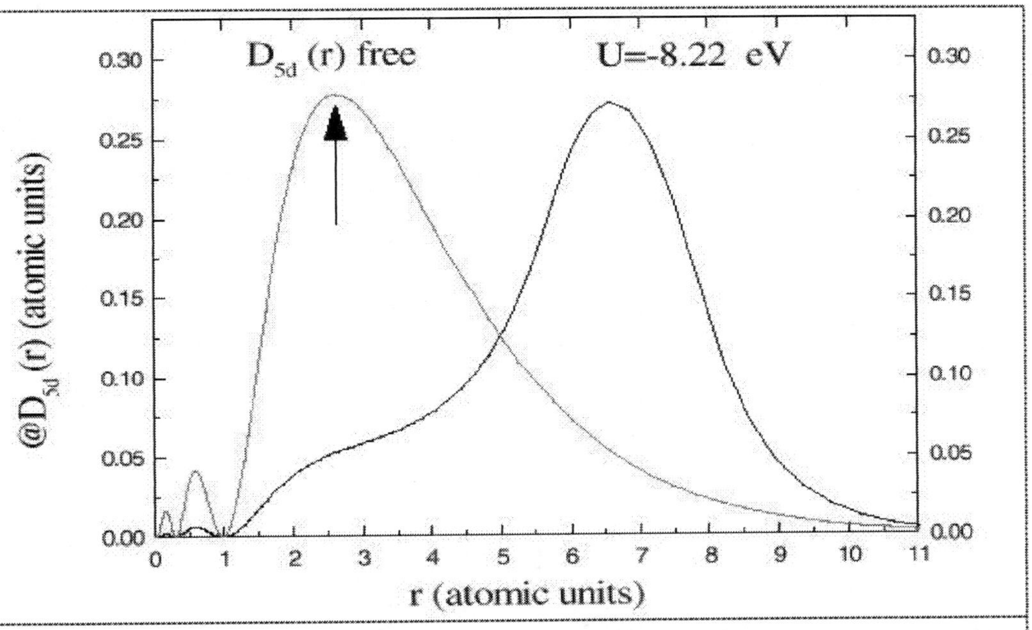

Figure 12. Change in the radial 5*d* wavefunction of La between the free atom and the atom confined by an attractive shell of depth 8.22 eV according to relativistic Dirac-Fock calculations [25].

14. The Periodic Table Under Pressure

The examples just given show that many-electron atoms can be profoundly affects by confinement, to the extent that the filling of their outer shells is modified. The shell structure of course controls ther valence and thus their chemistry. It is well known in compounds (for example SmS) that the appearance and physical properties of compounds can, under high pressure, undergo discontinuous changes and localisation / delocalisation transitions (termed first order Mott transitions) in solids. These fluctuations of valence or changes in physical properties occur for atoms located along the diagonal of Smith and Kmetko's [24] Quasi-Periodic Table for transition elements and rare-earths, and the underlying physics is discussed in ref [19]. It is clearly of atomic origin.

For confined atoms, the same or similar behaviour occurs, and for the same reasons: the immediate environment of the atom is capable of tipping the balance between the inner and outer reaches of the atomic potential for the *d* and *f* electrons. Broadly speaking, elements which do not follow the order of filling predicted by the Bohr-Stoner *aufbau* or 'building-up' principle can be made to follow it by placing them under high pressure. Thus, as explained in ref [28], the Periodic Table of certain atoms under high pressure is not the same as the Periodic Table of free atoms. This has a variety of interesting implications, including the fact that their chemistry is altered.

15. Confinement and reversible storage

The subject of reversible Li+ ion storage [4,5] is a vast one, with huge industrial and commercial implications, and it is therefore rather difficult to review it. Much of the research is unpublished. It would anyway be beyond the scope of the present article to attempt such a review, but some connections deserve mention here.

As is well-known, the rechargeable battery is the Achilles' heel of many commercial products, including in particular the electric vehicle. From the power to weight ratio of batteries, it turns out that Lithium rather than lead provides the ideal ion for a rechargeable cell. It is also a non-polluting element. However, for reasons of safety, one cannot make a lithium battery by using just the metal as the electrode as such a cell would readily catch fire and explode. Therefore, it is necessary to insert the lithium ion inside the electrode material, where it is no longer possible that it comes into direct contact with air. Obviously, if this process fails after a number of cycles and metallic lithium is deposited on the surface of the electrode, this would be a safety hazard.

The insertion process must be reversible, which means that the electrode must not undergo any phase transition such as recrystallisation during the charging / discharging cycle. It is also necessary that the Li+ ions should be able to migrate freely within the electrode material and find spaces within which to be encapsulated and from which they can readily be recovered during the cycle. Nor should these ions form permanent chemical bonds which would have to be broken to recover them. It is in practice very difficult to achieve a high level of reversible storage, and finding an appropriate electrode material for a high power Li+ reversible or 'rocking chair' battery remains a challenging problem. Much of the progress in this area is shrouded in commercial secrecy.

It is clear, however, that the mechanism involved in quantum confinement is of central interest: if the host material which temporarily holds the Li+ ions in place is made from 'flexible' atoms of the type just described and if there are spaces within it of sufficient size to pack ions inside, one has a basis for a good electrode material. It has also been found that the materials which have a sheet structure and which expand only in one dimension when *lithiated* (for example, some of the transition metal oxides) tend to weaken on repetitive cycling and become fragile. Thus, it is desirable to pick materials which expand in 3D on lithiation as well as being porous to the ions.

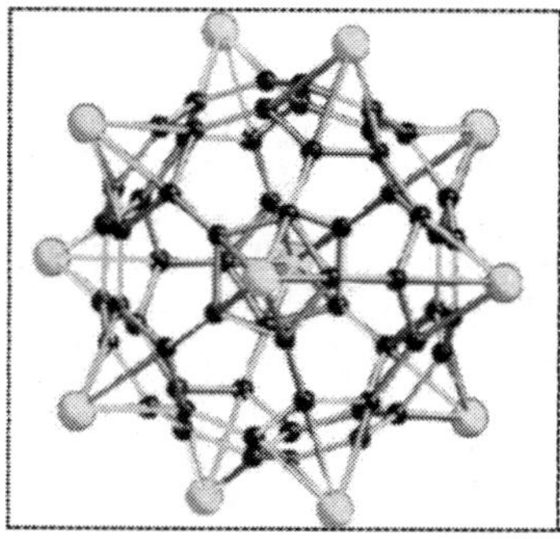

Figure 13. The Li-coated C_{60} molecule whose structure is shown alongside has been proposed as a medium for hydrogen storage in fuel cells. Similar molecules are also under investigation in connection with reversible lithium storage. Molecules suitable for this purpose must be readily penetrable by the ions and also provide cavities in which the ion can temporarily be stored. (see text)

Physically, the problems of storing Li + reversibly and the problem of storing hydrogen reversibly are quite closely related, and both of them provide options in the general quest for non-polluting and reversible energy storage. Quite often, similar host materials are involved in both cases. One seeks materials which can adapt to contain Li+ ions by what is called *polaronic distortion* (no change of structural symmetry) so that the process remains reversible.

Much research today is devoted to Li+ storage for future electric cars. In principle, the properties required and the processes by which higher current densities can be achieved for Li storage are well understood, but finding the best materials for the purpose is the main difficulty. Many different molecules have been suggested. They all seem to involve some at least of the principles described in the present article. For example, the role of $LiFePO_4$ as a suitable electrode material has been investigated. This again is a molecule which, under packing to form a material, creates a fairly open structure to enhance Li+ mobility and which also contains the atom Fe, which is one of the 'soft' atoms near the diagonal of Smith and Kmetko's Quasi-Periodic Table. It thus possesses properties very suitable to enhance lithium ion storage (see Fig. 14). Readers interested in the application of Li batteries to electric vehicle propulsion can refer to some general papers on the subject e.g. ref. [29]

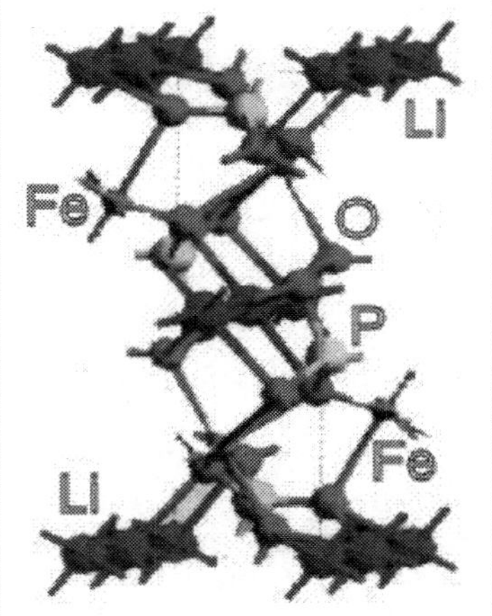

Figure 14. Structure of the $LiFePO_4$ molecules, proposed as the building block of a reversible lithium ion battery electrode material for high current applications. The molecule has an open structure favourable to lithium ion mobility and also contains the 'soft' atom Fe which facilitates polaronic distortion for storage. (see text).

Carbon is by no means the only atom from which to make hollow microspheres. Indeed, there are many alternatives. In 2008, $Li_4Ti_5 O_{12}$ microspheres were made in Dalian (China) and also turned out to possess excellent properties as anode materials [30]. Over 500 charge/discharge cycles, their capacity is very stable, with a loss of only 0.01% per cycle. Besides ordinary hollow spheres of this anode material, 3D hierarchical hollow microspheres assembled in thin nano-sheets have since been created. Hollow structured nanomaterials of other insertion-type oxide materials have equally been reported. The deposition of microspheres on graphene surfaces seems very promising.

16. Confinement versus ionisation

At the outset of the present paper, it was observed that, in the quest to control and probe atomic properties by external means, compression by a cavity provides a very useful and in some ways more subtle alternative to the rather more violent process of ionisation. So, it is perhaps interesting to conclude by comparing the levels of atomic compression achievable by both methods.

In order to do this, the best approach is to start with a rather compressible or soft ion, which displays the effects more clearly. The lightest one available is Ca+, which possesses a highly compressible *3d* wavefunction analogous to the *5d* wavefunction of La or of Cs discussed above. It experiences a strong centrifugal barrier which results in fairly flat external reaches of the effective atomic potential. Figure 15 illustrates the consequences of this (a) for confinement by an external barrier and (b) for incremental changes in the atomic number, which simulates progressive excitation / ionization.

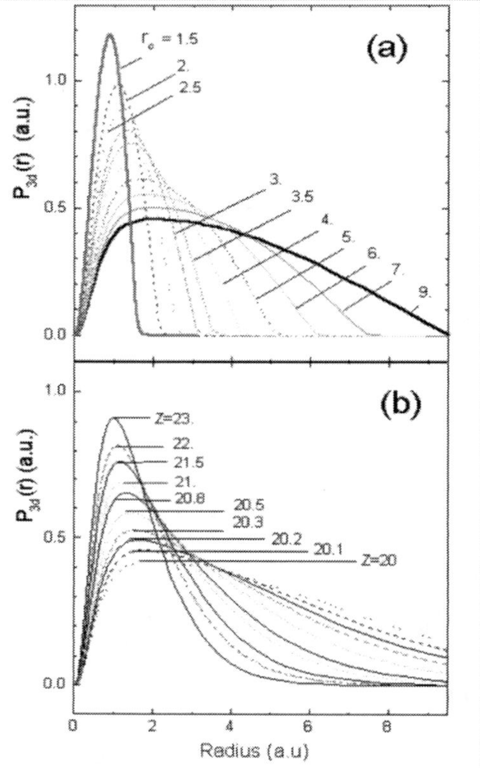

Figure 15. (a) represents compression by a shell acting like a spherical piston of decreasing radius, and (b) represents progressive ionisation, simulated by slowly increasing the nuclear charge state. Note that non-integral charge states do not exist in nature, whereas confining spheres can be created with different sizes and well depths. (the calculations are by VK Dolmatov)

Of course, increasing the charge state in this way is not achievable for free ions, although it may resemble what occurs for ions in certain solids. On the other hand, confining ions in spherical shells of different depths is in principle possible, for example by exciting or ionizing the outer electrons of a fullerene shell. Thus, confinement could be used as a tool to probe electron-electron interactions in a confined atom by exercising direct control on the compression. This would provide a new and very interesting technique for the study of correlations in multi-electron systems.

17. Conclusions

In the present review, I have not discussed a number of other interesting issues which concern quantum confinement and which may prove important for the future.

One of these is to explore effects similar to those which occur in thermodynamics when the compressibility of gases is discussed. It will not have escaped the reader that atomic volume and a quantum (or Hellman Feynman) pressure were both introduced in the context of quantum confinement, and that the effect of confinement compressing the atom was likened to an 'atomic piston'. This of course is a language similar to that of thermodynamics, except that the atom, considered here as the compressed medium, obeys the Schroedinger equation rather than the usual equation of state. In principle, it is possible to consider and set up an ideal compression cycle (involving, for example, the compression of a two-level atom (as done by Einstein to study radiation field effects) and thus to introduce a pseudo-temperature into the problem. One then has all the thermodynamic ingredients and one can set up some kind of idealised theory.

Amongst the problems of potential interest using this approach is the possibility to set up an ideal reversible Li ion storage cycle analogous in some sense to the Carnot cycle of thermodynamics, which could serve as a useful measure for an 'ideal battery' against which the performance of real batteries could be assessed.

Among experimental developments for the future is the fact that, as mentioned in the present paper, metallofullerenes can be transported in beams and, most probably, will soon be trapped. It is likely therefore that the laser spectroscopy of metallofullerenes will be developed to enable much more detailed studies than have been achieved so far. In this connection, it will become important to compute, not only the single-photon cross section, but also three-photon and eventually multiphoton properties.

Obviously, what I have described represents work by many collaborators over a long period, and I take this opportunity to thank them all. There are plenty of other important subjects I did not touch on which also involve quantum confinement, such as, e.g., high pressure bubbles in solids (in the walls of nuclear reactors, for example), zeolites, quantum dots, etc, etc. t would be difficult to cover all of these in a single review, which demonstrates that atomic confinement really is a growing field, and presents many new opportunities for research

18. References

[1] Sommerfeld A. and Welker H 1938 *Ann Phys* **32** 56
[2] Michels A, de Boer J and Bijl A 1938 *Physica* **4** S991
[3] Ley Koo E and Cruz SA 1981 *J Chem Phys* 74 4603
[4] Connerade JP 1996 *J of Alloys and Compounds* **255** 79
[5] Connerade JP Jumas JC and Olivier-Fourcade J 200 *J. Solid State Chemistry* **152** 533
[6] Varshni YP 1998 *J Phys B At Mol Opt Phys* **31** 2849
[7] Shi T-Y Qiao H-X and Li B-W 2000 *J Phys B At Mol Opt Phys* **33** L349
[8] Campbell E 1996 *Nature* **382** 407
[9] Xu YB Tan MQ and Becker U 196 *Phys Rev Lett* **76**
[10] Huang J Carman HS and Compton RN 1995 *J of Physical Chemistry* **99** 1719
[11] Pupyshev VI (2000) *J Phys B At Mol Opt Phys* **33** 961
[12] Connerade JP Dolmatov VK Lakshmi PA and Manson ST 1999 *J Phys B At Mol Opt Phys* **32** L239
[13] Kengkan P Seehamart K and Connerade JP (unpublished)
[14] Connerade JP Dolmatov VK and Manson ST 2000 *J Phys B At Mol Opt Phys* **33** 2279
[15] Korol AV and Solov'ov AV 20.. ArXiv 0912.2690 Physics Atm.Clus

[16] Kilkoyne ALD Aguilar A Muller A Schippers S Cisneros C Alma'Washi G Aryal NB Baral KK Esteves DA Thomas CM and Phaneuf RA 2010 *Phys Rev Lett* **105** 213001
[17] Zimkina TM and Gribovskii (1971) *J Phys (Pars) Colloq* **32** C4 282
[18] Connerade JP and Mansfield MWD 1974 *Proc Roy Soc (London)* **A341** 267
[19] Connerade JP *Highly Excited Atoms* 1998 Cambridge University Press Cambridge UK
[20] Amusia M Ya Dolmatov VK and Chernysheva L 2011 *Phys Rev* **A84** 063201
[21] Harneit W 2002 *Phys Rev* **A 65** 032322
[22] Connerade JP and Solov'yov AV 2005 *J Phys B At Mol Opt Phys* **38** L1
[23] Connerade JP Lyalin AG Semaoune R Semenov SK and Solov'yov AV 2001 *J Phys B At Mol Opt Phys* **34** 2505
[24] Boring AM and Smith JL 1987 in *Giant Resonances in Atoms Molecules and Solids* Connerade JP Esteva JM and Karnatak RC Eds NATO ASI Series B vol 151 Plenum Press New York page 311
[25] Connerade JP and Semaoune R (2000) *J Phys B At Mol Opt Phys* **33** 1
[26] Alonso V and De Vincenzo S (1997) *J Phys A Math Gen* **30** 8573
[27] Connerade JP and Semaoune R 2000 *J Phys B At Mol Opt Phys* **33** 3467
[28] Connerade JP Dolmatov VK and Lakshmi PA (2000) *J Phys B At Mol Opt Phys* **33** 251
[29] Stafford N *'Electric cars hoping for lithium ion revolution'* 2007 *Chemistry world* **17** October
[30] Liu J and Xue D 2010 *Nanoscale Research Letters* **5** 1525

Journal of Physics: Conference Series **438** (2013) 012002 doi:10.1088/1742-6596/438/1/012002

The Oblate Structure and Unexpected Resistance in Reactivity of Ag_{15}^+ with O_2.

Arthur C. Reber, Gabriel U. Gamboa, and Shiv N. Khanna[1]

Department of Physics, Virginia Commonwealth University, Richmond, VA, USA 23284.

E-mail: Snkhanna@vcu.edu

Abstract. Theoretical studies have been carried out to investigate the unusual reactivity of Ag_{15}^+ cations with oxygen. Our previous work has shown that the reactivity of free metal clusters with oxygen entails a spin excitation that causes reduced reactivity in clusters with filled electronic shells and large HOMO-LUMO gaps. Earlier experiments on Ag_{15}^+ have shown that the cluster exhibits remarkable resistance to reactivity with oxygen despite having a valence electron count that is not expected to result in a filled electronic shell within the spherical jellium model. It is shown that Ag_{15}^+, Ag_{14}, and Ag_{13}^- clusters with 14 valence electrons, all have oblate bilayer atomic structures that lead to a splitting of the superatomic D-shell in a manner analogous to crystal field splitting of d-states in transition metals. The importance of the oblate structure is demonstrated by showing a correlation between an oblate deformation coefficient and the HOMO-LUMO gap of the different structures of Ag_{15}^+. The oblate deformation results in the splitting of the electronic subshells and leads to an unusually large HOMO-LUMO gap and the unusual resistance of the cationic species to oxidation.

1. Introduction

The interaction of oxygen with metal clusters has attracted considerable attention as the reactivity pattern in small clusters can be different from the bulk and strongly dependent on the size.[1-19] These differences arise primarily due to two reasons. Firstly, the electronic states in small symmetric clusters are grouped into electronic shells such as 1S, 1P, 1D, 2S, 1F, 2P, much in the same way as in atoms, and clusters containing 2, 8, 18, 20, 34, and 40 valence electrons correspond to filled shells.[20-22] The shell structure was first seen through the experimental mass spectra in beams that showed marked peaks, usually referred to as magic numbers, at these sizes.[20] Later experiments also confirmed that the electronic shells govern the chemical reactivity and clusters having filled shells are generally less reactive to a variety of gases.[24] Secondly, the reactivity with oxygen is particularly interesting as the ground state of oxygen is a spin triplet with lowest unoccupied orbitals having a spin minority character. Activation of the O-O bond to promote reactivity requires a filling of the minority spin states. As suggested previously by the present authors, the spin conservation in free reacting clusters requires a spin excitation of the cluster to conserve the total spin, as the O-O bond is activated. Consequently, for a variety of anionic clusters, the reactivity with oxygen in even electron systems is found to be less than in odd electron systems.[3,4,9-11,19] More significantly, the reactivity in even electron clusters with singlet ground states is found to be governed by the difference between the highest

[1] To whom any correspondence should be addressed.

occupied molecular orbital (HOMO) and the lowest unoccupied molecular orbital (LUMO).This is because a spin excitation of clusters can be looked upon as a promotion from the singlet to the triplet configuration via the transfer of an electron in HOMO to the LUMO of opposite spin. Hence, the reactivity with oxygen not only probes the filled shells but also provides an indirect measure of the gap between the filled and unfilled electronic spectrum.

We have previously examined the reactivity of silver cluster anions with oxygen in a joint theory and experimental undertaking.[3] The experiments revealed variable reactivity with clusters containing an even number of electrons generally exhibiting lower reactivity than those with an odd number of electrons. In particular, Ag_{13}^- appeared as a stable species both in the mass spectra of unreacted species and after exposure to molecular oxygen. This was surprising since Ag is monovalent and Ag_{13}^- has 14 valence electrons and that does not correspond to a filled electronic shell in a spherically confined nearly free electron gas. We demonstrated that this behavior was rooted in the atomic structure of the clusters that was significantly different from a spherical shape. In fact, the cluster had a bi-layer arrangement. Our studies showed that the large deformation of the spherical shape to a bilayer atomic structure induced a splitting of the delocalized valence electron orbitals into $1S^2$ | $1P^4$ | $1P^2$ $2S^2$ 1D4 || 1D6 | shells where the vertical lines indicate the gaps between the filled orbitals and the double vertical line indicates the gap between filled and unfilled orbitals. Such a deformation induced splitting resembles the crystal field splitting of the atomic orbitals and shows thatthe geometry of the cluster may have a profound effect on the reactivity of a cluster with other oxidants.[25-28] These and other previous studies have all focused on anionic species.

In this work, we extend our studies to include cationic silver clusters. Brechignac and co-workers[1,2] have examined the reactivity of nitrogen and oxygen to silver cluster cations containing up to 27 atoms in order to examine the physio- and chemisorptions of the oxygen. Their results indicate that while molecular absorption is favored below 77K, above 105 K, the chemisorptions changes into oxidation of the cluster. One of the striking findings was the lack of reactivity of Ag_{15}^+ cluster, which had the lowest O_2 absorption rates in Ag_n^+ for n=10-25. Since the cluster has the same number of valence electrons as Ag_{13}^-, it raises the question if the lack of reactivity could also be due to atomic structure. In this work we perform theoretical investigations into silver clusters containing 14 valence electrons to examine if the strong oblate distortion, similar to that of Ag_{13}^-, may also explain the stability of Ag_{15}^+. Our theoretical studies focus on the role of the geometric deformation of the Ag_{15}^+ cluster and its effect on the electronic structure including a comparison with Ag_{14} and Ag_{13}^-. The key issue is to examine the atomic structure and to see if the electronic states in all three 14 valence electron systems are marked by large deformations that make all these clusters non-reactive towards oxidation.

2. Methods

A first-principles molecular orbital approach was used wherein the cluster wave function was expressed as a linear combination of atomic orbitals centered at the atomic sites and the exchange correlation effects were included within a generalized gradient density functional theory (DFT) formalism. The actual calculations were carried out using deMon2k set of computer codes.[29] For the exchange and correlation functionals, we used a generalized gradient approximation (GGA) functional as proposed by Perdew, Burke, and Ernzerhof (PBE).[30] The silver atoms were described using a 19 electron quasi-relativistic effective core potential (QECP) with a corresponding valence basis set as proposed by Andrae et al.[31] The auxiliary density was expressed in a combination of primitive Hermite Gaussian functions by using the GEN-A2* auxiliary function set.[32] To determine the geometry and spin multiplicity of the ground states, the configuration space was sampled by starting from several initial configurations and spin multiplicities and then optimizing the geometric structures via the quasi-Newton Levenberg-Marquardt method.[33] All structures were fully optimized in delocalized redundant coordinates without imposing any symmetry constrains to allow for full variational freedom. Previous studies have produced a large library of geometries for neutral silver clusters, and these methods have been successful in identifying the structure of free and ligand protected silver clusters.[34-41]

To parameterize the geometry of the clusters, a deformation coefficient, Δ, was calculated to determine the planarity of the structures. Note that this is a variation on previous coefficients that were used for primarily prolate clusters.[42] For a given set of coordinates, Δ is defined as

$$\Delta = \frac{2Q_x}{Q_y + Q_z} \qquad (1)$$

and Q_x, Q_y, and Q_z are the eigenvalues of the deformation tensor

$$Q_{ij} = \sum_I R_{Ii} R_{Ij} \qquad (2)$$

in which I runs over every ion, and R_{Ii} is the ith coordinate of ion I, and R_{Ij} is the jth coordinate of ion I relative to the center of mass. This produces a 3×3 matrix, and the eigenvalues are ordered so that $Q_z > Q_y > Q_x$. A roughly spherical cluster will have $Q_z \approx Q_y \approx Q_x$ and a Δ=1. An oblate structure will have $Q_z \approx Q_y > Q_x$, and as the oblate distortion becomes completely planar Δ=0, while a prolate cluster will have $Q_z > Q_y \approx Q_x$.

3. Results

To understand the observed resistance to oxidation of the Ag_{15}^+ cluster, we first performed theoretical studies to identify the cluster's ground state geometry. The structure, relative energy and HOMO-LUMO gap of a number of local minima of Ag_{15}^+ are shown in Figure 1. The lowest energy structure, **1**, has a HOMO-LUMO gap of 1.20 eV, significantly larger than that of the other structures, and is quite large for cluster that has a non-magic number of valence electrons. The structure of **1** is oblate, and the eigenvalues of the deformation tensor are given in Table 1. For comparison, we also list the relevant values for the higher energy structures of Ag_{15}^+ as well as for Ag_{14} and Ag_{13}^-. Note that the global minimum geometry of Ag_{15}^+ and Ag_{14} are remarkably oblate, with their Q_z being nearly identical to Q_y in both structures. The next three most stable geometries are also oblate with geometry **5** being the first with a prolate geometry. These results suggest that these clusters have a preference for geometries that are oblate. The most likely origin for the preference for oblate

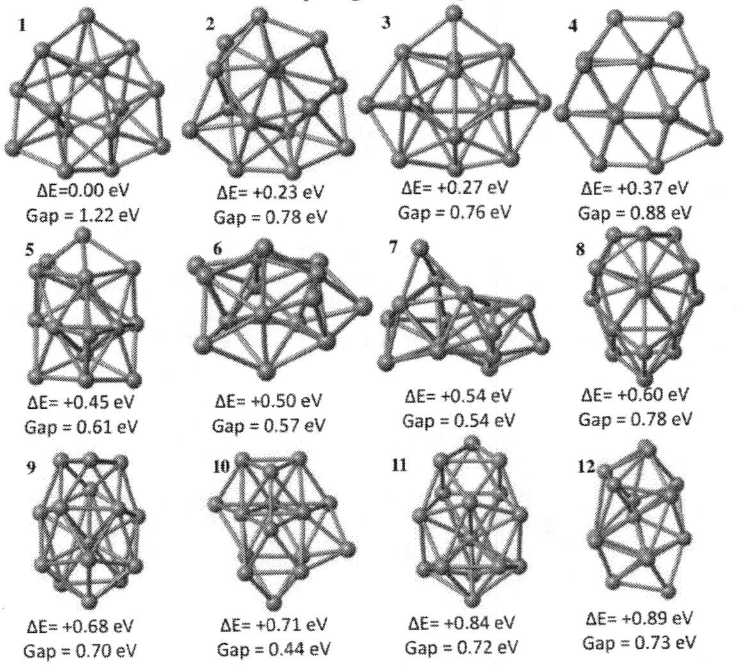

Figure 1. Geometry, relative energy, and HOMO-LUMO gap of different structures of the Ag_{15}^+ cluster.

geometries is that the deformation splits the subshell of delocalized orbitals so that the HOMO-LUMO gap is enhanced, resulting in enhanced stability.

To test our hypothesis that geometric factors play an important role in the energetic and electronic stability of Ag_{15}^+, we have plotted the relative stability of the Ag_{15}^+ clusters versus their HOMO-LUMO gaps in Figure 2. The result reveals that the most stable cluster, **1**, has the largest HOMO-LUMO gap by a wide margin, and a general trend in that clusters with larger HOMO-LUMO gaps are generally correlated with higher stability. While this agrees with conventional wisdom, that a large HOMO-LUMO gap results in enhanced stability, in the case of clusters which have a nonmagic number of electrons, there is potentially a tradeoff between the energy of cost of the deformation versus the enhanced stability caused by the increased HOMO-LUMO gap. In some systems, such as $CuAl_{12}^-$,[10,43] the cluster is icosahedral with a quite small gap because the reconstruction energy is too large despite other isomers having a larger HOMO-LUMO gap.

Table 1. Eigenvalues of the deformation tensor, deformation coefficient Δ, and HOMO-LUMO Gap.

Cluster	Q_x	Q_y	Q_z	Δ	Gap (eV)
Ag_{15}^+ (1)	14.50	64.63	64.66	0.22	1.22
Ag_{14}	13.78	54.88	56.61	0.25	1.12
Ag_{13}^-	12.90	45.77	69.53	0.22	1.28
Ag_{15}^+ (2)	17.97	58.08	64.74	0.29	0.78
Ag_{15}^+ (3)	15.22	55.92	71.54	0.24	0.76
Ag_{15}^+ (4)	15.98	59.97	66.54	0.25	0.88
Ag_{15}^+ (5)	29.67	39.04	68.74	0.55	0.61
Ag_{15}^+ (6)	28.56	45.78	60.87	0.54	0.57
Ag_{15}^+ (7)	28.28	43.26	68.9	0.50	0.54
Ag_{15}^+ (8)	32.15	35.33	70.34	0.61	0.78
Ag_{15}^+ (9)	29.21	34.01	72.54	0.55	0.70
Ag_{15}^+ (10)	30.02	42.37	63.72	0.57	0.44
Ag_{15}^+ (11)	28.81	34.17	75.68	0.52	0.72
Ag_{15}^+ (12)	31.31	31.42	74.35	0.59	0.73

We next plotted the HOMO-LUMO gap versus Δ, the deformation coefficient defined in equation 1, in Fig 3. A notable correlation is seen between the deformation coefficient and the HOMO-LUMO gap. Isomer **1** has both the lowest Δ and the largest HOMO-LUMO gap, further demonstrating that a geometric effect has a significant effect on the electronic stability of the Ag_{15}^+ clusters. A low value of Δ indicates that the clusters are more oblate, with the spread along the shortest coordinate axis being much lower than the spread along the other two axes. This shows that clusters whose geometry are closest to planar with a strong oblate deformation are the one's that have the largest HOMO-LUMO gap. This offers evidence that the large electronic stability of Ag_{15}^+ lies in its geometry.

We next analyze the electronic structure of the Ag_{15}^+ cluster, along with that of Ag_{14} and Ag_{13}^- for comparison. For the purpose of a shell model analysis, we consider Ag_{15}^+ as having 14 valence electrons. Each Ag atom contributes its $5s^1$ electron, and the localized 4d orbitals are excluded. Therefore, all three clusters are considered to have 14 valence electrons. Ag_{13}^- has been found to have an unusually large HOMO-LUMO gap among the anion clusters, and an enhanced resistance to oxygen etching.[4] Ag_{14} is found to have a relatively large HOMO-LUMO gap for a cluster with a non-magic number of electrons.[35] Fig 4 shows the geometry and electronic structure of the three clusters with 14 valence electrons. The side view of the geometry shows that all three clusters have an extreme oblate distortion. All three clusters have 4 orbitals with delocalized character within 0.6 eV of the HOMO, and all four orbitals may be assigned as a 2S orbital, a 1P orbital, and 2 1D orbitals. These orbitals are inconsistent with the ordering predicted by the Jellium model, which should order as $|1S^2|\ 1P^6|1D^{10}|2S^2|$. Instead the electronic structure is found to be $|1S^2|1P^4|1D^4 1P^2 2S^2||1D^4 1F^4|$, where the observed gaps are indicated by vertical lines, and the gap between HOMO and LUMO is indicated by a double vertical line. These results indicate that a splitting of the subshells is observed in Ag_{15}^+, Ag_{14}, and Ag_{13}^-. We have found that the subshells of the spherical jellium model have experienced significant splitting, and the cluster geometry has a strongly oblate distortion for all three electronically stable Silver clusters with 14 valence electrons.

The lowest energy structure of Ag_{15}^+ may be described as an oblate structure which is severely squashed along one axis. The effect of such a deformation on the electronic structure may be qualitatively predicted using concepts from crystal field theory. As shown in Fig. 5, we begin with the spherical jellium model, and we will only consider the 1P, 1D and 2S orbital whose splitting are likely to play a role in a 14 valence electron cluster. We will next consider a strong oblate distortion along the z axis, destabilizing all orbitals which extend along that axis. First, we find that the $1P_x$ and $1P_y$ orbitals are slightly stabilized, but the $1P_z$ orbital is destabilized and is now nearly a frontier orbital. Next we consider the splitting of the 1D orbitals and find that the D_{xy} and $D_{x^2-y^2}$ orbitals are most stable. The D_{xz} and D_{yz} orbitals are pushed up in energy due to their extension along the z axis, and the D_{z^2} orbital is pushed up most in energy and is almost completely destabilized. Lastly, the $2S^2$

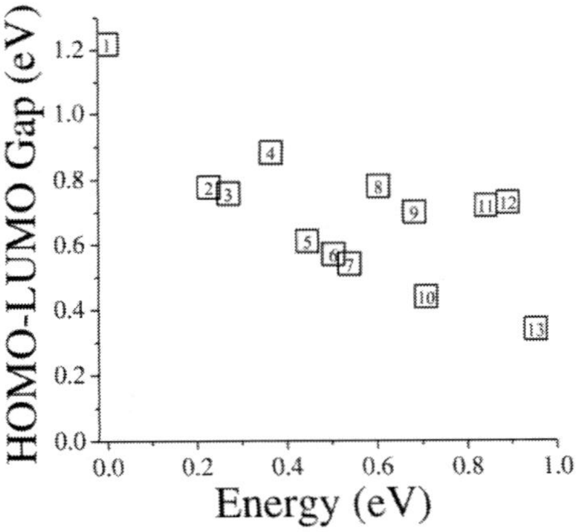

Figure 2. The HOMO-LUMO Gap and relative total energy of various structures of Ag_{15}^+.

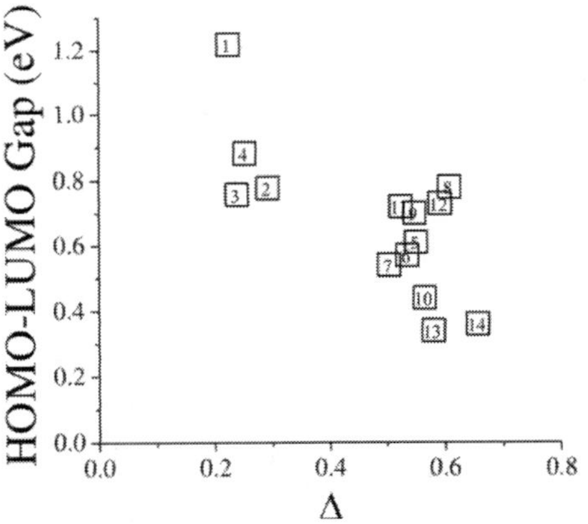

Figure 3. The HOMO-LUMO gap and a deformation coefficient (Δ) plotted against each other.

orbital is only weakly affected by the strong oblate distortion and is expected to be a frontier or near frontier orbital. We note here that the 2S and the D_z^2 orbital are both of the same symmetry group in an oblate distortion such that they hybridize. Next we compare our predicted splitting with the actual molecular orbitals in Ag_{15}^+. We find that our intuitive model gives effectively the same electronic structure as is found in Ag_{15}^+. The $1P_z$ orbital is pushed up to be 0.42 eV below the HOMO, consistent with the expected increase in energy. The D_{xy} and $D_{x^2-y^2}$ orbitals both lie 0.48 eV below the HOMO and are the most stable 1D orbitals. The HOMO is found to be a 2S orbital. The LUMOs are found to be the $1D_{xz}$ and $1D_{yz}$ orbitals at 1.22 and 1.23 eV above the HOMO. The destabilized $1D_z^2$ is pushed up to 2.68 eV above the HOMO as predicted by the cluster crystal field theory.

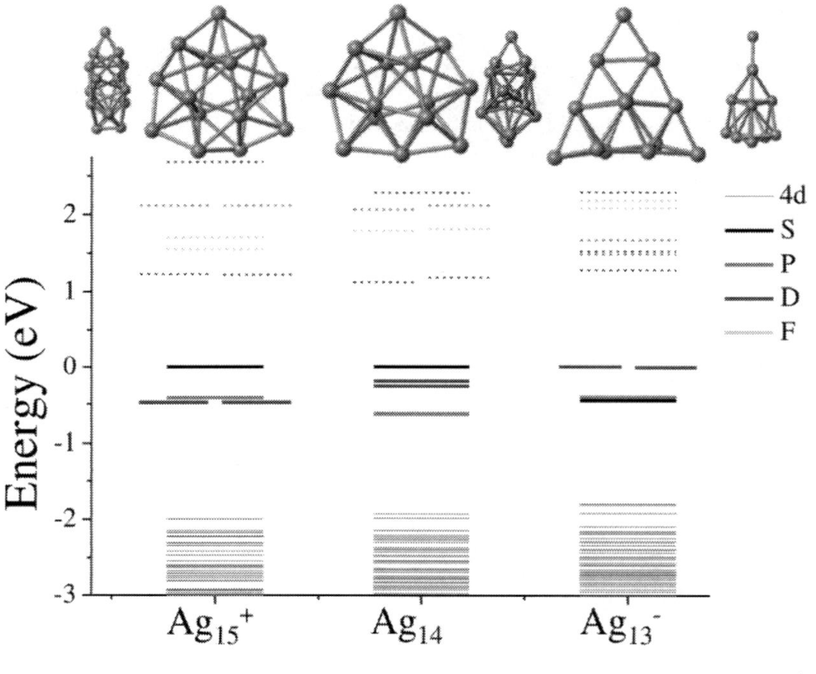

Figure 4. Geometry (Top view and side view) and electronic structure of Ag_{15}^+, Ag_{14}, and Ag_{13}^-. The orbitals with significant contributions from the Ag 5s orbitals are assigned subshells, while 4d orbitals are marked in gray.

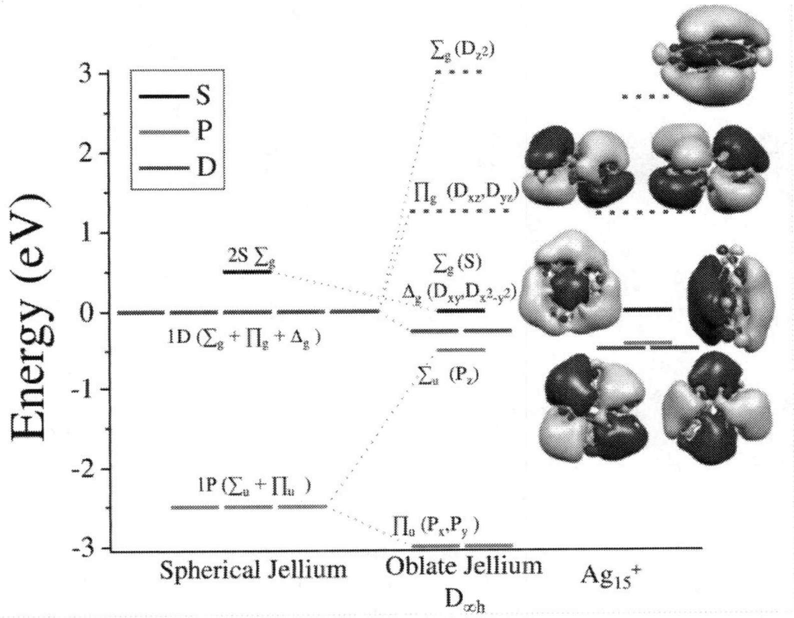

Figure 5. 1P, 1D and 2S orbitals predicted by the spherical Jellium model, their predicted splitting based on a strong oblate distortion, and the isosurfaces for the molecular orbitals for Ag_{15}^+. Note that the 1F and 2P orbitals have been left out for clarity.

To further test our hypothesis that a strong oblate distortion will result in silver clusters with 14 valence electrons having a large HOMO-LUMO gap through the similar splitting , we have extended our molecular orbital analysis to Ag_{14} and Ag_{13}^- in Fig. 6. We find that the electronic structure of Ag_{14} and Ag_{13}^- is nearly the same as that of Ag_{15}^+. The $1P_z$ orbital is pushed up to be 0.63 eV below the HOMO in Ag_{14} and 0.41 eV below the HOMO of Ag_{13}^-, both of which are consistent with the splitting caused by the strong oblate distortion. The D_{xy} and $D_{x^2-y^2}$ orbitals lie 0.27 eV and 0.20 eV below the HOMO in Ag_{14}, and is the HOMO and a second orbital 0.01 eV below the HOMO in Ag_{13}^-. The HOMO of Ag_{14} is found to be a 2S orbital, while in Ag_{13}^- it lies 0.46 eV below the HOMO. The LUMOs are found to be the $1D_{xz}$ and $1D_{yz}$ orbitals at 1.12 and 1.18 eV above the HOMO for Ag_{14}, and 1.28 and 1.52 eV above the HOMO in Ag_{13}^-. The destabilized $1D_z^2$ is pushed up to 2.28 eV above the HOMO in both the Ag_{14} and Ag_{13}^- clusters. The splitting of the 1P, 1D, and 2S orbitals in the Ag_{15}^+, Ag_{14}, and Ag_{13}^- cluster may all thus be explained by the strong oblate distortion which opens up a large HOMO-LUMO gap, resulting in an unusually electronicall stable cluster with a nonmagic number of electrons.

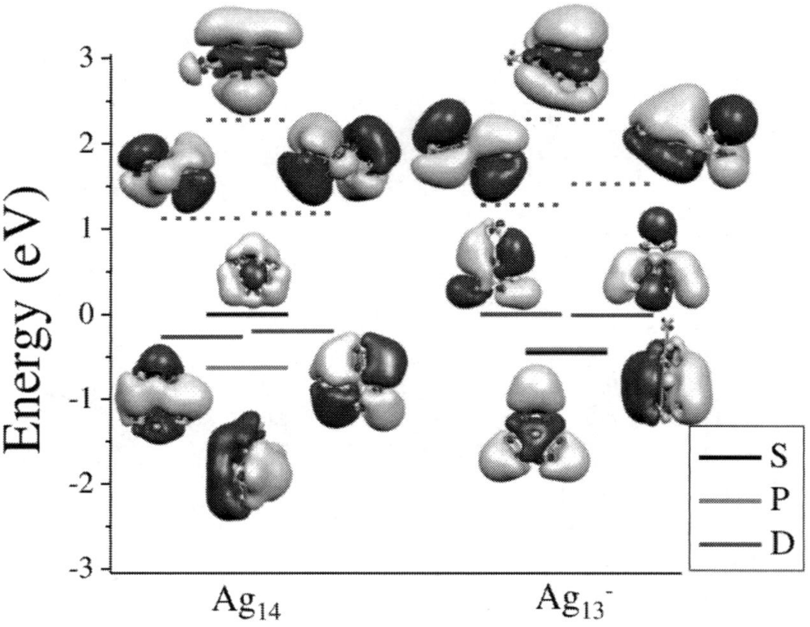

Figure 6. The observed position and isosurfaces of the molecular orbitals for Ag_{14} and Ag_{13}^{-} with the color denoting the assigned subshell. Note that the 1F and 2P orbitals have been left out for clarity.

4. Conclusions

We have demonstrated that Ag_{15}^{+} is resistant to reactivity with oxygen due to its large HOMO-LUMO gap, despite not having a magic number of electrons. The cluster is found to have a highly oblate structure, and the importance of this distortion is demonstrated by showing a correlation between an oblate deformation coefficient and the HOMO-LUMO gap of the different structures of Ag_{15}^{+}. The unusually large HOMO-LUMO gap may be understood qualitatively as a crystal field like splitting of the jellium orbitals, in which the $1P_z$, 2S, D_{xy} and $D_{x^2-y^2}$ orbitals are occupied as near frontier orbitals. An analysis of the molecular orbitals of Ag_{15}^{+}, Ag_{14} and Ag_{13}^{-} all reveal a similar electronic structure and a strongly oblate geometry. This work further demonstrates the interplay between the electronic structure and electronic structure and reveals how simple modifications of the spherical jellium model allow the intuitive understanding of cluster phenomenon. The present results also support our earlier finding that the spin accommodation plays a determining role in the reactivity of silver clusters with oxygen.

Acknowledgments
This material is based upon work supported by the Air Force Office of Scientific Research through a MURI Grant award No. FA9550-08-1-0400. SNK is grateful to Dr. C. Bréchignac for drawing his attention to this problem and for all the discussions.

References

[1] Schmidt M, Masson A and Bréchignac C 2003 *Phys. Rev. Lett.* **91** 243401
[2] Schmidt M, Masson A and Bréchignac C 2005 *J. Chem. Phys.* **122** 134712
[3] Luo Z, Gamboa GU, Smith JC, Reber AC, Reveles JU, Khanna SN and Castleman Jr. AW 2012

J. Amer. Chem. Soc. **134** 18973-78
[4] Reber AC, Khanna SN, Roach PJ, Woodward WH and Castleman Jr. AW 2007 *J. Amer. Chem. Soc.* **129** 16098-101
[5] Hagen J, Socaciu LD, Le Roux J, Popolan D, Bernhardt TM, Wöste L, Mitrić R, Noack H and Bonačić-Koutecký V 2004 *J. Amer. Chem. Soc.* **126**, 3442-43
[6] Bernhardt TM, Hagen J, Lang SM, Popolan DM, Socaciu-Siebert LD and Wöste L **2009** *J. Phys. Chem. A* **113** 2724-33
[7] Klacar S, Hellman A, Panas I and Grönbeck H 2010 *J Phys. Chem. C* **114** 12610-17.
[8] Reber AC, Claybourne PA, Reveles JU, Khanna SN, Castleman Jr. AW and Ali A 2006 *Nano Letters* **6** 1190-5
[9] Bergeron DE, Roach PJ, Castleman Jr. AW, Jones NO and Khanna SN 2005 *Science* **307** 231-235.
[10] Roach PJ, Woodward WH, Reber AC, Khanna SN and Castleman Jr. AW 2010 *Phys. Rev. B* **81** 195404
[11] Luo Z, Grover CJ, Reber AC, Khanna SN and Castleman Jr. AW 2013 *J. Amer. Chem. Soc.* **135** 4307-13
[12] Schmidt M, Cahuzac Ph and Bréchignac C 2003 *J. Chem. Phys.* **118** 10956-62
[13] Wu YN, Schmidt M, Leyrnier J, Cheng H-P, Masson A, and Bréchignac C 2012 *J. Chem. Phys.* **136** 024314
[14] Yuan X, Liu, L, Wang X, Yang M, Jackson KA and Jellinek J 2011 *J. Phys. Chem. A* **115** 8705-12.
[15] Socaciu-Siebert LD, Hagen J, Le Roux J, Popolan D, Vaida M, Vadja S, Bernhardt TM and Wöste L 2005 *Phys. Chem. Chem. Phys.* **7** 2706-9
[16] Socaciu LD, Hagen J, Heiz U, Bernhardt TM, Leisner T and Wöste L 2001 *Chem. Phys. Lett.* **340** 282-8
[17] Bernhardt TM 2005 *Int. J. Mass Spectrom.* **243** 1-29.
[18] Socaciu LD, Hagen J, Le Roux J, Popolan D, Bernhardt TM, Wöste L and Vajda SJ 2004 *Chem. Phys.* **120** 2078-81
[19] Burgert R, Schnockel H, Grubisic A, Li X, Stokes ST, Bowen KH, Gantefor GF, Kiran B and Jena P 2008 *Science* **319** 438-442
[20] Knight W, Clemenger K, de Heer WA, Saunders WA, Chou MY and Cohen ML. 1984 *Phys Rev. Lett.* 52 2141-43
[21] Clemenger K 1985 *Phys. Rev. B* **32** 1359-62
[22] Khanna SN and Jena P. 1995 *Phys. Rev. B* **51** 13705-16
[23] Castleman Jr. AW and Khanna SN. 2009 J. Phys. Chem. C 113 2664-75
[24] Leuchtner RE, Harms AC and Castleman Jr AW. 1991 J Chem Phys **94** 1093-1101
[25] Roach PJ, Woodward WH, Castleman Jr. AW, Reber AC and Khanna SN. 2009 *Science* **323** 492-5
[26] Reber AC, Khanna SN, Roach PJ, Woodward WH and Castleman Jr. AW 2010 *J. Phys Chem. A* **114** 6071-81
[27] Reber AC, Roach PJ, Woodward WH, Khanna SN and Castleman Jr. AW. 2012 *J. Phys. Chem. A* **116** 8085-91
[28] Abreu MB, Powell C, Reber AC and Khanna SN. 2012 J. Amer. Chem. Soc.**134** 20507-20512.
[29] Köster AM, Geudtner G, Calaminici P, Casida ME, Dominguez VD, Flores-Moreno R, Gamboa GU, Goursot A, Heine T, Ipatov A, Janetzko F, del Campo JM, Reveles JU, Vela A, Zuniga-Gutierrez B and Salahub DR 2011 deMon2k, Version 3; The deMon Developers: Cinvestav, Mexico City
[30] Perdew JP, Burke K and Enzerhof M, 1996 *Phys. Rev. Lett* **77** 3865-68
[31] Andrae D, Haeussermann U, Dolg M, Stoll H and Preuss H 1990 *Theor. Chim. Acta* **77** 123-41
[32] Calaminici P, Janetzko F, Köster AM, Mejia-Olvera R and Zuniga-Gutierrez BJ 2007 *Chem. Phys.* **126** 044108

[33] Reveles JU and Köster AM 2004 J. Comput. Chem. **25** 1109-16
[34] Fernández EM, Soler JM, Garzón IL and Balbás LC 2004 *Phys. Rev. B* **70** 165403
[35] Tian D, Zhang H and Zhao J. 2007 *Solid State Comm.* **144** 174-179
[36] Fournier R. 2001 *J Chem Phys* **115** 2165-2177
[37] Weis P, Bierweiler T, Gilb S and Kappes MM. 2002 Chem Phys Lett 355 355-364
[38] Blom MN, Schooss D and Kappes MM. 2006 J Chem Phys **124** 244308
[39] Xiang H, Wei SH and Gong X 2010 *J. Amer. Chem. Soc.* **132** 7355-60
[40] Biltek SR, Mandal S, Sen A, Reber AC, Pedicini AF, and Khanna SN. 2013 *J. Amer. Chem. Soc.* **135** 26-29
[41] Sun Y, Balasubramanian K, Rao TUB and Pradeep T 2011 *J Phys. Chem. C* **115** 20380-87
[42] Krishnamurty S, Joshi, K, Zorriasatein, S and Kanahere DG. 2007 *J. Chem. Phys.* **127** 054308
[43] Pal R, Cui L-F, Bulusu S, Zhai H-J, Wang L-S and Zeng XC 2008 *J. Chem. Phys.* **128** 024305

Journal of Physics: Conference Series **438** (2013) 012003 doi:10.1088/1742-6596/438/1/012003

The h-BN surface effect on CO oxidation reaction catalyzed by supported gold atom

Min Gao[1], **Andrey Lyalin**[1,2,3,4] **and Tetsuya Taketsugu**[1,2,3]

[1] Department of Chemistry, Faculty of Science, Hokkaido University, Sapporo 060-0810, Japan
[2] Center for Strategic Utilization of Elements, Graduate School of Science, Hokkaido University, Sapporo 060-0810, Japan
[3] Elements Strategy Initiative for Catalysts and Batteries (ESICB), Kyoto University, Katsura, Kyoto 615-8520, Japan

E-mail: lyalin@mail.sci.hokudai.ac.jp

Abstract. The mechanism of CO oxidation by O_2 on Au atoms supported on the pristine and defected hexagonal boron nitride (h-BN) surface has been studied theoretically using density functional theory. Two possible routes for catalytic oxidation are considered. The first route consists in a preliminary dissociation of the adsorbed O_2 followed by consequential oxidation of a reactant molecule by atomic oxygen. Although the presence of h-BN surface can change the O_2 dissociation barrier, it remains relatively high. The second route is a direct oxidation reaction between the activated molecular oxygen and the reactant. We have found two different pathways for CO oxidation: a two-step pathway where two CO_2 molecules are formed independently, and a self-promotion pathway where oxidation of the first CO molecule is promoted by the second CO molecule. Interaction of Au with the defect-free and defected h-BN surface considerably affects the CO oxidation reaction pathways and barriers. Therefore, Au supported on the h-BN surface (pristine or defected) cannot be considered as pseudo-free atom and support effects have to be taken into account, even when the interaction of Au with the support is weak.

1. Introduction

Since the pioneering work of Haruta on oxidation of carbon monoxide by small gold nanoparticles supported by metal oxides [1] it has become clear that gold can exhibit high catalytic activity in various oxidation reactions including the oxidation of carbon monoxide, aerobic alcohol oxidation, the direct synthesis of hydrogen peroxide, alkene epoxidation, etc. [2–10]. The catalytic properties of gold are strongly size-dependent and emerge when the size of particles decreases up to 1-5 nm, while larger particles and the bulk form of gold are catalytically inert [1, 2, 5]. The enormous interest in gold nanocatalysis is stipulated by the fact that gold nanoparticles are active and selective even at room temperatures that makes them unique catalysts for many industrial applications.

It is well established that there are several factors controlling catalytic activity of gold [10–14]. Apart from the strong size dependence of the catalytic activity of gold clusters one of the most important factors in gold nanocatalysis is the support effect [15, 16]. Thus, it was demonstrated that the so-called active metal oxide supports (such as, MgO, TiO_2, Al_2O_3, Fe_2O_3,

[4] On leave from: V. A. Fock Institute of Physics, St Petersburg State University, 198504 St Petersburg, Petrodvorez, Russia

etc.) can considerably influence the catalytic properties of gold clusters and modify (enhance or suppress) their reactivity [2, 9, 12, 17]. Such supports usually stabilize gold particles and can modify considerably their geometric structure and morphology [18–22], which in turn can affect catalytic properties of the supported particles. Charge transfer from the metal oxide support to the gold results in formation of the highly reactive charged gold clusters [17, 23]. Catalytic activity of the supported clusters can also be promoted by defects in the support. Defects can trap the metal cluster and enhance charge transfer between the support and the cluster. It was demonstrated that Au_8 clusters supported on the MgO(100) surface rich of F-center defects show high catalytic activity, while Au_8 deposited on the defect-poor MgO(100) surface are inert [17, 24].

One of the simplest and the most studied catalytic reaction on gold clusters is a low-temperature CO oxidation by molecular oxygen [1, 2, 10, 12, 17, 24–31]. It has been demonstrated that dispersed gold supported on Co_3O_4, NiO, TiO_2, and Fe_2O_3 possesses high catalytic activity for CO oxidation [32–35]. These supports affect dramatically the catalytic activity of gold clusters. The choice of support, the method of preparation and the pretreatment before use are very important factors controlling the reactivity [36]. The contact structure between the supported cluster and the surface is another important factor in the CO oxidation reaction [9, 32]. It has been also suggested that adsorption of O_2 and CO to the gold particles of finite size becomes stronger at low-coordinated sites [17, 24, 29, 37–42].

Activation of the adsorbed O_2 is a crucial step in CO oxidation reaction. A negative charge accumulated on the gold can weaken the O–O bond and activate oxygen molecule for further catalytic reaction; see, *e.g.*, Refs. [43–49] and references therein. On the other hand, the positive charge accumulated on the gold can promote adsorption of CO and hydrocarbons [7, 50]. There are two possible routes for catalytic oxidation on gold clusters. The first route consists in a preliminary dissociation of the adsorbed O_2 followed by consequential oxidation of a reactant molecule by atomic oxygen. The second route is a direct oxidation reaction between the activated molecular O_2 and the reactant. It has been demonstrated that CO oxidation reaction catalyzed by gold clusters can occur via Langmuir-Hinshelwood (LH) or Eley-Rideal (ER) mechanisms, depending on cluster charge. Thus the neutral Au_{10} can catalyze oxidation of CO via Langmuir-Hinshelwood mechanism even below room temperature [41]. CO oxidation on anionic gold clusters occurs via Eley-Rideal mechanism [38, 51, 52], while CO oxidation on cationic gold atoms can occur via both Eley-Rideal and Langmuir-Hinshelwood mechanisms [52]. Therefore, the charge state of gold particles (which can be controlled by the support) can considerably influence their reactivity and define the reaction mechanism [47, 48, 53].

On the other hand it is commonly accepted that the inert supports, such as SiO_2 or h-BN, do not affect the electronic and geometry structure of the supported clusters. Therefore in the absence of strong support interactions such clusters can be considered as pseudo-free [3, 54]. This suggestion is widely used to study catalytic properties of gold clusters free of support effects. However in our recent works we have demonstrated that adsorption, activation and dissociation of O_2 on h-BN supported gold atoms and dimers can be affected by the interaction with the inert h-BN support via electron pushing and donor/acceptor mechanisms [48, 55, 56]. We have shown that even weak interaction of Au and Au_2 with the defect-free inert h-BN surface can have an unusually strong influence on the binding and catalytic activation of the molecular oxygen [56]. This effect occurs due to the mixing of the 5d orbitals of the supported Au and Au_2 with the N-p_z orbitals. Although the defect-free h-BN surface does not act as a good electron donor for the supported O_2-Au, it promotes an electron transfer from the Au to O_2, pushing electrons from the gold to the adsorbed oxygen. In the case of defected h-BN surface Au and Au_2 can be trapped effectively by N or B vacancy and impurity point defects. Strong adsorption on the surface defects is accompanied by charge tranhsfer to/from the adsorbate which can affect catalytic activity of gold [48]. Very recently adsorption of small gold clusters on the h-BN/Rh(111) nanomesh has been studied theoretically [57]. It was noticed that all Au atoms in these clusters are negatively charged and thus might be catalytically active [57]. However to the best of our knowledge there were no attempts to study theoretically the specific catalytic reactions on h-BN supported Au clusters.

In the present paper we fill this gap and perform systematic theoretical investigation of the process of CO oxidation by O_2 on the h-BN supported gold atoms as the simplest model reaction. It is found that there are two different pathways for CO oxidation: a two-step pathway where two CO_2 molecules are formed independently, and a self-promotion pathway where oxidation of the first CO molecule is promoted by the second CO molecule. Interaction of Au with the defect-free and defected h-BN surface considerably affects the CO oxidation reaction pathways and barriers. Therefore, Au supported on the h-BN surface (pristine or defected) cannot be considered as pseudo-free atom and support effects have to be taken into account, even when the interaction of Au with the support is weak. It is demonstrated that the inert h-BN support can affect the process of CO oxidation and corresponding reaction barriers. It is shown that presence of h-BN support can affect the CO oxidation barriers and overall probability of oxidation reaction. Therefore h-BN cannot be considered as an inert support for Au.

2. Theoretical Methods

The calculations are carried out using density-functional theory (DFT) with the gradient-corrected exchange-correlation functional of Wu and Cohen (WC) [58] as implemented in the PWSCF package [59]. The choice of WC functional is stipulated by the fact that it allows to reproduce correctly the lattice constants, crystal structures and surface energies of solids with layered structures like graphite or h-BN, whose distances between the layers are determined by rather weak interactions [60]. The WC functional was successfully used in our previous works to describe adsorption of Au and Au_2 on the defect-free and defected h-BN surface as well as adsorption, activation and dissociation of molecular oxygen on h-BN supported Au particles [48, 55, 56, 61].

A plane-wave basis sets and Troullier-Martins norm-conserving pseudopotentials are used [62]. Relativistic effects are taken into account for Au via scalar-relativistic pseudopotentials. All calculations are spin polarized. The energy cutoff of 80 Ry is chosen to guarantee convergence of the total energies and forces. The self-consistency of the density matrix is achieved with a tolerance of 10^{-9}. A climbing image nudged elastic band (CI-NEB) method has been used for finding transition states and minimum energy pathways [63]. We have checked that the total spin of the whole system does not change at transition states.

Periodic boundary conditions are used for all systems, including free molecules and clusters. In the case of finite size systems the supercell was chosen to be large enough to make intermolecular interactions negligible. The h-BN surface is represented by the two-layer slab containing 6×6 unit cells (72 units of BN per slab). In all calculations the bottom layer in the slab is fixed, and all other atoms are fully relaxed. The periodically replicated slabs are separated by the vacuum region of 15 Å in the (001) direction. Only the Γ point is used for sampling the Brillouin zone due to the large size of the supercell.

In order to obtain the most stable configuration of the adsorbed O_2 and CO we have created a large number of initial geometries by adding adsorbants in different nonequivalent positions (up to 20 in each case) on the surface of the pristine and defected h-BN with the trapped Au atom. These structures have been optimized without any geometry constraints. The similar approach has been successfully used in our previous works to study structure of metal clusters [64–66] as well as adsorption and dissociation of O_2, H_2, and C_2H_4 molecules on free and supported gold clusters [46–48, 50, 55, 56, 67].

The binding energy of a molecule M (O_2 or CO) to a free or supported Au atom is defined as

$$E_b = E_{tot}(\mathrm{M}) + E_{tot}(\mathrm{Au}) - E_{tot}(\mathrm{M-Au}), \qquad (1)$$

where $E_{tot}(\mathrm{M}-\mathrm{Au})$ denotes the total energy of the complex system M − Au, while $E_{tot}(\mathrm{M})$ and $E_{tot}(\mathrm{Au})$ are the total energy of non-interacting M and Au species, respectively. Here M denotes O_2 or CO molecule, while Au corresponds to a free or supported Au atom.

3. Theoretical Results

There are two possible routes for catalytic oxidation on gold clusters. The first route consists in a preliminary dissociation of the adsorbed O_2 followed by the consequential oxidation of a reactant molecule by the atomic oxygen. The second route is a direct oxidation reaction between the activated molecular oxygen and the reactant. In the present work both of the possible processes are considered.

3.1. Dissociation of O_2 on h-BN supported gold atoms

First, we have studied the O_2 dissociation on Au atoms trapped by pristine and defected h-BN surfaces. Figure 1 presents potential energy profiles for O_2 dissociation on supported h-BN supported Au. It can be seen that oxygen molecule adsorbs on top of the Au atom supported on the defect-free h-BN surface and trapped by the V_B defect, but bridges Au and the surface B atom in the case of O_2 adsorption on the Au/V_N@h-BN center. Interaction of Au with the defect-free h-BN support stabilizes the initial state of O_2 adsorbed molecularly on Au/h-BN. As a result, the barrier for O_2 dissociation on Au/h-BN become 2.92 eV. The barrier for O_2 dissociation on a defected-free h-BN supported Au atom is too high and the dissociated configuration of O_2 is not favorable energetically; thus O_2 dissociation cannot occur spontaneously. The barrier for O_2 dissociation on Au/V_B@h-BN adsorption center is 2.48 eV, which lower than those calculated for Au and Au/h-BN adsorption centers, however the final dissociated state of oxygen is not favorable energetically. Figure 1 demonstrates that the potential energy profile calculated for O_2 dissociation on the Au/V_N@h-BN adsorption center differs considerably from those obtained for Au, Au/h-BN, and Au/V_B@h-BN centers. It is seen from Fig. 1 that O_2 dissociation on Au/V_N@h-BN occurs in two steps with formation of one intermediate configuration. The energy of transition states TS1 and TS2 are much lower if compared with those obtained for free Au atom and Au adsorbed on h-BN V_B@h-BN. Figure 1 also demonstrates the final configuration of the dissociated oxygen on Au/V_N@h-BN is energetically favorable. Thus, dissociation of O_2 can occur even on a single Au atom if it is trapped by the V_N defect on h-BN surface. Although the presence of h-BN support can lower the barrier of O_2 dissociation, it remains relatively high and O_2 dissociation cannot occur spontaneously with an exception of Au/V_N adsorption center [55]. It has been shown that the barrier of O_2 dissociation on a single Au atom is too high. Therefore, in the next step, we focus on the mechanism of direct oxidation of CO by activated O_2.

3.2. CO oxidation with activated O_2 on h-BN supported Au atoms

The optimized geometries of CO adsorbed on Au/h-BN, Au/V_B@h-BN and Au/V_N@h-BN centers are shown in Fig. 2. It is seen from Fig. 2 that CO adsorption on Au/h-BN, Au/V_B@h-BN and Au/V_N@h-BN centers, CO always binds on top of Au atom, forming Au−C bond.

This result is consistent with the previous study on CO adsorption on small cationic, neutral, and anionic Au_n (n = 1-6) clusters where it was found that among the various possible CO adsorption sites, the on-top (one-fold coordinated) sites are the most favorable, irrespective of the cluster charge [68].

Figure 3 demonstrates possible reaction pathways for this process. The results of our calculations demonstrate that CO adsorbs on Au/h−BN with the binding energy of 1.03 eV, which is larger that the binding energy of O_2 to Au/h−BN. After adsorption of the first CO on Au/h−BN two different pathways for O_2 adsorption are possible, as it is shown in Fig. 3(a). The first is the coadsorption of O_2 on Au without direct interaction with CO (Langmuir-Hinshelwood

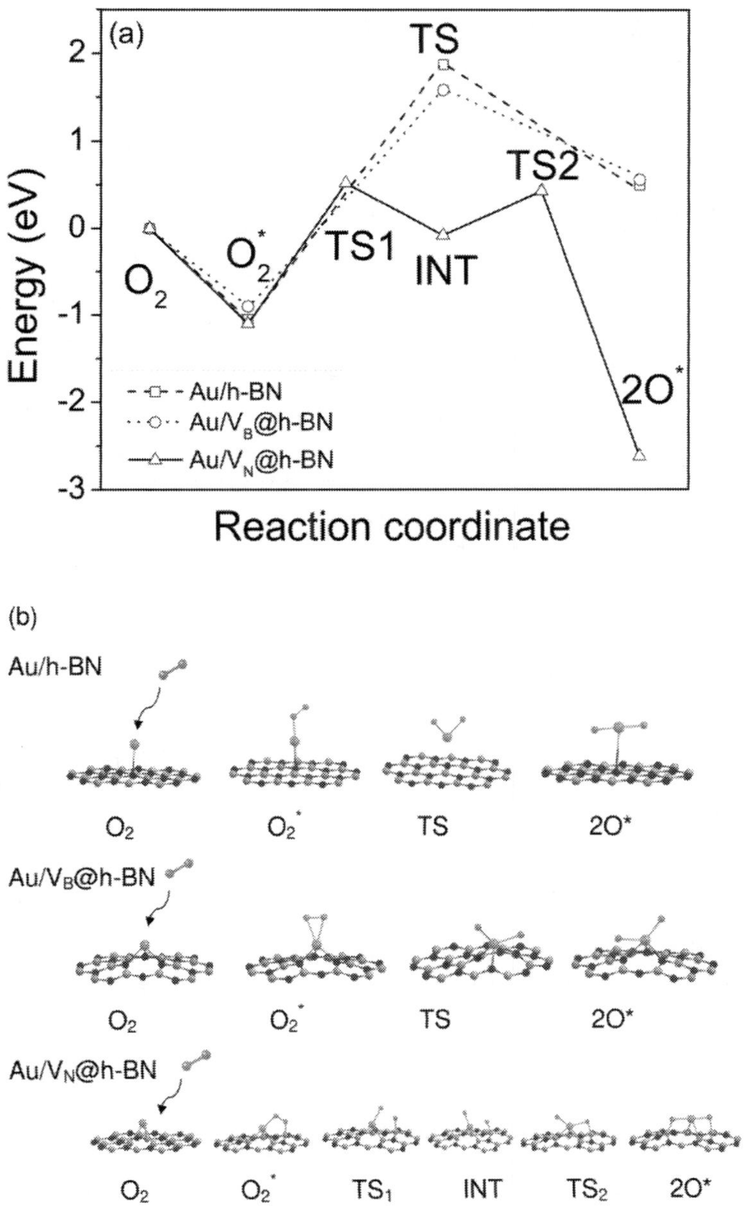

Figure 1. (a) Potential energy profiles for O_2 dissociation on Au/h-BN. The zero energy corresponds to the noninteracting O_2 and Au/h-BN, Au/V_B@h-BN, or Au/V_N@h-BN adsorption centers. The relative energies are given in eV. Adsorbed configurations of oxygen are marked by the star (*); (b) Optimized geometries of transition (TS) and intermediate (INT) states (only part of the slab is shown).

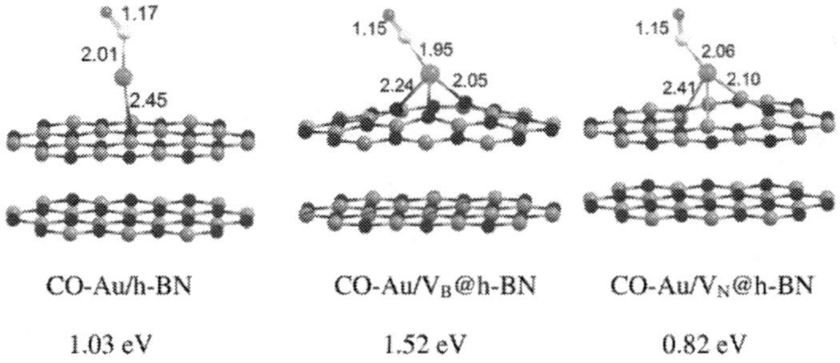

Figure 2. Adsorption of CO on free and supported Au atoms on defect-free h-BN surface and h-BN surface with boron (V_B) and nitrogen (V_N) vacancy defects (only part of the slab is shown).

mechanism) resulting in formation of the intermediate A_1. This is the barrierless process. After A_1 is formed it can be transformed to the intermediate A_2 via transition state TS_1 with a barrier of 0.98 eV. The second pathway consists in a direct O_2 attack of Au−CO and formation of the A_2 intermediate via transition state $TS_1^{'}$ with the barrier of 0.04 eV.

In order to obtain the final product CO_2 from the intermediate A_2 there are two possibilities: (i) a two-step process with formation of the first CO_2 product via transition state $TS_2^{'}$ with barrier of 0.67 eV, and consequent oxidation of the second CO by atomic oxygen on AuO/h-BN; (ii) a self-promoting path, when the second CO reacts with intermediate B_2 forming $2CO_2$ via transition state TS_2 with the barrier of 0.49 eV. It is important to note that in the latter pathway the oxidation of the first CO molecule is promoted by the second CO molecule (the self-promotion effect) [27, 30].

Let us consider CO oxidation on Au trapped by the nitrogen and boron point defects on h-BN surface. As it was discussed above, the boron vacancy V_B plays a role of the electron acceptor. Therefore, Au trapped by V_B@h-BN is positively charged. The positive charge localized on Au promotes CO adsorption. After the first adsorption of CO on Au/V_B@h-BN the intermediate C_1 can be formed directly without any barrier. After intermediate B_1 is formed there are two possible pathways to get $2CO_2$ as the final product. These pathways are similar to those obtained for the unsupported Au. However, in the case of Au trapped by V_B@h-BN the intermediate C_1 disintegrates, releasing one CO_2 with a barrier of 0.58 eV. After that the second CO attacks O-Au/V_B@h-BN and forms CO_2 overcoming the small barrier of 0.07 eV. This barrier is related to formation of CO-Au-O/V_B@h-BN complex, where both CO and O reactants are attached to the supported Au atom. After that step, oxidation of CO by O occurs spontaneously without any barrier. This path is energetically favorable in comparison with the self-promoting pathway. It is important to note that as the CO_2 product departs from the Au/V_B@h-BN center the total spin of the whole system changes from singlet to triplet. However this change occurs on the down-hill slope of the reaction path and does not affect the barrier.

For the CO oxidation reaction on the Au/V_N@h-BN center, Au is negatively charged, favoring O_2 adsorption and activation as a first step. After that CO can react with the dissociated O atom with a barrier of 1.21 eV, or with the activated O_2 molecule. Intermediate C_1 can be formed directly without any barrier after O_2 activation. The route from D_1 to the final products consists of two paths: a two-step pathway with a barrier of 1.98 eV and a self-promoting pathway with a barrier of 0.49 eV. Thus in the case of Au/V_N@h-BN center the self-promoting pathway is energetically favorable in comparison with a two-step pathway. The barrier of CO oxidation with dissociated O atom is higher than the case of reaction with activated O_2 molecule. The

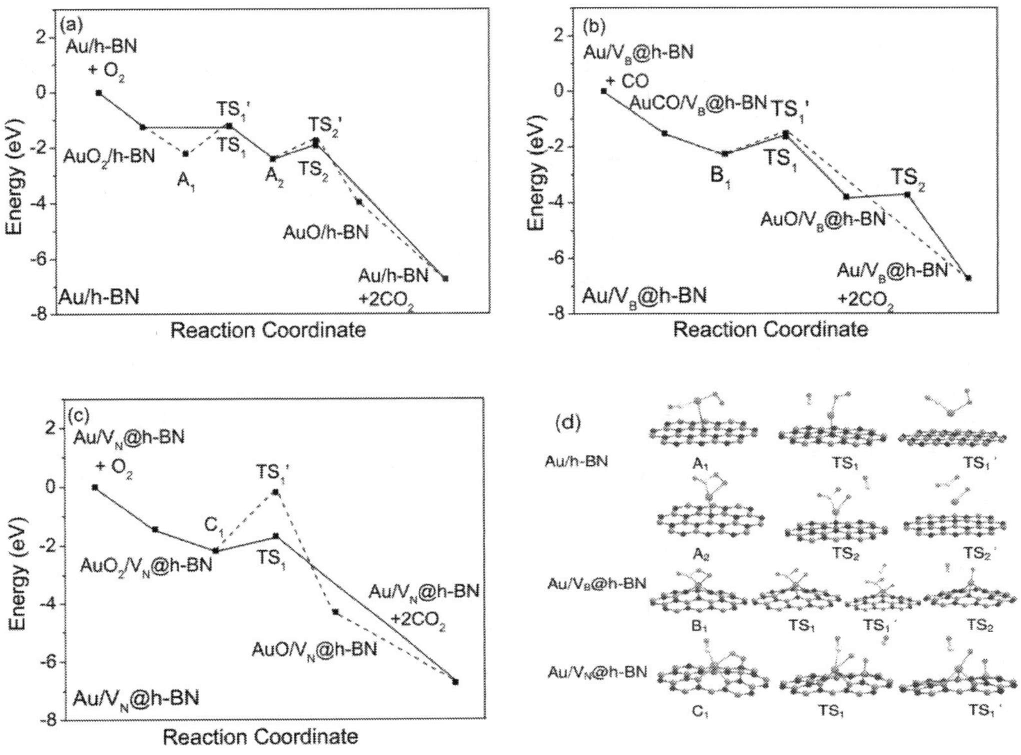

Figure 3. Reaction pathways for the CO oxidation by O_2 on Au/h-BN (a); Au/V_B@h-BN (b); Au/V_N@h-BN (c) adsorption centers. The key intermediate and transition state structures are shown in (d).

self-promoting pathway is energetically favorable except for the case of Au atom supported on V_N@h-BN.

4. Conclusion

In the present work, we have studied mechanism of CO oxidation reaction on free and h-BN supported gold atom. The pristine h-BN surface and h-BN surfaces with boron and nitrogen vacancies have been considered. Various pathways of CO oxidation reaction by molecular oxygen are studied. Although the presence of h-BN support can lower the barrier of O_2 dissociation, it remains relatively high and O_2 dissociation cannot occur spontaneously with an exception of Au/V_N adsorption center. It has been shown that the barrier of O_2 dissociation on a single Au atom is too high. The CO oxidation reaction follows the pathways that CO react with activated O_2 molecule. There are two possibilities: (i) the two-step pathway when two CO_2 molecules are formed independently; (ii) the self-promoting pathway when oxidation of the first CO molecule is promoted by the second CO molecule. CO oxidation reaction follows the self-promoting pathways except for the case of V_B@h-BN center. Interaction of Au with the defect-free and defected h-BN surface considerably affects the CO oxidation reaction pathways and barriers. Our finding leads to an important conclusion that Au supported on the h-BN surface (pristine or defected) cannot be considered as pseudo-free atom.

Acknowledgments
This work was supported by Elements Science and Technology Project of Ministry of Education, Culture, Sports, Science and Technology, Japan (MEXT) on "Nano-hybridized Precious-metal-free Catalysts for Chemical Energy Conversion"; MEXT program "Elements Strategy Initiative to Form Core Rersearch Center" (since 2012); and the JSPS Grant-in-Aid for Scientific Research C. The computations were partly performed using the Research Center for Computational Science, Okazaki, Japan.

References

[1] Haruta M, Kobayashi T, Sano H and Yamada N 1987 *Chem. Lett.* **16** 405–408
[2] Haruta M 1997 *Catal. Today* **36** 153–166
[3] Turner M, Golovko V B, Vaughan O P H, Abdulkin P, Berenguer-Murcia A, Tikhov M S, Johnson B F G and Lambert R M 2008 *Nature* **454** 981–984
[4] Hughes M D, Xu Y J, Jenkins P, McMorn P, Landon P, Enache D I, Carley A F, Attard G A, Hutchings G J, King F, Stitt E H, Johnston P, Griffin K and Kiely C J 2005 *Nature* **437** 1132–1135
[5] Tsunoyama H, Sakurai H, Negishi Y and Tsukuda T 2005 *J. Am. Chem. Soc.* **127** 9374–9375
[6] Landon P, Collier P J, Papworth A J, Kiely C J and Hutchings G 2002 *Chem. Commun.* 2058–2059
[7] Hutchings G J, Hall M S, Carley A F, Landon P, Solsona B E, Kiely C J, Herzing A, Makkee M, Moulijn J A, Overweg A, Fierro-Gonzalez J C, Guzman J and Gates B C 2006 *J. Catal.* **242** 71–81
[8] Chaki N K, Tsunoyama H, Negishi Y, Sakurai H and Tsukuda T 2007 *J. Phys. Chem. C* **111** 4885–4888
[9] Haruta M 2011 *Faraday Discuss.* **152** 11–32
[10] Heiz U and Landman U (eds) 2007 *Nanocatalysis* (Berlin, Heidelberg, New York: Springer)
[11] Bond G C, Louis C and Thompson D T 2006 *Catalysis by Gold* (Imperial College Press: London)
[12] Landman U, Yoon B, Zhang C, Heiz U and Arenz M 2007 *Topics in Catalisis* **44** 145–158
[13] Daté M, Okumura M, Tsubota S and Haruta M 2004 *Angew. Chem. Int. Ed.* **43** 2129–2132
[14] Zhang H, Watanabe T, Okumura M, Haruta M and Toshima N 2012 *Nat. Mater.* **11** 49–52
[15] Bond G C and Thompson D T 2009 *Gold Bull.* **42** 247–259
[16] Bond G C 2011 *Faraday Discuss.* **152** 277–291
[17] Yoon B, Häkkinen H, Landman U, Wörz A S, Antonietti J M, Abbet S, Judai K and Heiz U 2005 *Science* **307** 403–407
[18] Semenikhina V V, Lyalin A, Solov'yov A V and Greiner W 2008 *JETP* **106** 678–689
[19] Dick V V, Solov'yov I A and Solov'yov A V 2011 *Phys. Rev. B* **84**(11) 115408
[20] Mukhamedzyanova D F, Ratmanova N K, Pichugina D A and Kuz'menko N E 2012 *J. Phys. Chem. C* **116** 11507–11518
[21] Escamilla-Perea L, Nava R, Pawelec B, Rosmaninho M G, Peza-Ledesma C L and Fierro J L G 2010 *Appl. Catal. A* **381** 42–53
[22] Pillai U R and Deevi S 2006 *Appl. Catal. A* **299** 266–273

[23] Chen M, Cai Y, Yan Z and Goodman D W 2006 *J. Am. Chem. Soc.* **128** 6341–6346

[24] Sanchez A, Abbet S, Heiz U, Schneider W D, Hakkinen H, Barnett R N and Landman U 1999 *J. Phys. Chem. A* **103** 9573–9578

[25] Haruta M and Daté M 2001 *Appl. Catal. A* **222** 427–437

[26] Kung H H, Kung M C and Costello C K 2003 *J. Catal.* **216** 425–432

[27] Socaciu L D, Hagen J, Bernhardt T M, Wöste L, Heiz U, Häkkinen H and Landman U 2003 *J. Am. Chem. Soc.* **125** 10437–10445

[28] Soares J M C and Bowker M 2005 *Appl. Catal. A* **291** 136–144

[29] Molina L M and Hammer B 2004 *Phys. Rev. B* **69**(15) 155424

[30] Wang F, Zhang D, Xu X and Ding Y 2009 *J. Phys. Chem. C* **113** 18032–18039

[31] Stamatakis M, Christiansen M A, Vlachos D G and Mpourmpakis G 2012 *Nano Lett.* **12** 3621–3626

[32] Haruta M, Tsubota S, Kobayashi T, Kageyama H, Genet M and Delmon B 1993 *J. Catal.* **144** 175–192

[33] Yuan Y, Kozlova A P, Asakura K, Wan H, Tsai K and Iwasawa Y 1997 *J. Catal.* **170** 191–199 ISSN 0021-9517

[34] Grunwaldt J, Kiener C, Wögerbauer C and Baiker A 1999 *J. Catal.* **181** 223–232 ISSN 0021-9517

[35] Haruta M, Yamada N, Kobayashi T and Iijima S 1989 *J. Catal.* **115** 301–309 ISSN 0021-9517

[36] Bond G C and Thompson D T 2000 *Gold Bull.* **33** 41–50

[37] Molina L M and Hammer B 2005 *Appl. Catal. A* **291** 21–31

[38] Yoon B, Häkkinen H and Landman U 2003 *J. Phys. Chem. A* **107** 4066–4071

[39] Lopez N and Nørskov J K 2002 *J. Am. Chem. Soc.* **124** 11262–11263

[40] Mills G, Gordon M S and Metiu H 2002 *Chem. Phys.Lett.* **359** 493–499 ISSN 0009-2614

[41] Remediakis I N, Lopez N and Nørskov J K 2005 *Applied Catalysis A: General* **291** 13 – 20 ISSN 0926-860X

[42] Lemire C, Meyer R, Shaikhutdinov S and Freund H J 2004 *Angew. Chem. Int. Ed.* **43** 118–121 ISSN 1521-3773

[43] Coquet R, Howard K L and Willock D J 2008 *Chem. Soc. Rev.* **37** 2046–2076

[44] Ding X, Li Z, Yang J, Hou J G and Zhu Q 2004 *J. Chem. Phys.* **120** 9594–9600

[45] Fernández E, Ordejón P and Balbás L C 2005 *Chem. Phys. Lett.* **408** 252–257

[46] Lyalin A and Taketsugu T 2009 *J. Phys. Chem. C* **113** 12930–12934

[47] Lyalin A and Taketsugu T 2010 *J. Phys. Chem. Lett.* **1** 1752–1757

[48] Gao M, Lyalin A and Taketsugu T 2011 *Catalysts* **1** 18–39

[49] Okumura M, Kitagawa Y, Yabushita H, Saito T and Kawakami T 2009 *Catalysis Today* **143** 282 – 285

[50] Lyalin A and Taketsugu T 2010 *J. Phys. Chem. C* **114** 2484–2493

[51] Häkkinen H and Landman U 2001 *J. Am. Chem. Soc.* **123** 9704–9705

[52] Kimble M L, Castleman A W, Mitrić R, Bügel C and Bonačić-Koutecký V 2004 *J. Am. Chem. Soc.* **126** 2526–2535

[53] Bürgel C, Reilly N M, Johnson G E, Mitrić R, Kimble M L, Castleman Jr A W and Bonačić-Koutecký V 2008 *J. Am. Chem. Soc.* **130** 1694–1698

[54] Zheng X, Veith G M, Redekop E, Lo C S, Yablonsky G S and Gleaves J T 2010 *Ind. Eng. Chem. Res* **49** 10428–10437

[55] Gao M, Lyalin A and Taketsugu T 2012 *Int. J. Quantum Chem.* URL `http://dx.doi.org/10.1002/qua.24066`

[56] Gao M, Lyalin A and Taketsugu T 2012 *J. Phys. Chem. C* **116** 9054–9062

[57] Koch H P, Laskowski R, Blaha P and Schwarz K 2012 *Phys. Rev. B* **86**(15) 155404

[58] Wu Z and Cohen R E 2006 *Phys. Rev. B* **73** 235116/1–6

[59] Giannozzi P, Baroni S, Bonini N, Calandra M, Car R, Cavazzoni C, Ceresoli D, Chiarotti G L, Cococcioni M, Dabo I, Corso A D, de Gironcoli S, Fabris S, Fratesi G, Gebauer R, Gerstmann U, Gougoussis C, Kokalj A, Lazzeri M, Martin-Samos L, Marzari N, Mauri F, Mazzarello R, Paolini S, Pasquarello A, Paulatto L, Sbraccia C, Scandolo S, Sclauzero G, Seitsonen A P, Smogunov A, Umari P and Wentzcovitch R M 2009 *J. Phys.: Condens. Matter* **21** 395502

[60] Tran F, Laskowski R, Blaha P and Schwarz K 2007 *Phys. Rev. B* **75**(11) 115131

[61] Gao M, Lyalin A and Taketsugu T 2013 *J. Chem. Phys* **138** 034701

[62] Troullier N and Martins J L 1991 *Phys. Rev. B* **43** 1993–2006

[63] Henkelman G, Uberuaga B P and Jónsson H 2000 *J. Chem. Phys.* **113** 9901–9904

[64] Lyalin A, Solov'yov I A, Solov'yov A V and Greiner W 2003 *Phys. Rev. A* **67**(6) 063203

[65] Lyalin A, Solov'yov I A, Solov'yov A V and Greiner W 2007 *Phys. Rev. A* **75**(5) 053201

[66] Lyalin A, Solov'yov A V and Greiner W 2006 *Phys. Rev. A* **74**(4) 043201

[67] Lyalin A and Taketsugu T 2011 *Faraday Discuss.* **152** 185–201

[68] Wu X, Senapati L, Nayak S K, Selloni A and Hajaligol M 2002 *J. Chem. Phys.* **117** 4010–4015

Journal of Physics: Conference Series **438** (2013) 012004 doi:10.1088/1742-6596/438/1/012004

Ultrahigh-sensitive detection of molecules produced in catalytic reactions by uni-atomic-composition bi-element clusters supported on solid substrate

H Yasumatsu[1] and N Fukui[2]
[1]Cluster Research Laboratory, Toyota Technological Institute:
In [2]East Tokyo Laboratory, Genesis Research Institute, Inc.
717–86 Futamata, Ichikawa, Chiba 272-0001, Japan

E-mail: yasumatsu@clusterlab.jp

Abstract. An apparatus has been developed for measuring catalytic activities of uni-atomic-composition bi-element clusters supported on a solid substrate. The cluster sample is prepared by irradiating a cluster-ion beam having the uni-atomic composition onto the substrate on a soft-landing condition in an ultra-high vacuum. The catalytic activity is measured by temperature-programmed desorption (TPD) mass analysis. Molecules at a density as low as $<10^3$ cm^{-3} have been detected with an ultrahigh-sensitive TPD mass spectrometer consisting of a cylindrical electron gun, a quadrupole mass filter and a micro-channel-plate ion-detector. The high reproducibility has been achieved by careful calibration of the TPD mass spectrometer. As a benchmark example, thermal oxidation of CO catalysed on Pt_{30} disks supported on a silicon surface was studied. The CO_2 products have been successfully observed at the Pt_{30} density as low as 3×10^{12} clusters in a circular area of 8 mm in diameter at the ramping rate of the sample temperature as low as 0.3 K s^{-1}.

1. Introduction

Clusters, which are aggregates of a few to hundred atoms, can be promoted as functional materials, considering a number of research results achieved so far [1–6]. It is sure that the mechanisms of the cluster-based functions are different from those of the corresponding bulk materials and even nanoparticles because of a number of specific properties originating from both many-body and small natures. Another advantage of the cluster materials is precise controllability of their performances in a wide dynamic range with the size and the atomic composition of the clusters. Indeed, a variety of studies on the size-dependent specific properties of the clusters have been pursued; for instance, odd-even alternation in electronic and magnetic properties, metal-insulator transition due to quantization of the electronic energy levels, magic numbers of the cluster size for stability and reactivity due to closing of electronic and atomic shells, splitting a chemical bond with a molecular wedge, etc.

Supporting the clusters on a solid substrate [5,6] is a promising way for cluster-based functional materials due to not only the practical reason in their real application but promotion of the cluster-specific properties through electronic/chemical interaction with the substrate. They are tuned further with internal species such as electric charges, spins, atomic dopants and ligands to control the internal

[1] To whom any correspondence should be addressed.

electric, magnetic and chemical fields as well as the corresponding external fields. Indeed, we have shown successfully in our experimental [7] and theoretical [8] studies that two-dimensional charge polarization is created at a sub-nanometer interface between a monatomic-layered platinum cluster disk [9,10] and a silicon surface. This originates from electron accumulation at the interface between the cluster disk and the substrate surface in a similar manner to the Schottky barrier junction, and the resulting positive charging in the center part of the cluster disk due to the two-dimensional metallic interaction among the Pt atoms arranged in a close-packed manner [9]. It is straightforward that the accumulated chares can be adjusted by doping another element of an electron donor or acceptor into the cluster. We have successfully produced uni-atomic-composition bi-element cluster ions of Pt and Ag with sufficient intensities [11]. Furthermore, we have found that two- to three-dimensional geometric transition occurs in tungsten clusters anchored on a graphite surface between the cluster size of 10 and 11 due to competition between the cluster-substrate and intracluster interactions [12,13]

We are aiming to educe novel catalytic functions from clusters. It is obvious from the definitions of reduction and oxidation as capture and release of electrons, respectively, that one can obtain high catalytic activities of the reduction and the oxidation by utilizing the rich electric charges. It is highly expected that the subnano-space charge accumulation also leads to an efficient advantage of higher molecular selectivity because of the charged area as small as a molecule.

Vigorous research studies have been prosecuted on catalytic activities driven by supported clusters [6,14–30]. For instance, experimental and theoretical studies have been reported that a planar gold cluster supported on the MgO(100) thin film prepared on a metal surface is charged due to electron transfer from the metal support to the cluster [14–17]. Furthermore, the charging state relates closely to catalytic activities [18–20]; the electron transfer from an oxygen-vacancy defect of the MgO(100) film to a three-dimensional Au_8 induces a high catalytic activity in the CO oxidation, but that supported on a defect-free MgO(001) film has no catalytic activity because of much less electron transfer between them [18,19]. On the other hand, a planner Au_{20} supported on a defect-free MgO(100) substrate has the catalytic activity due to efficient charging [20]. Distinct cluster-size dependence has been reported in the catalytic CO oxidation by platinum clusters on a thin MgO(100) film, indicating the way to tuning the heterogeneous catalyst with the cluster size [21]. Clusters, Pt_7, supported on a slightly reduced $TiO_2(110)$ substrate reveal a high catalytic activity in the CO oxidation, while those on a strongly reduced one quenches the CO oxidation [22].

The size- and composition-selected cluster catalyst samples can be prepared by soft-landing of the uni-size cluster ions onto the substrate [4–6]. Considering a typical intensity (<100 pA) of the most intense bi-element cluster-ion sources [11], the number of the clusters on a substrate is limited as low as $\sim 10^{12}$ in a realistic time required for the sample preparation. Therefore, the critical procedure in this kind of experimental studies is high-sensitive and reproducible detection of the reaction products. In this report, we present our study to develop an apparatus for the detection of low-density molecules produced in a catalytic reaction on the supported-cluster sample. The thermal oxidation of CO by O_2 catalysed by the Pt_{30} disks supported on a silicon substrate [7–9,31] was employed as a benchmark system of heterogeneous catalyses [32].

2. Experimental apparatus and method

The apparatus consists of a bi-element cluster ions source, a mass filter for the selection of the size and atomic composition, a cluster-cooling device, a reaction device and a detection device [30].

2.1. Preparation of cluster-catalyst sample

The samples were prepared by the soft-landing of the uni-atomic-composition cluster ions onto the substrate. The details have been described in elsewhere [11,30], and several modifications were made thereafter. The bi-element cluster ions were produced by means of dual magnetron-sputtering devices having different-material targets equipped in a gas-aggregation chamber and the cluster cations were selected out of the neutral and anionic species in a quadrupole ion-deflector. A desired atomic composition was selected out according to the mass difference by passing through the quadrupole mass filter (Extrel MAX-16000 operated at 440 kHz, rod diameter is 9.5 mm, maximum mass is 16,000 dalton), and both their kinetic energies and beam diameter were reduced in collision with cold He gas at 100 K in the cooling device. The cluster ions were always guided with octopole ion guides. The low-temperature cluster ions with a uni-atomic composition were allowed to collide onto the (111)–7×7 reconstructed surface of a silicon substrate placed in the reaction chamber (see figure 1) at 300 K at a collision energy of 1 eV per atom so as to fix the uni-atomic-composition clusters on it. The ambient pressure of the reaction chamber was maintained less than 5×10^{-8} Pa during the cluster deposition due to two differential-pumping chambers, where the most of the residual gas in the reaction chamber was helium leaked from the upstream cooling device. The number density of the clusters was set at 4×10^{12} cm^{-2} in order to avoid undesired direct interaction between the clusters on the substrate; the density is obtained by time-integration of the cluster-ion current (500 nC) and the known diameter (8 mm) of the cluster-ion beam [30]. Then, the surface was terminated with hydrogen by irradiating an atomic hydrogen beam from a thermal hydrogen cracker (Tectra H-flux) in order to prevent the silicon surface from oxidation during the TPD measurements; the hydrogen-terminated silicon surface is stable below 750 K [33], while the H atoms desorb from the Pt clusters at ~300 K by analogy with the H desorption from the Pt(111) surface [34]. It has been confirmed by STM observation that the Pt_{30} disks on the silicon substrate are stable below 673 K [31].

The (111)–7×7 reconstructed surface of the silicon substrate was prepared from a Si wafer (Nilaco n-type, specific resistance less than 0.01 Ωcm, 10×20×0.5 mm^3) by repeating a series of Ar^+ sputtering (10 μA for one hours at 5×10^{-4} Pa of Ar), heating resistively at 1200 K for 50 s and successive annealing down to 850 K at a rate of –0.5 K s^{-1} at an ambient pressure less than 1×10^{-7} Pa, where the temperature of the silicon surface was measured with a pyrometer (Chino IR-AH1S). Each end of stacked two silicon wafers was placed on a copper electrode through a tantalum thin plate (0.2-mm thick), and fixed with a molybdenum plate (1-mm thick) and two molybdenum bolts; the wafer stacking is employed for the accurate measurement of the substrate temperature as described in the next subsection. The reaction chamber was evacuated with a turbo-molecular pump (Mitsubishi PT-500).

2.2. *Observation of catalytic reaction*

The temperature-programmed desorption (TPD) mass-spectroscopy was carried out repeatedly for a given cluster catalyst sample in a series of sample cleaning, reactant adsorption and the TPD measurement. Figure 1 shows a sectional top view of the sample stage and the TPD mass spectrometer installed in the reaction and detection chambers, respectively [30]. The sample was cooled with liquid nitrogen through the copper electrodes and a copper block attached at the end of a liquid-nitrogen tank (310-mm long, 45 mm in diameter), where the liquid-nitrogen tank was placed on an x-y-z-θ manipulating stage. On the other hand, the sample was heated resistively with dc current though the silicon substrates programed for the linear temperature ramping, where the electric cables were thermally anchored on the copper block. The product molecules desorbed from the cluster catalyst were introduced into a cylindrical electron gun (Extrel axial ionizer) to be ionized through a home-made skimmer having a 6-mm hole placed at less than 1 mm far from the sample substrate. Thoriated iridium filaments, which emit electrons at lower temperature than tungsten one, were employed in order to suppress emission of contaminating species from the filaments, where they were, if produced, evacuated readily through mesh surrounding the electron gun. The ions were focused onto the entrance lens of a high-transmission quadrupole mass filter (Extrel MAX-300 operated at 1.2 MHz, rod diameter is 19 mm) with the cylindrical ion lenses and detected with a micro-channel plate (Hamamatsu MCP F4655) placed in a shielding metal box in the counting method at a time constant of 0.3–3 s. The MCP faced to 90 degrees with respective to the mass-filter axis and was placed at an off-axis position in order to suppress the back ground level by reducing the frequency of electronically-excited neutral species in metastable and Rydberg states hitting the detector. The sample surface was faced to the skimmer only during the TPD measurement in order for minimizing undesired sample heating and contamination by the electron gun.

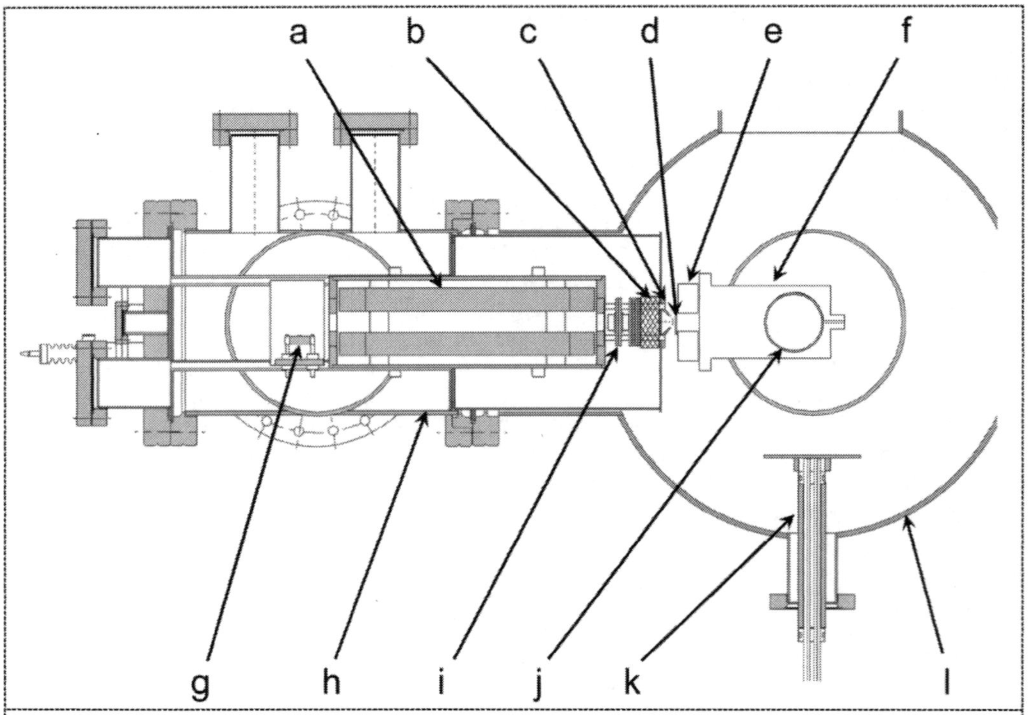

Figure 1. Sectional top view of the reaction and detection chambers. (a) quadrupole mass filter, (b) cylindrical electron gun, (c) skimmer, (d) sample substrate, (e) copper electrode, (f) copper block, (g) micro-channel plate, (h) detection chamber, (i) cylindrical ion lenses, (j) liquid-nitrogen tank, (k) guide for the uni-size cluster ions (from bottom to top of the view) and (l) reaction chamber.

The isotope-labelled reactants, $^{18}O_2$ and $^{13}C^{16}O$, were employed in order to rule out contributions of the major residual $^{12}C^{16}O$ in the reaction chamber. The dose amounts of the reactants were regulated by their partial pressures measured with a calibrated cold-cathode pressure gauge (Pfeiffer IKR270) and the exposure time. The cluster catalyst was exposed to the reactant gases one by one by leaking one of the gasses through the corresponding variable leaking valve into the reaction chamber at 6×10^{-7} Pa for 200 s (exposure at 0.45 L) and at the substrate temperature of 105 K. The cold-cathode gauge was calibrated independently for $^{18}O_2$ and $^{13}C^{16}O$ in the pressure range of $3–15\times10^{-7}$ Pa everyday before and after all the series of the TPD measurements with a calibrated commercial residual-gas analyzer (SRS RGA-100) installed in the reaction chamber. However, the residual gas analyser was turned off during the TPD measurements in order to avoid sample contamination due to species produced by its electron gun. The cluster catalyst was cleaned within 20 minutes before every set of the reactant dose by burning the impurity adsorbates at 650 K after exposure to O_2 at more than 1 L at 105 K.

The intensity of the reaction product, $^{13}C^{16}O^{18}O$, was measured as a function of the temperature of the cluster catalyst whilst heating it from 110 K to 550 K. The temperature of the substrate was measured as that of the cluster catalyst with a chromel-alumel thermocouple (0.05 mm in diameter for each) referred to liquid nitrogen out of the vacuum chamber. Good thermal contact between the substrate and the thermocouple was achieved by tight insertion of the contact end of the thermocouple wrapped with tantalum foil (0.075-mm thick) between the stacked silicon wafers. The temperature uniformity was measured with the pyrometer (spatial resolution of 3 mm at the substrate) to be as good as <10 K in a circular area of 10 mm in diameter at the center of the substrate at the substrate temperature of 850 K. The ramping rate as low as 0.3 K s^{-1} was employed in the TPD measurements also for the uniform heating of the rather large substrate.

The pressure of the reaction chamber was less than 1×10^{-8} Pa during the TPD measurement by filling the tank with liquid nitrogen, while that of the detection chamber was 3×10^{-8} Pa by differential pumping with a turbo-molecular pump (Mitsubishi PT-500).

3. Results and discussion

3.1. Production of intense platinum cluster cations

Figure 2 shows mass spectra optimized for the production of Pt_{30}^+ (5,853 dalton) and Pt_{82}^+ (15,997 dalton); the discharging voltage and current and the flow rates for He and Ar for the Pt magnetron were (238 V, 83 mA, 53 and 21.6 sccm, standard cc per minute) and (278 V, 216 mA, 46 and 15.6 sccm) for panels (a) and (b), respectively. The series of mass peaks are assignable to Pt_N^+ (N=1–82) and Ar^+. In the entire size range, Pt_N^+ are successfully produced as intense as or more intense than ~100 pA, which is high enough to prepare the cluster catalysis sample at the desired number density within 2 hours .

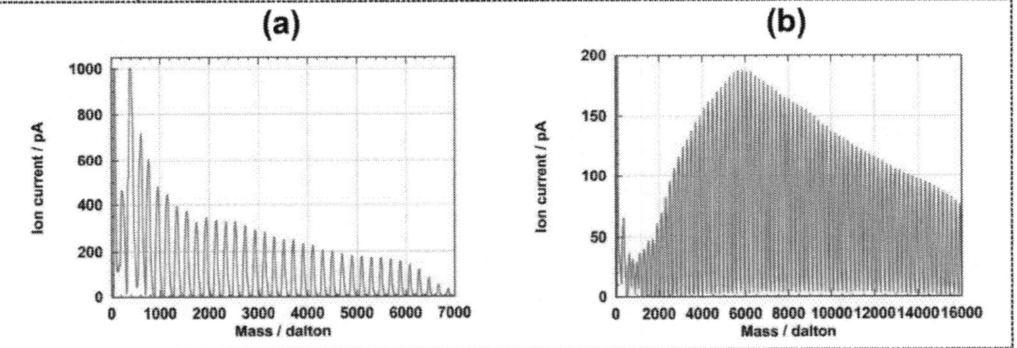

Figure 2. Mass spectra of platinum cluster cations optimized at (a) Pt_{30}^+ and (b) Pt_{82}^+.

3.2. Detection efficiency of product molecules

The detection efficiency of the TPD products was calibrated also everyday before and after all the series of the TPD measurements with the residual-gas analyzer. Leaking a trace amount of argon into the reaction chamber, the intensity of Ar^+ and the partial pressure of Ar were measured simultaneously with the TPD mass spectrometer and the residual gas analyser, respectively. Figure 3 shows mass spectra measured with the TPD mass spectrometer at the time constant of 0.3 s and the residual gas analyser at the Ar partial pressure of 4×10^{-10} Pa as shown in panel (c) of this figure. It is apparent that the TPD mass spectrometer has much higher sensitivity than the residual gas analyser. The mass resolution of the TPD mass spectrometer was intentionally lowered at 1 dalton in order to obtain the high-ion transmittance through the mass filter, leading to the high sensitivity. All the back ground (~300 cps) at 47 dalton ($^{13}C^{16}O^{18}O$) is due to the excited neutral species passing through the quadrupole mass filter to hit the detector, which cannot be distinguished from the ion signals with the discriminator because their pulse heights from the MCP detector are similar. These mass spectra show that the vacuum of the reaction and the detection chambers are sufficiently clean as the partial pressures of the hydrocarbons are smaller than 10^{-11} Pa, plausibly supposing almost the same detection efficiencies for Ar and the hydrocarbons. All the mass peaks are accompanied with a small shoulder at their higher-mass ends due to nonstandard tuning of the TPD mass spectrometer for the high detection efficiency. This is located sufficiently far from the adjacent peak, so that no influence is given to the TPD measurements.

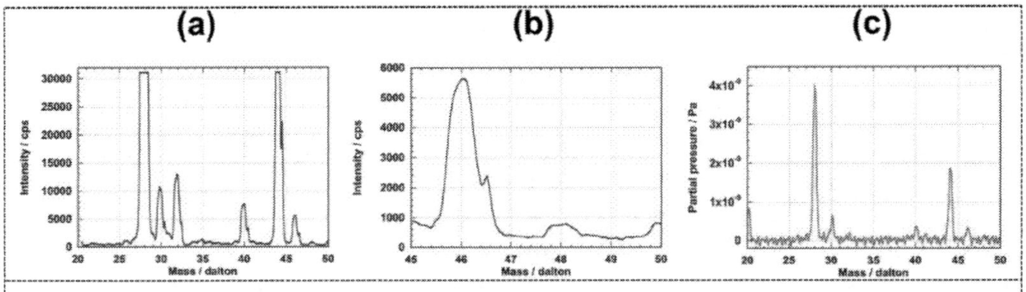

Figure 3. Mass spectra of residual gases and Ar leaked intentionally into the reaction chamber measured with the TPD mass spectrometer (a and b) and the residual gas analyser (c), where panel (b) shows the spectrum shown in panel (a) in the expanded scale.

Figure 4 compares the Ar^+ intensity, I_{Ar+}, measured at the time constant of 0.3 s and the Ar partial pressure, P_{Ar}, as a function of time with reducing the leaking rate of Ar. As shown in panel (b), they are correlated in the following equation.

$$P_{Ar} / \mathrm{Pa} = 5.2 \times 10^{-14} (I_{Ar+} / \mathrm{cps} - 1550). \qquad (1)$$

The conversion coefficients before and after the TPD measurements in a day were within the experimental errors, but changed by <20% day by day. As shown in the expanded plots in panel (b) of figure 4, the detection limit of the TPD mass spectrometer is lower than 4×10^{-12} Pa (number density at 10^3 cm^{-3}) for Ar in the reaction chamber. This high sensitivity has been achieved by the high ion-transmittance through the quadrupole mass filter and the optimization of the electron-gun filament at 4.3 A and 5 V so as to obtain the electron emission current as high as 7 mA. The electron gun operated at a higher current resulted in lower sensitivity probably due to charging up in the electron gun. The off-axis arrangement of the MCP detector was also effective for the high sensitivity by suppressing the back ground noise level originating from the excited neutral species.

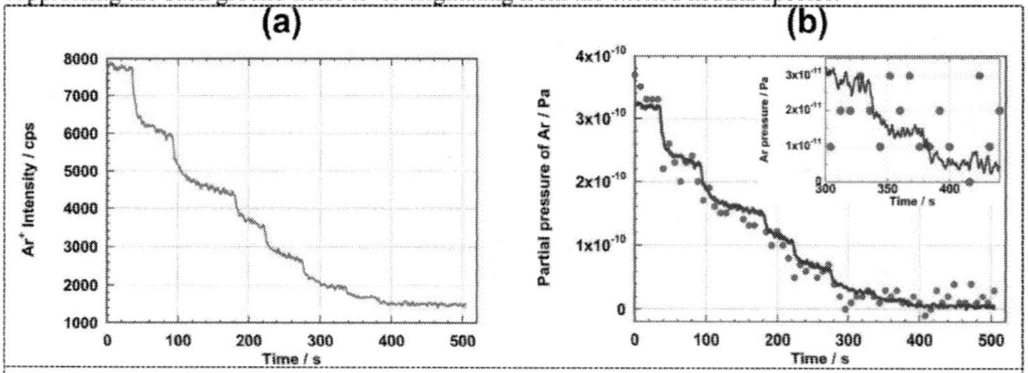

Figure 4. Time propagations of (a) Ar^+ intensity measured with the TPD mass spectrometer and (b) Ar partial pressure simultaneously measured with the residual gas analyser (solid circles) with reducing the Ar leaking rate into the reaction chamber. In panel (b), the Ar partial pressure calculated from the Ar^+ intensity according to equation (1) (see text) is also plotted as the solid curve. The insert shows the plots in the expanded scale.

3.3. TPD measurements of CO oxidation

Figure 5 shows the TPD spectra of the $^{13}C^{16}O^{18}O$ production catalysed on the Pt_{30} disk supported on the silicon substrate at the exposure of 0.45 L for both the reactants of $^{18}O_2$ and $^{13}C^{16}O$ exposed in this order, where T is the temperature of the Pt_{30} disk catalyst measured with the thermocouple. Displayed are the spectra of the single TPD run measured at the time constant of 3 s and averaged for the five sets of the TPD data measured for the same catalyst in different days, where the averaging was made by factoring in the detection efficiency measured in a manner described in the previous subsection. The single-run signal intensities are not integer, because an analogue voltage from the counting-rate meter, which is proportional to the counting number, was measured as a signal. The back-ground level (280 cps for the spectrum shown in panel (a) of figure 5) was obtained by referring the mass spectra observed 5 minutes before each single TPD run, and this was subtracted from the single-run TPD data before averaging. In the averaging, T was smoothed in the following manner; all T measured in a range of $T_{int} \pm 0.5$ K were smoothed into T_{int}, where T_{int} is an integer. The TPD spectrum has been successfully measured with high sensitivity and good reproducibility. Even the single-run spectrum reveals a sufficient quality at the temperature ramping rate as low as 0.3 K s^{-1}, where the lower the ramping rate, the lower the counting number in a unit time.

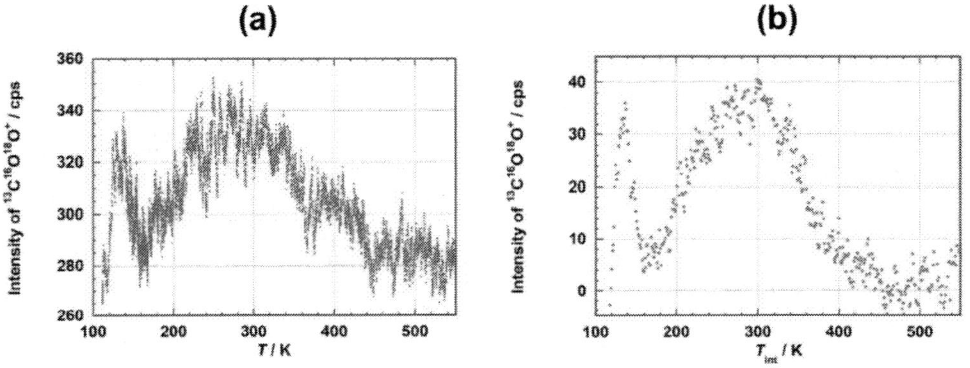

Figure 5. TPD spectra of the $^{13}C^{16}O^{18}O$ production from $^{18}O_2$ and $^{13}C^{16}O$ catalysed on the Pt_{30} disk supported on the silicon substrate. (a) Single-run TPD spectrum and (b) averaged one. T and T_{int} are the measured and smoothed temperatures of the catalyst, respectively

A sharp peak and a broad one appear at 130 K and 300 K, respectively. Considering the close-packed atomic arrangement of the Pt_{30} disk [9], it is a good starting point to compare the present results with the same reaction on the bulk Pt(111) surface having the close-packed Pt arrangement [32]. Therefore, the sharp TPD peak is assigned to the CO oxidation by the molecular O_2, which is activated by electron capture in its anti-bonding molecular orbitals from the catalyst, so that the internuclear distance of O_2 is elongated. On the other hand, the broader one is assigned to the catalytic action by the dissociatively adsorbed O .

4. References

[1] de Heer, W A 1993 *Rev. Mod. Phys.* **65** 611
[2] Jena P and Castleman Jr A W 2006 *Proc. Natl. Acad. Sci. U.S.A.* **103** 10560
[3] ed Sattler K 2010 *Handbook of Nanophysics* (Boca Raton : CRC Press)
[4] Yasumatsu H and Kondow T 2003 *Rep. Prog. Phys.* **66** 1783
[5] ed Meiwes-Broer K H 2000 *Metal Clusters at Surfaces* (Heidelberg : Springer)
[6] ed Heiz U and Landman U 2007 *Nanocatalysis* (Berlin: Springer)
[7] Yasumatsu H, Hayakawa T and Kondow T 2010 *Chem. Phys. Lett.* **487** 279
[8] Yasumatsu H, Murugan P and Kawazoe Y 2012 *Phys. Stat. Solidi B* **249** 1193
[9] Yasumatsu H, Hayakawa T, Koizumi S and Kondow T 2005 *J. Chem. Phys.* **123** 124709
[10] Yasumatsu H, Hayakawa T and Kondow T 2006 *J. Chem. Phys.* **124** 014701
[11] Yasumatsu H 2011 *Euro. Phys. J. D* **63** 195
[12] Hayakawa T and Yasumatsu H 2012 *J. Nanopart. Res.* **14** 1022
[13] Hayakawa T, Yasumatsu H and Kondow T 2009 *Euro. Phys. J. D* **52** 95

[14] Ricci D, Bongiorno A, Pacchioni G and Landman U 2006 *Phys. Rev. Lett.* **97** 036106
[15] Walter M, Frondelius P, Honkala K and Häkkinen H 2007 *Phys. Rev. Lett.* **99** 096102
[16] Lin X, Nilius N, Freund H J, Walter M, Frondelius P, Honkala K and Häkkinen H 2009 *Phys. Rev. Lett.* **102** 206801
[17] Lin X, Nilius N, Sterrer M, Koskinen P, Häkkinen H and Freund H J 2010 *Phys. Rev. B* **81** 153406
[18] Sanchez A, Abbet S, Heiz U, Schneider W D, Häkkinen H, Branett R N and Landman U 1999 *J. Chem. Phys. A* **103** 9573
[19] Yoon B, Häkkinen H, Landman U, Wörz A S, Antonietti J M, Abbet S, Judai K and Heiz U 2005 *Science* **307** 403
[20 Zhang C, Yoon B, and Landman U 2007 *J. Am. Chem. Soc.* **129** 2228
[21] Heiz U, Sanchez A, Abbet S and Schneider W D 1999 *J. Am. Chem. Soc.* **121** 3214
[22] Bonanni S, Aït-Mansour K, Harbich W and Brune H 2012 *J. Am. Chem. Soc.* **134** 3445
[23] Heiz U, Vanolli F, Sanchez A and Schneider W D 1998 *J. Am. Chem. Soc.* **120** 9668
[24] Yoon B, Landman U, Habibpour V, Harding C, Kunz S, Heiz U, Moseler M and Walter M 2012 *J. Phys. Chem. C* **116** 9594
[25] Röttgen M, Abbet S, Judai K, Mntonietti J M, Wörz A S, Arenz M, Henry C R and Heiz U 2007 *J. Am. Chem. Soc.* **129** 9635
[26] Kunz S, Schweinberger F F, Habibpour V, Röttgen M, Harding C, Arenz M and Heiz U 2010 *J. Chem. Phys. C* **114** 1651
[27] Kaden W E, Wu T, Kunkel W A and Anderson S L 2009 *Science* **326** 826
[28] Watanabe Y, Wu X, Hirata H and Isomura N 2011 *Catal. Scie. Technol.* **1** 1490
[29] Wyrzgol S A, Schäfer S, Lee S, Lee B, Vece M D, Li X, Seifert S, Winans R E, Stutzmann M, Lercher J A and Vajda S 2010 *Phys. Chem. Chem. Phys.* **12** 5585
[30] Yasumatsu H, Fuyuki M, Hayakawa T and Kondow T 2009 *J. Phys. Conf. Ser.* **185** 012057
[31] Fukui N and Yasumatsu H 2012 submitted to *Euro. Phys. J. D*
[32] Allers K H, Pfnür H, Feulner P and Menzel D 1994 *J. Chem. Phys.* **100** 3985
[33] Dinger A, Lutterloh C and Küppers J 2001 *J. Chem. Phys.* **114** 5338
[34] Kim J Y and Lee J J 2000 *Chem. Phys.* **113** 2856

Acknowledgments
We are deeply indebted to the late Professor Tamotsu Kondow for his guidance and advice to start this study. Mr. Tomonori Ito is also acknowledged for his assistance in the tuning of the apparatus. This research was supported by the Special Cluster Research Project of Genesis Research Institute, Inc.

Size Selected Clusters on Surfaces

X. Tang, X. Li, Y. Wang, K. Wepasnick, A. Lim, D. H. Fairbrother, K. H. Bowen[1],

Departments of Chemistry and Materials Science, Johns Hopkins University, Baltimore, MD USA

T. Mangler, S. Noessner, C. Wolke, M. Grossmann, A. Koop, G. Gantefoer

Department of Physics, University of Konstanz, Konstanz, Germany

B. Kiran

Departments of Chemistry and Physics, McNeese University, Lake Charles, LA, USA

A. K. Kandalam

Deptartment of Physics, West Chester University, West Chester, PA

[1] E-mail : kbowen@jhu.edu

Abstract. Size selected metal oxide, sulfide and oxynitride clusters, soft-landed onto highly ordered pyrolytic graphite (HOPG) at room temperature, have been studied in this work. Based on their size and chemical compositions, the deposited clusters have exhibited various surface structures as illustrated by *in situ* Scanning Tunnelling Microscope (STM) and *ex situ* Atomic Force Microscope (AFM). In contrast to pure metal clusters, size selected metal compound clusters have shown different surface behaviours due to their different cluster-surface interactions.

1. Introduction

Clusters contain an ensemble of bound atoms whose size is intermediate between an individual molecule and a bulk solid. As such, clusters are a unique class of material which continues to attract intense scientific interest due to the fact that they often exhibit extraordinary electronic, optical, catalytic properties, etc. [1-3] Many of these properties are strongly size-dependent, e.g. gold clusters in the size regime of 2-4 nm catalyze CO oxidization at low temperatures, whereas bulk gold is catalytically inert [4]. Metal oxides have also been shown to exhibit size-dependent reactivity. For example, in a study by Castleman *et al.* [5], the reactivity of group V transition metal oxide cluster cations with *n*-butane has been shown to be strongly dependent on the number of metal atoms in the cluster. This underscores the need for means that allow one to control the size of deposited clusters as a way of tailoring the properties and functions of these nanostructures. Amongst the various methods of cluster deposition, which include conventional molecular beam [6] and wet-chemistry [7] techniques, the soft-landing of size selected gas phase clusters onto supported surfaces is a promising

[1] To whom any correspondence should be addressed.

approach due to its ability to utilize gas phase clusters that occur in a wide range of sizes and chemical compositions. However, controlling the ultimate size, structure, and therefore functionality of deposited clusters created by this means relies on understanding the relationships between the size of the incident gas phase clusters and the nature of the deposited clusters, determined by cluster-cluster and cluster-surface interactions [8; 9].

Our knowledge of the structures that gas phase clusters adopt on solid surfaces comes predominantly from studies on metal clusters deposited by soft-landing, or, in a few cases, low energy ion beam deposition on atomically smooth surfaces such as highly ordered pyrolytic graphite (HOPG) and mica [10-17]. In general, the surface structures formed have been found to depend on the size of the incident gas phase clusters. For example, antimony clusters deposited on graphite surface coalesce into larger clusters when the incident cluster size is small (Sb_4). In contrast, cluster-cluster interactions become less favorable when larger incident antimony clusters (Sb_{350}, Sb_{2200}) are deposited [11]. These size-dependent effects are ascribed to the lower surface mobility of larger clusters, coupled with their greater thermodynamic stability due to the lower fraction of surface atoms. The observations are, in turn, consistent with experiments on gold clusters. In a series of experiments, small gold clusters (Au_1, Au_3, Au_7) were observed to be mobile on HOPG before coalescing into larger clusters with a much larger diameter [18], while large gold clusters (Au_{250}) were less mobile and only partially coalesced into fractal structures [19]. However, in experiments performed using size selected silver clusters, it was observed that silver clusters containing 50 to 250 atoms deposited onto HOPG diffuse and coalesce into larger nanoparticles with a diameter of about 14 nm, irrespective of the incident gas phase cluster size [13]. In contrast, our knowledge of the size-dependent surface behavior of metal compound clusters (e.g, metal oxides[20-22], sulfides[23], nitrides), is extremely limited despite their importance in catalysis and material science [24].

From these discussions, understanding how the size of incident metal compound clusters affects their surface structures and properties is clearly important. In this proceeding, we have focused on three different types of clusters: molybdenum oxides, lead sulfides, and titanium oxynitrides. Results obtained by local probe microscopes reveal that these metal compounds sometimes result in significantly different surface structures from their pure metal clusters. We attribute these differences to their different cluster-cluster and cluster-surface interactions.

2. Apparatus

Cluster anions were produced by the cluster deposition apparatus shown in Figure 1. Briefly, a metal target was placed in a cylindrical magnetic field and biased to -500 V. A mixture of argon and helium was then introduced into the source. Argon ions created within the source sputtered the metal target, producing metal clusters. To produce different types of clusters, different reactive gases were introduced to react with the metal cluster vapors: Oxygen (O_2) was used to produce oxide clusters; metal sulfides were prepared by reacting metal ions with hydrogen sulfide (H_2S); finally, for titanium oxynitride clusters, nitrogen was provided by ammonia gas (NH_3), and oxygen came from residual gases such as water (H_2O). The added helium served to cool and transport the ions downstream.

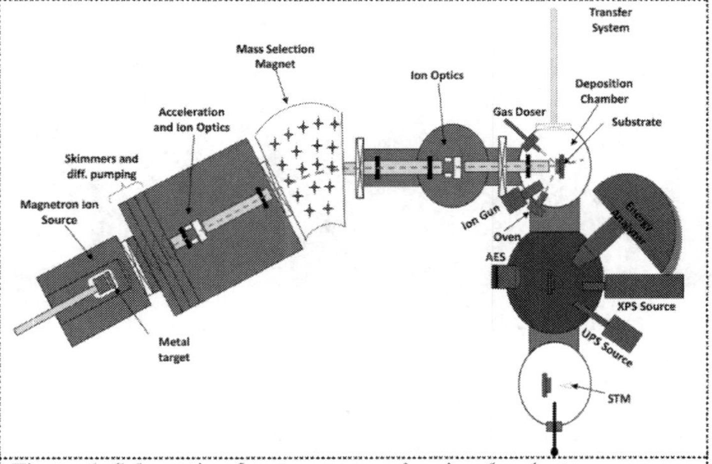

Figure 1. Schematic of our apparatus, showing the cluster deposition, beam-line, and the surface analytical instrumentation.

After the clusters anions were produced, they were accelerated before entering a magnetic sector mass spectrometer (25° sector magnet with resolution of $m/\Delta m = 20$). By tuning the magnetic field strength and/or the ion energy, clusters of various sizes were mass-selected and then passed through a series of ion optics before entering a Ultra-High Vacuum (UHV) deposition chamber where they were soft-landed (kinetic energy < 0.1eV/atom) onto a 1 cm^2 piece of freshly peeled HOPG.

Once clusters had been deposited onto the HOPG substrate, samples could be internally transferred to the adjunct UHV chamber, where they could be characterized by *in situ* Auger electron spectroscopy (AES), X-ray Photoelectron Spectroscopy (XPS), and Scanning Tunneling Microscopy (STM). *Ex situ* Atomic force microscopy (AFM) was also employed to characterize the structures of deposited clusters.

3. Structures of Size Selected of Clusters on Surfaces

3.1. Metal /Metal Oxide Clusters on HOPG

The mobility and aggregation behaviors of metal versus metal oxide clusters on HOPG surfaces have been examined as a function of their coverage. To accomplish this, mass-selected, cluster anion beams of Mo_{100}^- and $(MoO_3)_{67}^-$ were produced in a magnetron sputter source and soft-landed onto HOPG under UHV conditions. These two clusters were chosen because they possess essentially the same masses and consequently could be soft-landed with the same low kinetic energies, resulting in deposition energies of ~0.1 eV/ atom. The *in situ* AES confirmed the chemical composition of molybdenum metal and metal oxide clusters, respectively. Their surface morphologies were then characterized as a function of cluster coverage by using *in situ* STM and *ex situ* AFM. Both STM and AFM results indicated high mobility for the metal atom clusters on HOPG at room temperature as shown Figure 2(a) and Figure 3.

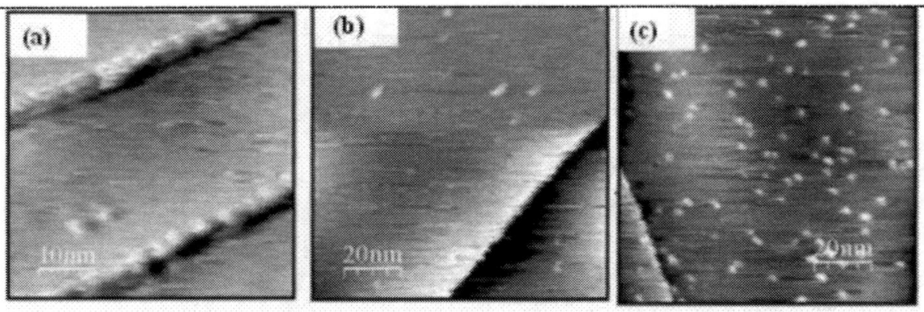

Figure 2. Deposited (a) Mo_{100} clusters and (b)(c) $(MoO_3)_{67}$ clusters on HOPG imaged by *in situ* STM .

At low coverages, Mo_{100} clusters migrated over the surface and nucleated preferentially at step edges as shown in Figure 2(a) and Figure 3(a), but at higher coverages they aggregated on the terraces until a fully saturated over-layer was eventually created, as shown in Figure 3(b). This behavior by metal clusters on HOPG surfaces was in accordance with prior studies on other metal cluster systems due to the high mobility of metal clusters on HOPG substrates.[10] With the same low kinetic energies and the same masses, one might have expected the metal oxide clusters to disperse over HOPG in the same way. In contrast with Mo_{100} metal clusters, the $(MoO_3)_{67}$ oxide clusters, instead of showing high mobility, produced a random array of adsorbed clusters at all coverages as shown in Figure 2(b)(c) . Consequently, on the same type of surface and with the same energy per atom, these metal oxide clusters were comparatively immobile relative to metal clusters, and differences in the mobility and aggregation behaviors of Mo_{100} and $(MoO_3)_{67}$ clusters at room temperature were attributed to differences in the interactions of metal and metal oxide clusters with carbonaceous surfaces.

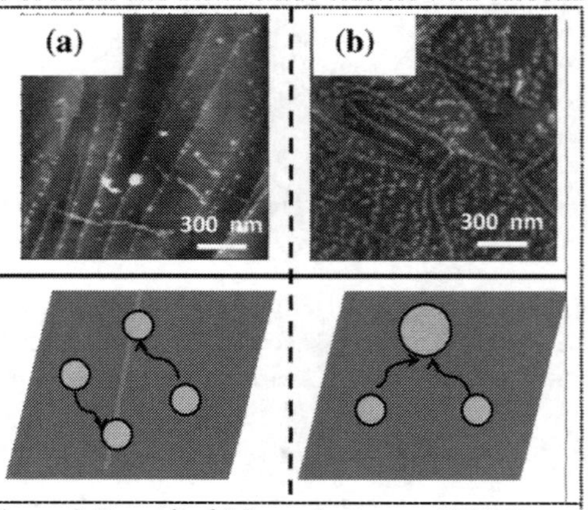

Figure 3 Deposited Mo_{100} clusters on HOPG imaged by *ex situ* AFM: (a) low coverage (b) high coverage

3.2. Lead Sulfide: Its Clusters and its Embryonic Crystal

It has been a goal of cluster science to connect the microscopic, molecular level world to the macroscopic, bulk level world.[1] In particular, at what size do clusters (aggregates) first take on properties which are characteristic of the bulk? Here, we are interested in identifying the smallest, viable crystalline unit of the lead sulfide rock salt crystal, i.e., its "baby crystal."[23]

To set the stage for the STM studies of "baby crystals," our previous work combined both experiments and theory to explore the geometric structures and electron affinities of small, gas-phase, lead sulfide clusters, $(PbS)_n$, ranging in size from $n = 2$-15. [25]

First, photoelectron spectroscopy was conducted on size-selected cluster anions to acquire both ground and excited state electronic structure information as well as the electron affinity values of the cluster anions' neutral counterparts. Figure 4 and Figure 5 display the calculated geometric structures of neutral and anionic $(PbS)_n$ clusters with n=1-10, respectively. Further comparison of the calculated electron affinities and vertical detachment energies with measured photoelectron spectra has authenticated the calculated structures. Among these calculated structures, the $(PbS)_4$ cube was found to be the most stable of the lead sulfide clusters studied due to a large, calculated HOMO-LUMO gap (2.96 eV), and it was also suggested to be the preferred product of cluster fragmentation processes. These findings are consistent with the fact that the primitive cell in crystalline lead sulfide is made up of four lead sulfide molecules. Notably, a structural growth pattern for the lowest energy isomers in this cluster size range has been proposed here based on a two-dimensional stacking of face-sharing (side-by-side) lead sulfide cubical units. For example, the face sharing of $(PbS)_4$ cube would lead to the lowest energy isomer of $(PbS)_8$. However, over the studied size range, the lead and sulfur atoms in these structures possessed a maximum of five-fold coordination, rather than the six-fold coordination found in bulk, lead sulfide crystals. As a general conclusion for these calculated structures, lead sulfide clusters in this size range form slightly distorted, fused cuboids.

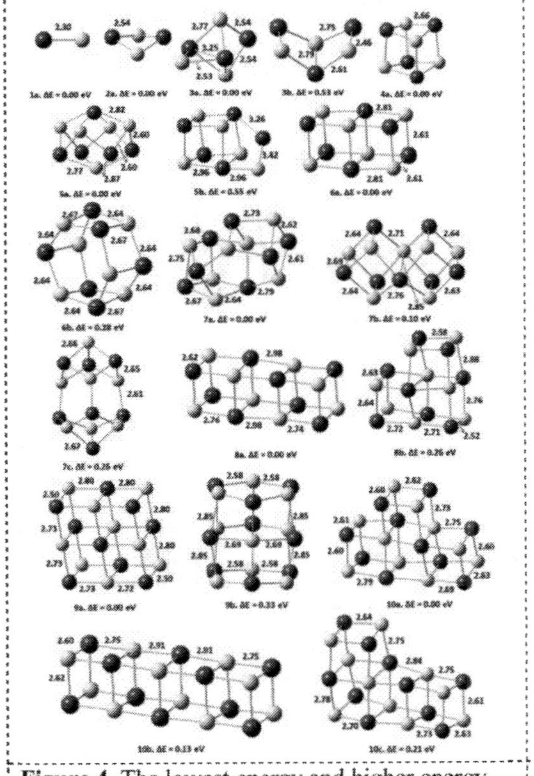

Figure 4. The lowest energy and higher energy isomers of neutral $(PbS)_n$ (n = 1–10) clusters.

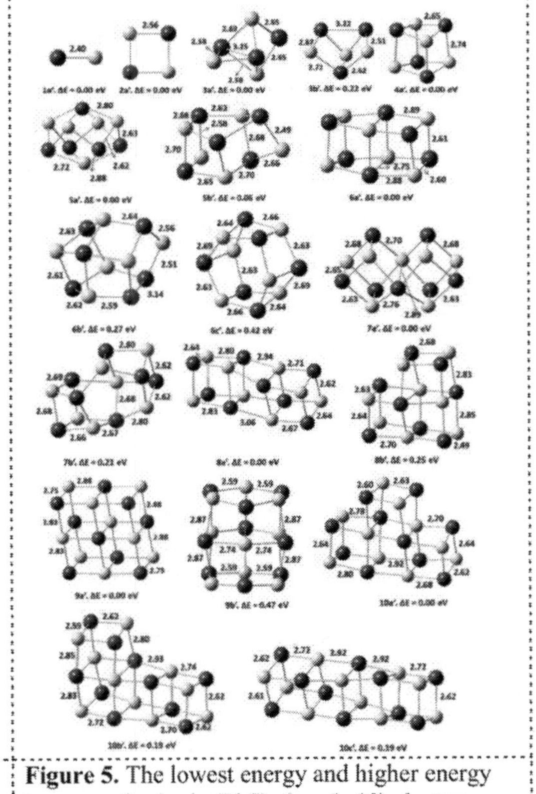

Figure 5. The lowest energy and higher energy isomers of anionic $(PbS)_n$ (n = 1–10) clusters.

Since lead sulfide clusters in this size range were too small to adopt bulk-like six-fold coordination, this left open the question: what is the smallest $(PbS)_n$ cluster size that can do so and that can also be replicated to form the bulk crystalline material? The successive dimerization processes from the $(PbS)_4$ cluster were used to search for this minimum cluster size, as illustrated in Figure 6. The dimerization leads to the formation of $(PbS)_n$ clusters from n=4 to 8, 16, and further on until 32, where the inner $(PbS)_4$ core of $(PbS)_{32}$ enjoys bulk-like six-fold coordination. On other hand, the calculated structures of $(PbS)_n$ cluster with n=12, 20, 24 did not have six-fold coordination in their bulk. According to the calculations, the lowest energy isomer of $(PbS)_{28}$ did have bulk-like six-fold coordination. However, the replication of $(PbS)_{28}$ failed to lead to the bulk rock-salt structure while $(PbS)_{32}$ did. Thus, cubic $(PbS)_{32}$ is a "*baby crystal*," i.e., the smallest cluster exhibiting six-fold coordination that can be replicated to obtain the bulk crystal. Additionally, the calculated dimensions of the $(PbS)_{32}$ cluster further provide a rubric for understanding how the successive fusion of multiple $(PbS)_{32}$ building blocks results in the formation of square and rectangular, crystalline nano-blocks, e.g., $(PbS)_{64}$, $(PbS)_{128}$, and $(PbS)_{256}$, with specific dimensions as shown in Figure 7.

We then conducted cluster deposition to test the above theoretical predictions. Herein. mass-selected $(PbS)_{32}$ clusters were soft-landed onto HOPG surfaces and imaged by STM to gauge their dimensions. These dimensions were then compared with those of the fused $(PbS)_{32}$ nano-blocks predicted by theory.

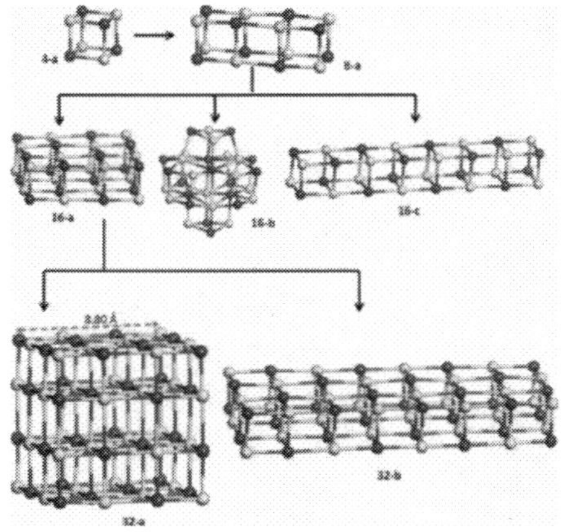

Figure 6. The growth pattern of lead sulfide cluster structures via dimerization of $(PbS)_{4n}$ units. For $(PbS)_{16}$ and $(PbS)_{32}$, both, the lowest, (a), and higher, ((b) and (c)), energy isomers are shown.

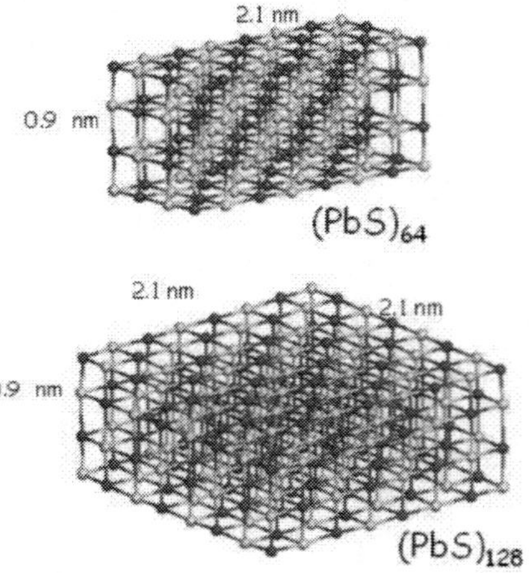

Figure 7. Calculated lowest energy structures for the $(PbS)_{64}$ and $(PbS)_{128}$ nano-blocks.

The chemical compositions of lead sulfide clusters were confirmed by *in situ* XPS measurements. The XP spectra of Pb (4f) and S (2p) regions are presented in Figure 8. Both the peak transitions of $Pb(4f_{7/2})$ centered at 137.4 eV and S(2p) centered at 160.6 eV indicated the unoxidized nature of the deposited lead sulfide clusters.[26] A further quantitative analysis of XPS results revealed that the ratio of Pb to S was found to be about 1:1, which doubly confirmed the deposition of PbS clusters.

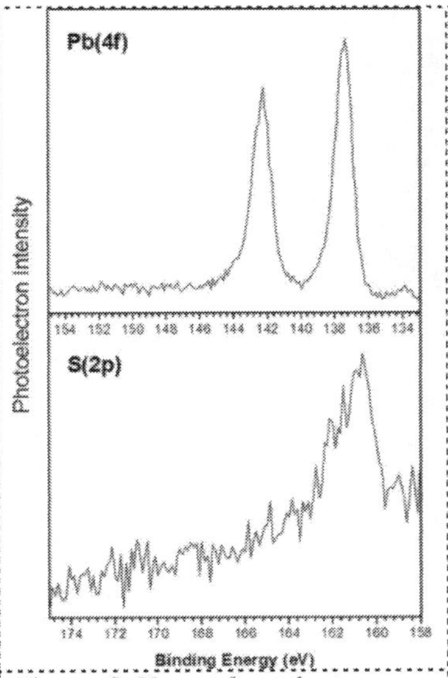

Figure 8. X-ray photoelectron spectra [Pb(4f) and S(2p) regions] of lead sulfide aggregates on HOPG.

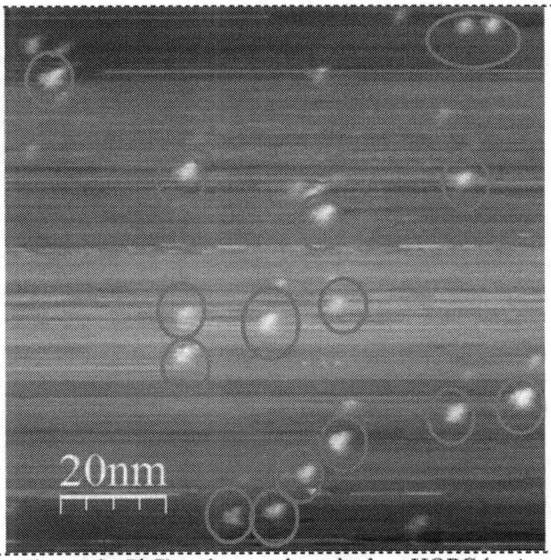

Figure 9. $(PbS)_{32}$ clusters deposited on HOPG by *in situ* STM

Given the presence of size-defined $(PbS)_{32}$ clusters on the surface as the starting point, we were assured that the observed STM images were the result of two-dimensional nucleation between $(PbS)_{32}$ clusters and/or between nano-blocks composed of multiple $(PbS)_{32}$ units. A typical STM image of the as deposited $(PbS)_{32}$ clusters at low coverage were displayed in Figure 9, in which many of these clusters exhibited non-spherical shapes, consistent with structures formed by the growth of cuboids. Two additional figures in Figure 10 are presented showing another two cluster that $(PbS)_{32}$ clusters have been deposited at low coverage. One of these figures shows an image of a square nano-block, while the other shows an image of a rectangular nano-block. Note that none of the clusters in the STM image have the dimensions of the originally deposited $(PbS)_{32}$ clusters; instead, they are significantly larger. This is because small $(PbS)_{32}$ clusters are mobile on HOPG; thus, they have aggregated into larger objects. Beyond qualitative assessments of shapes, STM imaging was also used to determine approximate lateral and vertical dimensions of the various aggregated clusters on the surface. We have observed STM images of these clusters, where the lateral dimensions are consistent with the predictions of theory. For example, the one labeled A has measured lateral dimensions, ~2 nm × ~2 nm corresponding to $(PbS)_{128}$. Likewise, the one labeled B has measured lateral dimensions,~4 nm × ~2 nm corresponding to $(PbS)_{256}$. Moreover, line-scans of adjacent nano-blocks, (e.g. $(PbS)_{128}$ and $(PbS)_{256}$) showed that both aggregates had heights in the range of 0.8 – 1 nm, consistent with the theoretical prediction of 0.88 nm for the heights of nano-blocks formed by the side-by-side aggregation of $(PbS)_{32}$ units. A further height analysis for a larger population of clusters was conducted to analyse the size distribution of the clusters. Consistent with the previous results, the height distribution of these clusters ranged from 0.8 to 1.2 nm, resulting from the two dimensional aggregation behaviours of $(PbS)_{32}$ clusters.

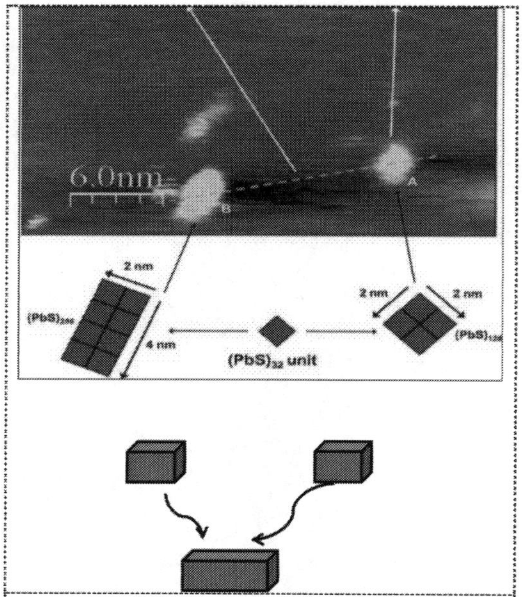

Figure 10. A STM image showing aggregates resulting from the soft-landing of mass-selected $(PbS)_{32}$ clusters onto a HOPG surface.

Therefore our STM images show shapes and dimensions that are consistent with the predictions of the theoretical part of this study. Thus, this study has provided a pathway for better understanding the mechanisms involved in the formation of this class of solids.

3.3. Titanium Oxynitride Snowflakes (Fractals)

It is known that titanium nitride clusters, $(TiN)_n$, can be prepared by ablating or sputtering titanium metal in the presence of NH_3 or N_2 gas[27; 28]. In our work, we magnetron sputtered titanium metal in the presence of ammonia and then size-selected and soft-landed $(TiN)_{141}$ onto HOPG. While still at 10^{-9} Torr, *in situ* XPS spectra of the deposited clusters was recorded to verify the chemical compositions of the as-deposited clusters.

Figure 11(a) showed the presence of XPS peak features in both the Ti (2p) and N (1s) regions, with an addition peak feature in O (1s) region that suggested the presence of oxygen within the clusters. While this was unwelcome, titanium (an excellent getter) had incorporated any available oxygen into the putative titanium nitride clusters.

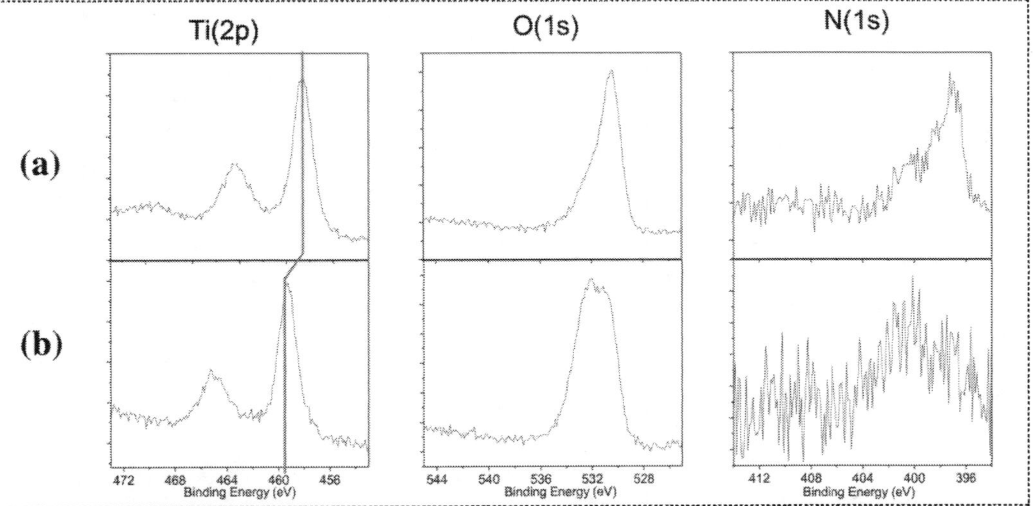

Figure 11. X-ray photoelectron spectra [Ti(2p) ,O(1s) and N(1s) regions] of (a) as-deposited and (b) air-exposed titanium oxynitride cluster aggregates on HOPG

While the sample was clearly contaminated by oxygen, thus making the as-deposited clusters essentially titanium oxynitride clusters instead of titanium nitride clusters, the clusters' surface structures were further characterized by *ex situ* AFM. This procedure necessitated exposing the sample to air. To our surprise, the resulting image in Figure 12 showed beautiful fractal (snowflake) patterns. The successive XPS characterization as shown Figure 11(b) showed the evidence for titanium and oxygen as expected, but much less nitrogen for the air-exposed clusters.

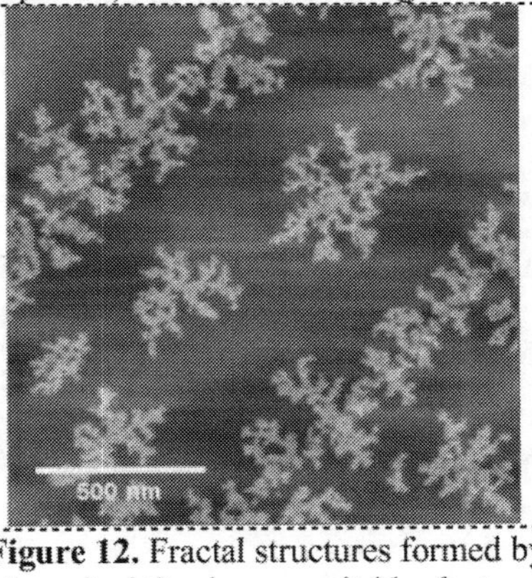

Figure 12. Fractal structures formed by deposited titanium oxynitride clusters

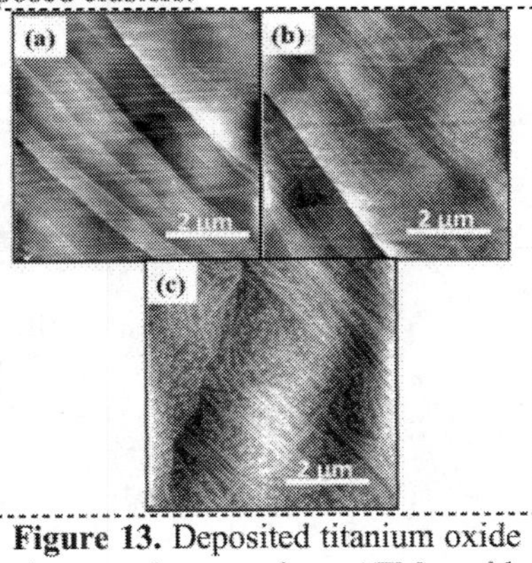

Figure 13. Deposited titanium oxide clusters by ex-situ AFM with coverage increasing from (a) to (c)

To study the effect of nitrogen on the structures, pure titanium oxide clusters with the same mass were produced by magnetron sputtering titanium metal with pure O_2 gas and with no NH_3 or N_2 gas. As expected, XPS confirmed that the cluster sample consisted only of titanium and oxygen. Then, the sample was exposed to air and recorded with *ex situ* AFM. No snowflakes had appeared for aggregated titanium dioxide clusters at low and medium coverages(Figure 13(a) and Figure 13(b), respectively). At much higher coverage, which corresponds to Figure 13(c), very small fractal structures begin to appear on the surface, indicating that, indeed, the presence of nitrogen had mattered.

The coverage of the titanium oxynitride clusters was varied systematically to investigate the formation of the fractal structures. The AFM images are shown in Figure 14, where coverage increases from (a) through (f). At the lowest coverage, nucleation along step edges is evident. As coverage increases, one observes randomly distributed aggregates on the terraces, which serve as feeds to further formation of fractal patterns. At the highest coverage, snowflakes are seen to form and to arrange themselves into intricate, elaborate patterns. Fractal patterns due to aggregation of atoms/molecules on surfaces have often been seen in the past, but only rarely due to mass-selected cluster deposition [10; 29; 30] and, to our knowledge, never under these circumstances. Usually, a diffusion-limited aggregation mechanism[31] is used to understand the growth behavior of the fractal structures. Within the growth model, the fractal formation is strongly correlated to the flux intensity of the incoming clusters from the gas phase. For low coverage, due to low flux intensity, there are not enough cluster-cluster interactions to immobilize the clusters on the terrace. Consequently, they diffuse on the terrace until they are trapped by the step edges. In contrast, as coverage increases, more cluster-cluster interactions are favored due to higher flux intensity. The abundance of cluster-cluster aggregations will then lead to the formation of nucleation seeds on the terrace as starting points for further fractal formations on the HOPG surface.

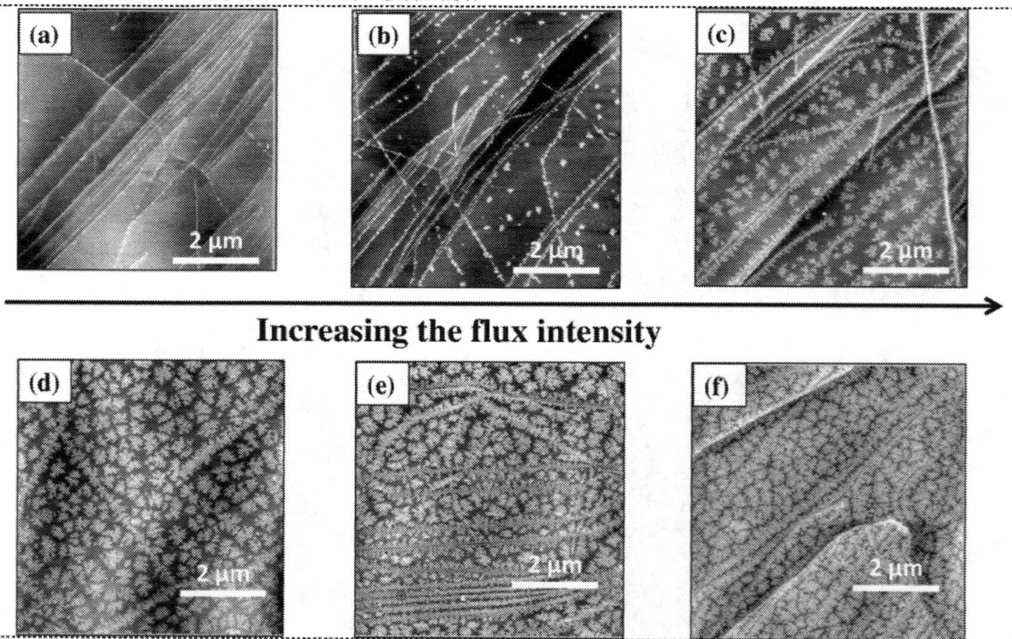

Figure 14. Deposited titanium oxynitride clusters by *ex situ* AFM with coverage increasing from (a) to (f)

A size-dependent study was also conducted by depositing the titanium oxynitride clusters with a different mass (3,700 amu) as presented in Figure 15(b). Compared to the same cluster with a higher mass (8,600 amu), the structures of clusters with lower mass showed much smaller and more compact fractal structures compared to Figure 15(a). The differences in structures of two clusters can be understood by an edge diffusion mechanism[32]; that is, clusters with a lower mass possess more active edge diffusion due to their higher mobility. Thus, once they are attached to the fractal island, they are more robust to diffuse along the edge in order to search for the adsorption sites corresponding to the global energy minimum. As a result, the deposition of smaller clusters will lead to more compact fractal patterns.

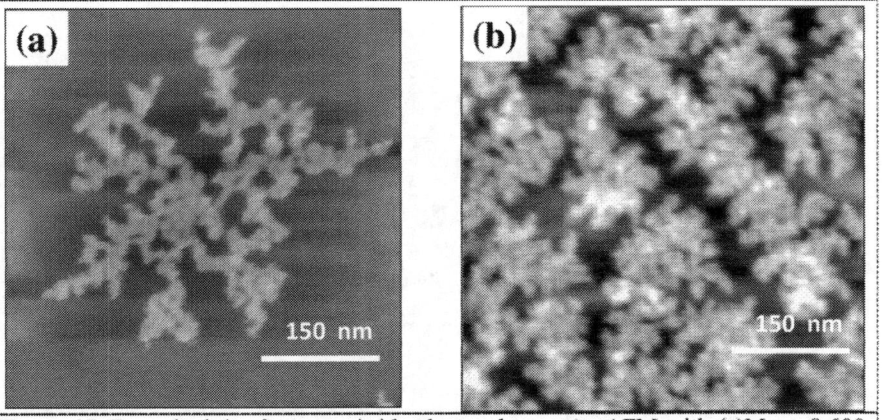

Figure 15. Deposited titanium oxynitride clusters by *ex situ* AFM with (a)Mass~8,600 amu and (b)Mass~3,700 amu

Herein, we have obtained fractal structures through the deposition of size-selected titanium oxynitride clusters on an HOPG surface. The nitrogen content was found to be responsible for the formation of fractal structures. Varying the flux intensity revealed the different stages of fractal growth. Furthermore, the fractal structures showed a strong size-dependent behavior, which provide possible pathways to regulating the fractal structure via size control.

4. Conclusion and Outlook

Three different types of systems of size-selected metal compound clusters were introduced in this proceeding. In all three systems, cluster behaviours significantly different from previously studied metal cluster systems were observed. Results from these studies illustrate how the surface morphologies of adsorbed clusters are influenced by their chemical composition, their size, and their coverage for metal compound clusters. The knowledge acquired here will benefit further applications of these metal compound clusters. For example, all three systems in this work have potential catalytic applications, i.e. molybdenum oxides are widely used in oxidative dehydrogenation, and lead sulfide is a good photocatalyst due its tunable band gap to adsorb solar light. Thus, these supported size-selected metal compound clusters may offer a new pathway to study their catalytic properties correlated to their size, structures, and chemical compositions.

Acknowledgments

This material is based upon work supported by the Division of Materials Science and Engineering, Basic Energy Sciences, U.S. Department of Energy, under Grant No. DE-FG02-09ER46558 (K.H.B).

References

[1] Castleman A W and Bowen K H 1996 *The Journal of Physical Chemistry* **100** 12911-44
[2] Morse M D, Geusic M E, Heath J R and Smalley R E 1985 *The Journal of Chemical Physics* **83** 2293-304
[3] de Heer W A 1993 *Reviews of Modern Physics* **65** 611-76
[4] Hashmi A S K and Hutchings G J 2006 *Angewandte Chemie International Edition* **45** 7896-936
[5] Zemski K A, Justes D R, Bell R C and Castleman A W 2001 *The Journal of Physical Chemistry A* **105** 4410-7
[6] Benten W, Nilius N, Ernst N and Freund H J 2005 *Physical Review B* **72** 045403
[7] Vossmeyer T, Katsikas L, Giersig M, Popovic I G, Diesner K, Chemseddine A, Eychmueller A and Weller H 1994 *The Journal of Physical Chemistry* **98** 7665-73

[8] Popok V N, Barke I, Campbell E E B and Meiwes-Broer K-H 2011 *Surface Science Reports* **66** 347-77
[9] Jensen P 1999 *Reviews of Modern Physics* **71** 1695-735
[10] Bardotti L, Jensen P, Hoareau A, Treilleux M, Cabaud B, Perez A and Aires F C S 1996 *Surface Science* **367** 276-92
[11] Bréchignac C, Cahuzac P, Carlier M d F, A. Masson F, Mory C, Colliex C and Yoon B 1998 *Physical Review B* **57** R2084-R7
[12] Fuchs G, Melinon P, Santos Aires F, Treilleux M, Cabaud B and Hoareau A 1991 *Physical Review B* **44** 3926-33
[13] Goldby I M, Kuipers L, von Issendorff B and Palmer R E 1996 *Applied Physics Letters* **69** 2819-21
[14] Tong X, Benz L, Kemper P, Metiu H, Bowers M T and Buratto S K 2005 *Journal of the American Chemical Society* **127** 13516-8
[15] Tournus F, Bardotti L and Dupuis V 2011 *Journal of Applied Physics* **109** 114309-4
[16] Di Vece M, Young N P, Li Z, Chen Y and Palmer R E 2006 *Small* **2** 1270-2
[17] Vandamme N, Janssens E, Vanhoutte F, Lievens P and Haesendonck C V 2003 *Journal of Physics: Condensed Matter* **15** S2983
[18] Saito Y, Murata K, Hamaguchi K, Fujita H, Kotake S, Suzuki Y, Senoo M, Hu C W, Kasuya A and Nishina Y 1998 *Journal of Cluster Science* **9** 123-30
[19] Bardotti L, Prével B, Treilleux M, Mélinon P and Perez A 2000 *Applied Surface Science* **164** 52-9
[20] Wepasnick K A, Li X, Mangler T, Noessner S, Wolke C, Grossmann M, Gantefoer G, Fairbrother D H and Bowen K H 2011 *J. Phys. Chem. C* **115** 12299-307
[21] Price S P, Tong X, Ridge C, Shapovalov V, Hu Z, Kemper P, Metiu H, Bowers M T and Buratto S K 2011 *Surface Science* **605** 972-6
[22] White J M, Bondarchuk O, Huang X, Kim J, Kay B D, Wang L S and Dohnalek Z 2006 *Angew. Chem.* **45** 4786-9
[23] Kiran B, Kandalam A K, Rallabandi R, Koirala P, Li X, Tang X, Wang Y, Fairbrother H, Gantefoer G and Bowen K 2012 *The Journal of Chemical Physics* **136** 024317-7
[24] Rao C N R and Raveau B 1998 *Transition metal oxides : structure, properties, and synthesis of ceramic oxides* (New York: Wiley-VCH)
[25] Koirala P, Kiran B, Kandalam A K, Fancher C A, de Clercq H L, Li X and Bowen K H 2011 *The Journal of Chemical Physics* **135** 134311-12
[26] Osherov A, Matmor M, Froumin N, Ashkenasy N and Golan Y 2011 *The Journal of Physical Chemistry C* **115** 16501-8
[27] Chen Z Y and Castleman J A W 1993 *The Journal of Chemical Physics* **98** 231-5
[28] Kraft J, Rattunde O, Rusu O, Häfele A and Haberland H 2002 *Surface and Coatings Technology* **158–159** 131-5
[29] Bréchignac C, Cahuzac P, Carlier F, Colliex C, Leroux J, Masson A, Yoon B and Landman U 2002 *Physical Review Letters* **88** 196103
[30] Bottcher A, Weis P, Jester S-S, Loffler D, Bihlmeier A, Klopper W and Kappes M M 2005 *Physical Chemistry Chemical Physics* **7** 2816-20
[31] Meakin P 1983 *Physical Review Letters* **51** 1119-22
[32] Röder H, Bromann K, Brune H and Kern K 1995 *Physical Review Letters* **74** 3217-20

International Conference on Dynamics of Systems on the Nanoscale (DySoN 2012) IOP Publishing
Journal of Physics: Conference Series **438** (2013) 012006 doi:10.1088/1742-6596/438/1/012006

Simulation of nanofractal dynamics with MBN Explorer

Ilia A. Solov'yov[1,2,3], **and Andrey V. Solov'yov**[1,3,4]

[1] Virtual Institute on Nano Films (VINF), Allée des Noisetiers, 2 bte 30 Angleur, Belgium
[2] Beckman Institute for Advanced Science and Technology, University of Illinois at Urbana-Champaign, 405 N. Mathews Ave, Urbana Illinois 61801, USA
[3] On leave from A.F. Ioffe Physical-Technical Institute, Politechnicheskaya Str. 26, 194021 St. Petersburg, Russia
[4] Frankfurt Institute for Advanced Studies, Ruth-Moufang-Str. 1, 60438 Frankfurt am Main, Germany

E-mail: `ilia@illinois.edu`

Abstract. One of the goals of nanotechnology is the development of controlled, reproducible, and industrially transposable nanostructured materials. In this context, controlling of the final architecture of such materials by tuneable parameters is one of the fundamental problems. Post-growth processes occurring in patterns grown on a surface were studied using a multi-purpose computer code MBN Explorer introduced in the present paper. The package allows to model molecular systems of varied level of complexity, and in the present paper was used, in particular, to study dynamics of silver nanofractal formation and fragmentation on graphite surface. We demonstrate that the detachment of particles from the fractal and their diffusion within the fractal and over the surface determines the shape of the islands remaining on a surface after the fractal fragmentation.

1. Introduction

In recent years, numerous nanosystems possessing unique structural, optical, electric and magnetic properties have been discovered [1–11]. The aggregation of atoms and small molecules into clusters, nanoparticles, micro-droplets is a process in which a wide range of complex bio-, nano- and mesoscopic objects can be created. Often, such systems are seen as building blocks for new nanostructured materials with specific tailored properties, and many of these systems have become a subject of intensive investigations because of the variety of potentially important applications [5, 8, 12–18].

The present paper features a multi-purpose computer package, MesoBioNano Explorer (MBN Explorer), which can be used for advanced theoretical characterization and analysis of a variety of bio- nano systems [1]. The package has been specifically developed to enable a unified approach for simulating of complex molecular structure and dynamics on different temporal and spatial scales, ranging from atoms to clusters, macromolecules, crystals and surface patterns. [1] The ultimate goal of MBN Explorer is to expand the understanding of mechanisms of stability,

[1] The binary files of MBN Explorer, the user's guide and a number of representative examples are available upon registration online at http://www.mbnexplorer.com/.

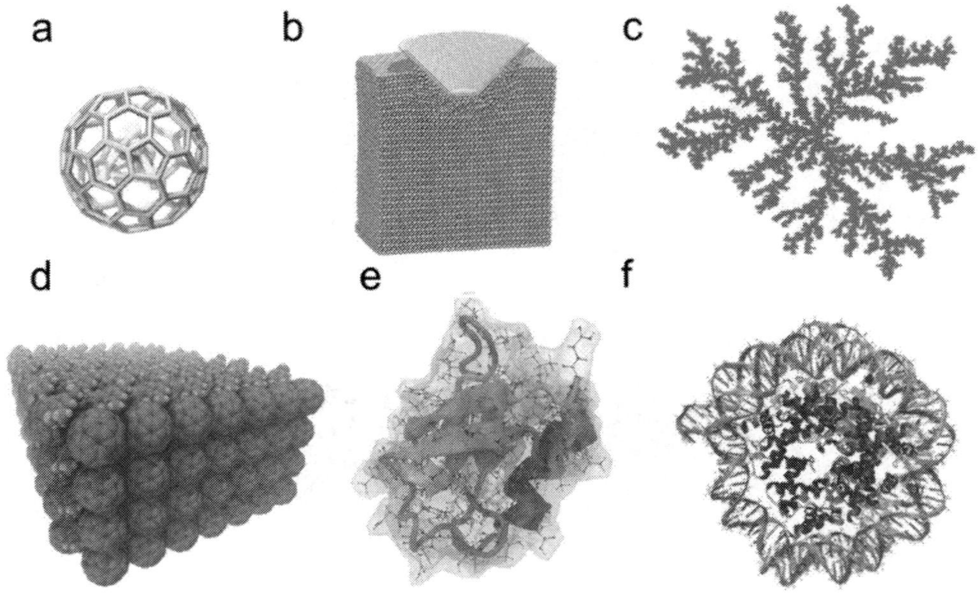

Figure 1. Variety of molecular systems which can be simulated using MBN EXPLORER: (a) encapsulated clusters, (b) nanoindentation of titanium crystal, (c) nanofractals, (e) proteins, (f) biomolecules, e.g., DNA on the histone. Structure and properties of any of these objects or any of their combinations can be studied using MBN EXPLORER. The figure has been adapted from Ref. [19].

self-organization and growth, as well as the ways of manipulation and control of bio-nano systems with potential applications in nanotechnology, microelectronics and medicine.

Figure 1 highlights a variety of molecular systems, which can be simulated using MBN EXPLORER. In particular, MBN EXPLORER is suited to compute the system's energy, to optimize molecular structures, as well as to explore the molecular and random walk dynamics. MBN EXPLORER allows to use a broad variety of interatomic potentials, to model different molecular systems, such as atomic clusters [20–23], fullerenes, nanotubes [24], polypeptides, proteins [25], composite systems [26–29], nanofractals [30–32], etc.

In this paper we overview the key functionality of the present version of MBN EXPLORER. Then we utilize the program for a concrete case-study. In particular, the investigation of the dendritic structures (fractals) has recently raised an increased interest [10, 33–40] and, therefore, is in the focus of the present investigation. The formation of fractals provides a natural framework for studying disordered structures on a surface because fractals are generally observed in far-from-equilibrium growth regime.

Here we make an important step towards understanding of the nanofractal stability. Using MBN EXPLORER we perform analysis of the post-growth processes occurring in a nanofractal on a surface. We demonstrate that the detachment of particles from the fractal and their diffusion within the fractal and over the surface determines the shape of the islands remaining on a surface after the fractal fragmentation. Additionally we advocate MBN EXPLORER as a universal tool for studying structure and dynamics, as it has also been succesfully used for studying such processes as nanoindentation of titanium alloys [41], conventional molecular dynamics [19], and other problems in bio-nano science [42].

2. MBN Explorer functionality

MBN Explorer is a multi-purpose computer code [1] availiable for download at http://www.mbnexplorer.com/. The essential functionality modules of MBN Explorer are summarized below, subdivided into several categories. In particular, MBN Explorer allows to perform the following *computational tasks*:

- Single point energy calculation;
- Molecular structure optimization;
- Classical molecular dynamics simulation;
- Relativistic molecular dynamics simulation;
- Random walk dynamics;

MBN Explorer has a rapidly expanding list of *interaction potentials*, that include

- Power potential;
- Exponential potential;
- Coulomb potential;
- Yukawa potential;
- Moliere potential;
- Pacios potential;
- Lennard-Jones potential;
- Morse potential;
- Girifalco potential;
- Dzugutov potential;
- Quasi Sutton-Chen potential;
- Sutton-Chen potential;
- Gupta potential;
- Brenner potential;
- Tersoff potential;
- Finnis-Sinclair potential;
- CHARMM molecular mechanics potential;
- Dissociative CHARMM molecular mechanics potential;
- External electric fields;
- External viscous fields;

The program is developed based on the following *basic algorithms*

- OpenMP parallelization for shared memory computers;
- Velocity Verlet and Leapfrog integrators;
- Velocity quenching and conjugate gradient optimization methods;
- Linked Cell algorithm;
- Periodic and reflective boundary conditions;
- Langevin and Berendsen thermostats;
- Ewald summation for long-range interactions;

- Rigid body Euler dynamics;
- Adaptive 4th order Runge-Kutta integrator for relativistic dynamics;
- External forces and user-defined particle manipulation;
- Momentum and angular momentum control;

MBN Explorer supports different standards of *input and output formats*, in particular:

- Cartesian XYZ;
- Binary DCD;
- Protein data bank, PDB;
- X-PLOR PSF;

More detailed description of algorithms and methods implemented in MBN Explorer is availiable in the key publication [1] and also on the website http://www.mbnexplorer.com/.

3. Stochastic Monte-Carlo-based dynamics in MBN Explorer

This section describes the basics of the stochastic Monte-Carlo-based dynamics algorithm implemented in MBN Explorer. In this case, the time-evolution of a coarse-grained molecular system is modeled stepwise in time, where at each step of the simulation, the system undergoes a structural transformation with a certain probability. The new configuration of the system is then used as the starting point for the next simulation step. The transformation of the system is governed by predefined kinetic rates which are determined by interparticle interactions. Due to its probabilistic nature, stochastic Monte-Carlo-based dynamics allows to study processes on time scales significantly exceeding time scales of conventional molecular dynamics and is ideally suited for simulations where details of individual kinetic processes are not important, and can be parameterized by a few kinetic rates.

MBN Explorer allows to study random walk dynamics in open molecular systems. The implemented algorithm is illustrated in Fig. 2. Consider a molecular system at time $t = 0$ which evolves in time. At periodic time intervals new particles are added to the system with a certain probability. At each step of the simulation the list of mobile particles is constructed, and these particles are then displaced in a random direction with a certain probability. The algorithm is repeated until the necessary number of simulation steps is performed.

To simplify calculations, the random walk motion of particles is simulated on a grid. A grid represents an artificial subdivision of the simulation space into cells which accommodate particles of the system. A particle in the grid occupies a single cell and can move from this cell only to a neighboring non-occupied cell. The number of possible directions for a particle displacement depends on the grid packing and on the number of occupied neighbor cells. The present version of MBN Explorer considers interaction between neighboring particles in Arrhenius approximation. Thus, two coarse-grained particles being in the stochastic Monte-Carlo-based dynamics are assumed to have a bond if they occupy two neighboring grid cells.

Monte-carlo-based dynamics module of MBN Explorer allows to study self-assembly, pattern formation and pattern fragmentation on a surface. in this case the diffusion rate of a free particle Γ_d is determined by the diffusivity of a particle on the surface. If a particle is bound to periphery of a group of particles (an island) it can experience diffusion along the periphery, thereby maintaining some of its bonds with neighboring particles, or detach from the island. The probability of particle detachment in MBN Explorer is given by:

$$\Gamma_{de} = \Gamma_d \exp\left[-\frac{lE_b}{k_\mathrm{B}T} - \frac{\Delta\mu}{k_\mathrm{B}T}\right], \quad (1)$$

where l is the number of bonds broken after particle detachment, E_b is the binding energy of two particles, $\Delta\mu$ is the chemical potential of particle detachment [36, 43–45], T is the temperature

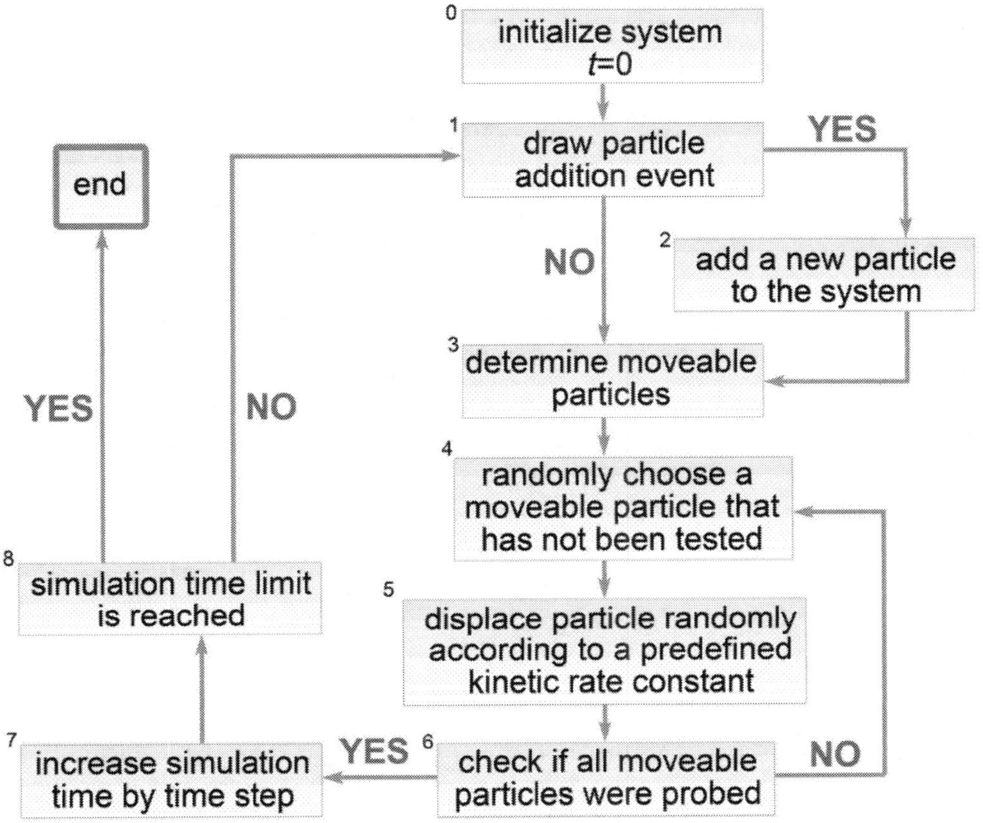

Figure 2. The principal scheme of stochastic Monte-Carlo algorithm implemented in MBN Explorer and used for random walk dynamics simulations. The figure has been adapted from Ref. [1].

of the system and k_B is the Boltzmann constant. The peripheral diffusion of a particle is parameterized as

$$\Gamma_{pd} = \Gamma_d \exp\left[-\frac{mE_b}{k_B T} - \frac{n\Delta\epsilon}{k_B T}\right], \tag{2}$$

where m is the number of bonds that are broken due to particle motion, n is the number of maintained neighbor bonds between two particles and $\Delta\epsilon$ is the diffusion energy barrier.

4. Results

Using the method outlined above we have performed analysis of the silver cluster fractal post-growth relaxation on a graphite surface. The fractal structure shown in Fig. 3 has been chosen for the further investigation of the post-growth relaxation processes. The diameter of the fractal in the considered example is 1.2 μm, which corresponds to the diameter of the experimentally grown structures [10, 36, 37], and consist of 38,012 Ag_{500} particles.

An important quantity in the stochastic Monte-Carlo-based dynamics method is the

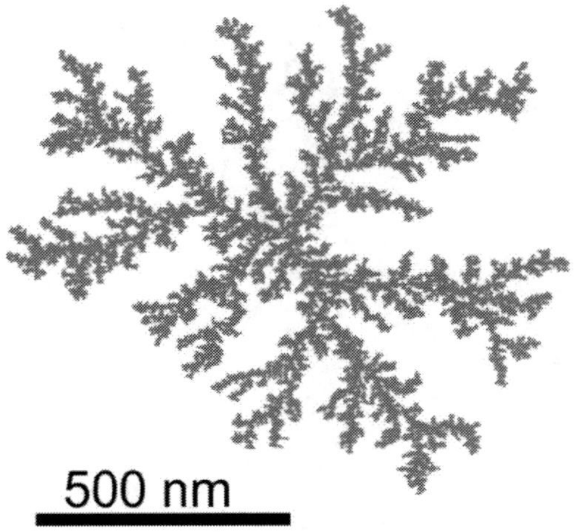

Figure 3. The fractal structure chosen for the further investigation of the post-growth relaxation processes in fractals. The figure has been adapted from Ref. [1].

simulation time step, Δt, which defines the characteristic time for particle diffusion over a surface as [30–32, 46, 47]

$$\Delta t(T) = \frac{d_0^2}{4D(T)}. \tag{3}$$

Here d_0 is the diameter of a particle and $D(T)$ is the diffusion coefficient of the particle at a given temperature, T.

Stability of the silver cluster fractals on graphite surface has been studied earlier in detail [30–32] using a prototype version of MBN Explorer for a smaller system, and here we illustrate it for a μm-large fractal which evolves on the timescale of seconds. Stability of silver cluster fractals on graphite surface is of great relevance to experiment, as the post-growth relaxation of silver fractals has been extensively studied over the last decade [10, 36, 37]. To make our simulations consistent with experimental observations we thus have used the model parameters derived from the experiment: the diameter of a particle has been taken equal to 2.5 nm, and the diffusion coefficient of a silver cluster on graphite at room temperature was reported [37] to be equal to $2 \cdot 10^{-7}$ cm^2/s. Substituting this value into Eq. (3), one obtains $\Delta t(300 \text{ K}) = 78$ ns, which allows one to simulate the process during the time period

$$t = N_{step}\Delta t, \tag{4}$$

where N_{step} is the number of simulation steps. The simulation stepsize Δt in stochastic Monte-Carlo-based dynamics is significantly larger than the stepsize of conventional molecular dynamics (where it is typically taken as 1-2 fs), and, therefore, allows to perform simulations on significantly larger time scales. In the present example we analyze several paths of fractal fragmentation. The rate of fractal decay depends on the interparticle interaction, and it defines the morphology of the fragments that are formed during the process. Snapshots of the structures arising at different stages of the fragmentation process simulated at different parameters of interparticle interactions are shown in Fig. 4. This example shows that fragmentation paths and the fragments morphology depends on the interparticle interaction energies. In particular we analyze the post-growth evolution of the fractal on surface depending on the particle interaction energy E_b and peripheral diffusion barrier $\Delta\epsilon$, see Eq. (1)-(2) for the definition of E_b and $\Delta\epsilon$.

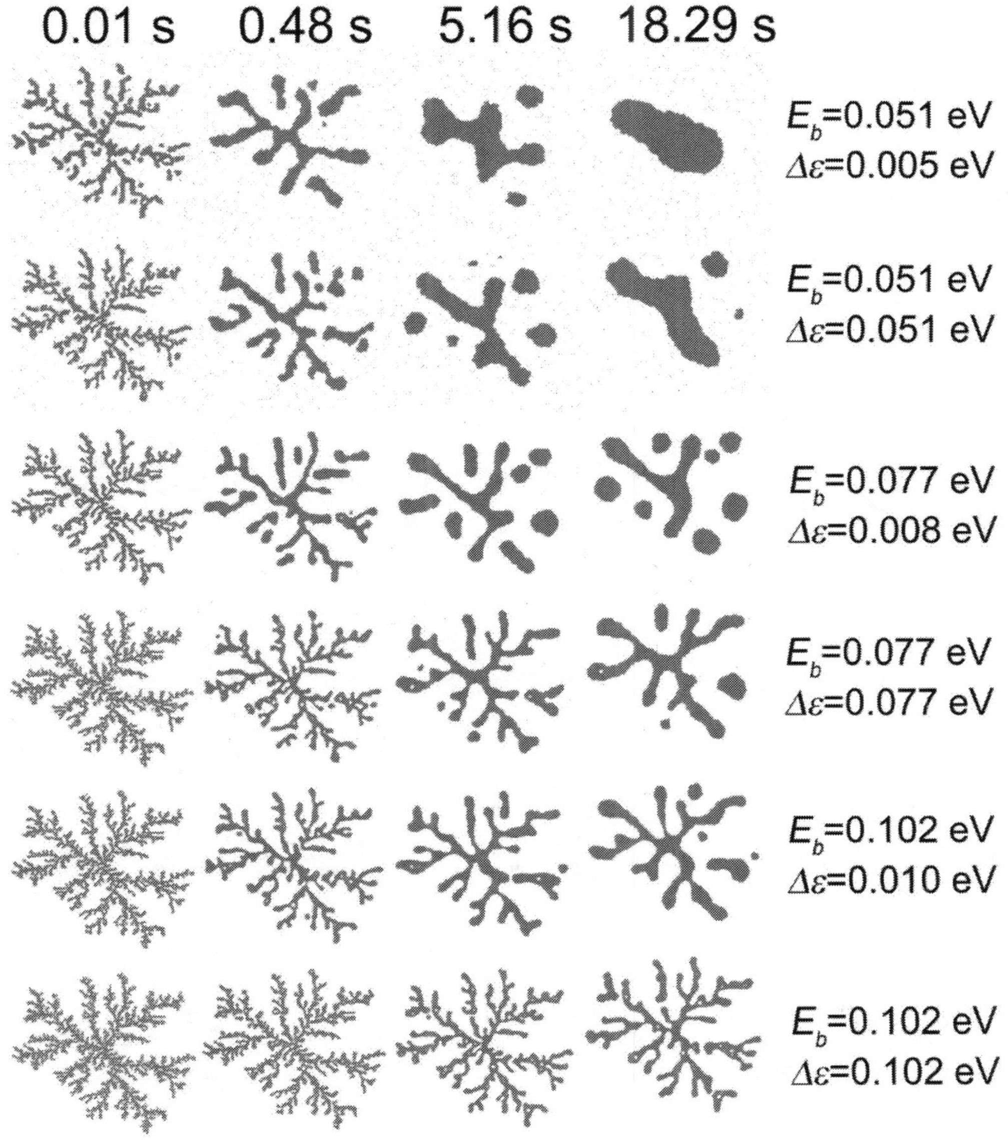

Figure 4. Evolution of a fractal structure on a 4.3×5.0 μm^2 substrate with periodic boundary conditions at temperature $T = 300$ K. The initial fractal structure, consisting of 38,012 particles (shown in Fig. 3) undergoes fragmentation in different final states depending on the interparticle interactions in the system (shown right). Numbers above the images indicate the simulation time t, see Eq. (4), at which the corresponding snapshots were rendered. The figure has been adapted from Ref. [1].

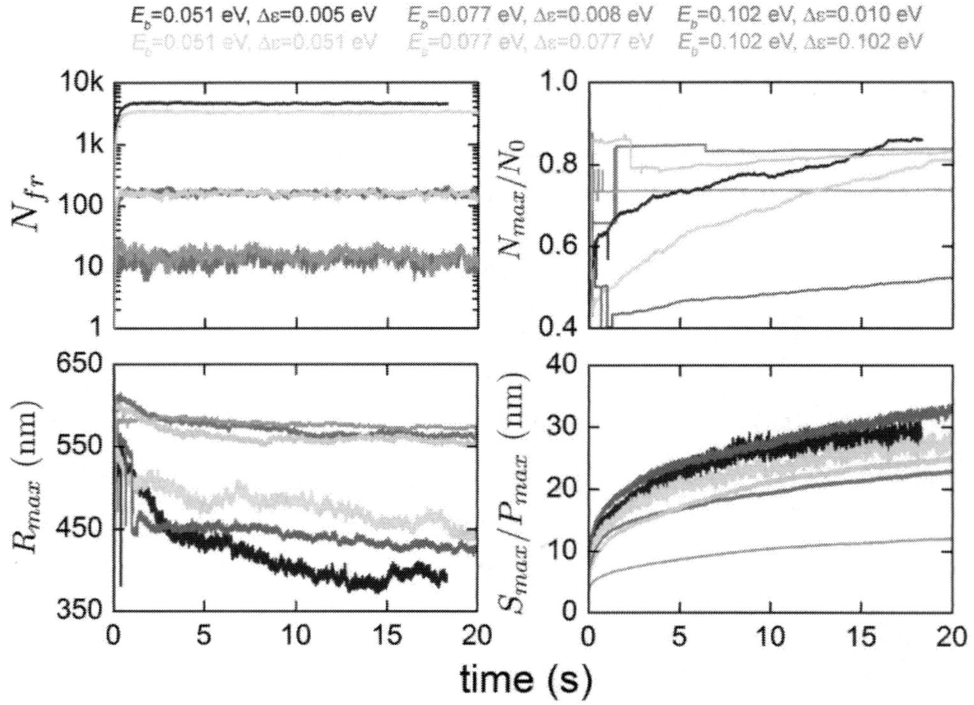

Figure 5. Time evolution of the number of fragments N_{fr}, normalized maximal island size N_{max}/N_0 (where $N_0 = 38,012$ is the number of particles in the initial fractal structure), maximal radius of an island R_{max}, and area to perimeter ratio for the largest island S_{max}/P_{max} calculated for the fractal structure shown in Fig. 4. The fractal fragmentation has been analyzed for different values of the binding energy E_b and the barrier energy $\Delta\epsilon$, as indicated in the figure legend. The figure has been adapted from Ref. [1].

Figure 4 shows that for $E_b = 0.051$ eV, $\Delta\epsilon = 0.005$ eV the entire fractal structure melts and forms a large compact droplet, while two additional smaller droplet are seen for $E_b = 0.051$ eV, $\Delta\epsilon = 0.051$ eV after 18.29 s of simulation. In both cases the binding energy E_b between the particles is small, allowing an easy detachment of particles, but at the same time it is large enough to make the characteristic particle detachment time comparable with the characteristic particle nucleation time, thereby preventing the system from entire defragmentation. Thus, the fragmentation path at $E_b = 0.051$ eV goes via the rearrangement of the entire system and results in the formation of large stable droplets.

An increase of the interparticle interaction energy leads to the change of the fractal fragmentation pattern. As seen in Fig. 4 at $E_b = 0.08 - 0.1$ eV the fractal fragments into several compact droplets, while the overal profile of the initial fractal structure is preserved even after t =18.29 s. The analysis of morphology of the created patterns leads us to the following main conclusions: (i) the growth of E_b leads to the decrease of the average fractal-core branch width during the fragmentation process. This happens because the detachment of particles from the fractal becomes energetically unfavorable process, and the fractal fragments slower. (ii) The increase of the peripheral diffusion barrier energy $\Delta\epsilon$ suppresses the diffusion of particles, resulting in a slower evolution and fragmentation of the fractal shape.

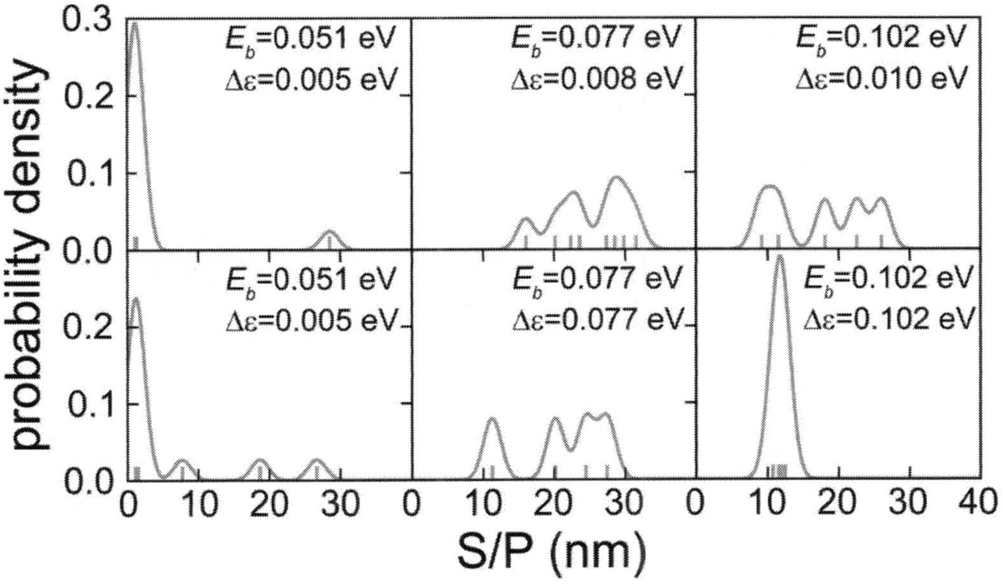

Figure 6. Probability density functions of the S/P ratio calculated for the fractal fragments after $t = 18.29$ s with different sets of the model parameters. The calculated distributions correspond to the final patterns shown in Fig. 4. The figure has been adapted from Ref. [1].

An important characteristic of the fractal fragmentation process is the number of fragments at a given time, N_{fr}. The smallest fragment is a single particle. The time evolution of the number of fragments calculated for different sets of model parameters is shown in Fig. 5. The figure shows that the number of fragments in the six considered cases rapidly approaches an asymptotic value which decreases with increasing the E_b value. It is interesting to note that at $E_b = 0.051$, 0.077 and 0.102 eV there are $\sim$1000, 100, and 10 fragments on the surface, respectively, i.e., increasing the binding energy between particles in the fractal on $k_{\rm B}T$ leads to an order of magnitude change of the number of fragments on the surface. For the largest island on the surface we have calculated the normalized maximal island size N_{max}/N_0, see Fig. 5, where N_0 is the number of particles in the initial fractal structure. This characteristic is useful to illustrate the dispersion of island sizes on the surface and demonstrates that except the $E_b = 0.1$ eV the largest fragment on the surface still continues to nucleate individual particles dispersed on the surface.

Useful quantity for the characterization of patterns on a surface are: (i) the ratio between the area and perimeter of the structures (S/P ratio) as this ratio characterizes the branch width of the fragmenting structures [10], and (ii) the maximal radius of a structure R_{max}. Figure 5 shows the area to perimeter ratio for the largest island on the surface, S_{max}/P_{max}. The dependencies calculated for all studied energies show a monotonous increase with slight fluctuations due to structural rearrangements in the system. Note that the smallest asymptotic value of S_{max}/P_{max} is observed for $E_b = 0.102$ eV and $\Delta\epsilon = 0.102$ eV, because the width of the fractal branches after 20 s of fragmentation are the smallest in this case.

Figure 6 shows the probability density functions of the S/P ratio calculated for the morphology on surface after $t = 18.29$ s of the fractal post-growth relaxation. The distributions plotted in Fig. 6 have been obtained with the model parameters $E_b = 0.05 - 0.1$ eV, $\Delta\epsilon = 0.1E_b$

and $\Delta\epsilon = 1.0E_b$ and correspond to the final patterns shown in Fig. 4. The position of the maxima in these distribution functions allows to determine the branch width of the probable fragments on the surface. Thus, for $E_b = 0.05$ eV fragments of smaller size appear on the surface with increased probability together with one large fragment having $S/P \sim 30$ nm. Contrary, for $E_b = 0.1$ eV, $\Delta\epsilon = 0.1$ eV the probability distribution of the S/P ratio appears to be symmetric, centered at $S/P = 11$ nm. Note that Fig. 6 did not account for the area to perimeter ratio of individual particles and particle dimers on surface, as such configurations are typically not visible in experiment [10, 36, 37].

In this example we have performed analysis of the post-growth processes occurring in a nanofractal on a surface using the method implemented in MBN Explorer, which models the internal dynamics of particles in a fractal and accounts for their diffusion and detachment. We have demonstrated that these kinetic processes are responsible for the shape of the islands created on a surface after the post-growth relaxation. In the present example we have studied particle dynamics in 2D, while MBN Explorer allows to perform stochastic Monte-Carlo-based dynamics in 3D. This is especially interesting to do, because there are many examples of three dimensional fractal systems in biology [48, 49], where the dendritic shapes are rather common.

5. Conclusion

MBN Explorer is a program with several important features, which make it potentially interesting to diverse communities including physicists, chemists, biologists and material scientists. MBN Explorer permits energy calculation, structure optimization, molecular and random walk dynamics simulations. The development of a new program for studying molecular structures and dynamics is motivated by the need of a universal extendable code, which would permit investigation of molecular processes on the nano- and meso-scales. Through the combination of advanced MD techniques and simplified coarse-grained models, MBN Explorer is suitable for the description of molecular systems on the basis of a multiscale approach, which is not provided by the most of existing standard molecular dynamics codes: the standard codes would require additional input, involving either plugin or module development, to permit the multiscale calculations, while MBN Explorer supports it intrinsically.

As an example of MBN Explorer extended functionality, the program allows for straightforward simulations using Monte-Carlo-based dynamics, classical molecular dynamics involving a large number of rigid molecules of arbitrary shapes, molecular dynamics simulations of bio-nano systems modeled through a combination of pairwise and many-body potentials (e.g., a gold nanoparticle marker linked to a DNA molecule). In the paper we have considered only one illustrative example, which outline the key features of MBN Explorer, but, of course, do not describe all of them, since MBN Explorer allows to combine most of its features in a user-friendly manner, permitting simulations of various molecular assemblies in special environments. It is worth noting that although most of the computational features of MBN Explorer can be found in other computer packages, MBN Explorer is unique because it combines a vast majority of these features into one computationally efficient code. The current version of MBN Explorer supports calculations on the multiple CPU cores with sheared memory by means of OpenMP parallelization technique.

MBN Explorer is made compatible with the popular molecular dynamics programs NAMD [50], CHARMM [51] and visualization program VMD [52]. This permits to easily swap files between the programs and prepare molecular systems for simulations in a straightforward way. The visualization support through the VMD program is necessary for analysis of MBN Explorer output and the build-in Tcl scripting language in VMD allows to develop specialized plugins to perform the extended analysis of the calculations results.

Acknowledgments
We acknowledge the Frankfurt Center for Scientific Computing for providing computer resources that have contributed to the research results reported within this paper. I.S. acknowledges support as a Beckman Fellow.

References

[1] Solov'yov I A, Yakubovich A V, Nikolaev P V, Volkovets I and Solov'yov A V 2012 *Journal of Computational Chemistry* **33** 2412–2439
[2] Neto A H C, Guinea F, Peres N M R, Novoselov K S and Geim A K 2009 *Rev. Mod. Phys.* **81** 109–162
[3] Ohkoshi S, Imoto K, Tsunobuchi Y, Takano S and Tokoro H 2011 *Nat. Chem.* **doi:10.1038/nchem.1067**
[4] Tan S J, Campolongo M J, Luo D and Cheng W 2011 *Nat. Nanotech.* **6** 268–276
[5] Nie Z, Petukhova A and Kumacheva E 2010 *Nat. Nanotech.* **5** 15–25
[6] Adler-Abramovich L, Aronov D, Beker P, Yevnin M, Stempler S, Buzhansky L, Rosenman G and Gazit E 2009 *Nat. Nanotech.* **4** 849–854
[7] Yan R, Gargas D and Yang P 2009 *Nat. Photon.* **3** 569–576
[8] Mann S 2009 *Nat. Mat.* **8** 781–792
[9] Withers N 2009 *Nat. Chem.* **doi:10.1038/nchem.476**
[10] Lando A, Kébaïli N, Cahuzac P, Masson A and Bréchignac C 2006 *Phys. Rev. Lett.* **97** 133402
[11] Carlier F, Benrezzak S, Cahuzac P, Kébaïli N, Masson A, Srivasta A K, Colliex C, de Frutos M and Bréchignac C 2006 *Nanoletters* **6** 1875–1879
[12] Stuart M A C, Huck W T S, Genzer J, Müller M, Ober C, Stamm M, Sukhorukov G B, Szleifer I, Tsukruk V V, Urban M, Winnik F, Zauscher S, Luzinov I and Minko S 2010 *Nat. Mat.* **9** 101–113
[13] Nie Z and Kumacheva E 2008 *Nat. Mat.* **7** 277–290
[14] Sun T and Qing G 2011 *Adv. Mat.* **23** H57–H77
[15] Ciofani G 2010 *Expert Opinion on Drug Delivery* **7** 889–893
[16] Yang P, Yan R and Fardy M 2010 *Nano Lett.* **10** 1529–1536
[17] Schneider G F, Kowalczyk S W, Calado V E, Pandraud G, Zandbergen H W, Vandersypen L M K and Dekker C 2010 *Nano Lett.* **10** 3163–3167
[18] Garaj S, Hubbard W, Reina A, Kong J, Branton D and Golovchenko J A 2010 *Nature* **467** 190–194
[19] Yakubovich A, Solov'yov I and Solov'yov A 2013 *Physics Procedia* **40** 93 – 99
[20] Solov'yov I A, Solov'yov A V and Greiner W 2003 *Int. J. Mod. Phys. E* **13** 697–736
[21] Obolensky O I, Solov'yov I A, Solov'yov A V and Greiner W 2005 *Comp. Lett.* **1** 313–318
[22] Koshelev A, Shutovich A, Solov'yov I A, Solov'yov A V and Greiner W 2003 *Proceedings of International Workshop "From Atomic to Nano-scale", Old Dominion University* 184–194
[23] Solov'yov I A, Solov'yov A V, Greiner W, Koshelev A and Shutovich A 2003 *Phys. Rev. Lett.* **90** 053401–(1–4)
[24] Solov'yov I A, Mathew M, Solov'yov A V and Greiner W 2008 *Phys. Rev. E* **78** 051601–(1–13)
[25] Solov'yov I A, Yakubovich A V, Solov'yov A V and Greiner W 2007 *Phys. Rev. E* **75** 051912–(1–9)
[26] Geng J, Solov'yov I A, Zhou W, Solov'yov A V and Johnson B F G 2009 *J. Phys. Chem. C* **113** 6390–6397
[27] Solov'yov I A, Geng J, Solov'yov A V and Johnson B F G 2009 *AIP Conf. Proc.* **1197** 89–102
[28] Solov'yov I A, Geng J, Solov'yov A V and Johnson B F G 2009 *Chem. Phys. Lett.* **472** 166–170
[29] Geng J, Solov'yov I A, Reid D G, Skelton P, Wheatley A E H, Solov'yov A V and Johnson B F G 2010 *Phys. Rev. B* **81** 214114–(1–13)
[30] Dick V V, Solov'yov I A and Solov'yov A V 2009 *AIP Conf. Proc.* **1197** 76–88
[31] Dick V V, Solov'yov I A and Solov'yov A V 2010 *J. Phys.: Conf. Ser.* **248** 012025–(1–11)
[32] Dick V V, Solov'yov I A and Solov'yov A V 2011 *Phys. Rev. B* **84** 115408–(1–14)
[33] Conti M, Meerson B and Sasorov P V 1998 *Phys. Rev. Lett.* **80** 4693
[34] Sharon E, Moore M G, McCormick W D and Swinney H L 2003 *Phys. Rev. Lett.* **91** 205504
[35] Sempéré R, Bourret D, Woignier T, Phalippou J and Jullien R 1993 *Phys. Rev. Lett.* **71** 3307
[36] Bréchignac C, Cahuzac P, Carlier F, Colliex C, de Frutos M, Kébaïli N, Roux J L, Masson A and Yoon B 2003 *Eur. Phys. J. D* **24** 265–268
[37] Lando A, Kébaïli N, Cahuzac P, Colliex C, Couillard M, Masson A, Schmidt M and Bréchignac C 2007 *Eur. Phys. J. D* **43** 151–154
[38] Li W, Zheng X and Fei W 2009 *Vacuum* **83** 949
[39] Jensen P, Barabási A L, Larralde H, Havlin S and Stanley H E 1994 *Phys. Rev. B* **50** 15316
[40] Jensen P 1999 *Rev. Mod. Phys.* **71** 1695
[41] Yakubovich A V, Verkhovtsev A V, Hanauske M and Solov'yov A V 2013 *Computational Materials Science* – ISSN 0927-0256 URL `http://www.sciencedirect.com/science/article/pii/S0927025613000037`
[42] Surdutovich E, Yakubovich A V and Solov'yov A V 2013 *Sci. Rep.* **3** 1289

[43] Irisawa T, Uwaha M and Saito Y 1995 *Europhys. Lett.* **30** 139–144
[44] Bréchignac C, Houdy P and Lahmani M (eds) 2007 *Nanometerials and Nanochemistry* (Cambridge University Press)
[45] Frenkel J (ed) 1955 *Kinetic Theory of Liquids* (Dover)
[46] Einstein A 1956 *Investigations on the Theory of Brownian Movement* (New York: Dover)
[47] Landau L D and Lifshitz E M 1987 *Fluid Mechanics* (Elsevier, Butterworth-Heinemann, Oxford)
[48] Mashiah A, Wolach O, Sandbank J, Uziel O, Raanani P and Lahav M 2008 *Acta Haematol* **119** 142–150
[49] Baish J W and Jain R K 2000 *Cancer Reasarch* **60** 3683–3688
[50] Phillips J C, Braun R, Wang W and *et al* 2005 *J. Comp. Chem.* **26** 1781–1802
[51] Brooks B, Bruccoleri R, Olafson B, States D, Swaminathan S and Karplus M 1983 *J. Comp. Chem.* **4** 187–217
[52] Humphrey W, Dalke A and Schulten K 1996 *J. Molec. Graphics* **14** 33–38

Ion interactions with pure and mixed water clusters

R. Maisonny[1], M. Capron[1,2], S. Maclot[1,2], J. C. Poully[1,2], A. Domaracka[1], A. Méry[1,2], L. Adoui[1,2], P. Rousseau[1,2] and B.A. Huber[1,*]

[1]Centre de Recherche sur les Ions, les Matériaux et la Photonique (CIMAP) – CEA/CNRS/ENSICAEN/UCBN, Bd Henri Becquerel, 14070 Caen, Cedex 5, France
[2]Université de Caen Basse-Normandie, Esplanade de la Paix, CS 14032, 14032 Caen cedex , France

E-mail: huber@ganil.fr

Abstract. We report on collisions of highly charged Xe^{20+} ions with weakly bound clusters of water molecules and water/adenine mixtures. Singly and doubly charged water clusters are observed in the size range of n=1 to 65 and n=49 to 61, respectively. In contrast to other comparable systems, the dominant monomer fragment (H_3O^+) is formed with very low kinetic energy, hence indicating that it is formed by evaporation processes. Larger fragments are produced with larger kinetic energies due to charge-separating processes. Furthermore, water clusters are found to be protonated, only a very small amount of the non protonated dimer $(H_2O)_2^+$ is observed. Excited mixed adenine/water clusters fragment by the loss of the surrounding water molecules, thus, adenine fragments are formed without water molecules attached. In addition, the adenine monomer is found to be partly protonated.

1. Introduction

Ion collisions in the gas-phase with weakly bound clusters (hydrogen- or van der Waals-bonded) may induce chemical reactions between the constituents of the cluster. This is possible due to the transfer of energy and charge during the collision which allows the system to overcome existing reaction barriers or which can create highly reactive species which interact immediately with other surrounding cluster constituents.

It has been shown experimentally, that coalescence and growth reactions occur when highly charged Xe ions collide with clusters of fullerene molecules, resulting in the formation of giant fullerenes [1]. More recently, measurements performed by the same group indicated that highly reactive C_{59}^+ fullerene ions are formed by kicking out one carbon atom from a C_{60} molecule in close He^{2+} / fullerene cluster collisions. As supported by quantum mechanical simulations the C_{59}^+ molecule reacts immediately (~100 fs) with one of the neighbored C_{60} molecules within the cluster forming covalently bound C_{119}^+ ions [2].

Important representatives of weakly bound clusters are water clusters which play a crucial role in different physical, chemical and biological processes on the Earth and its atmosphere. Liquid water consists of a macroscopically connected, random network of hydrogen bonds; therefore, these weakly bound aggregates of molecules have been suggested as possible transient intermediates [3]. Recently, the formation and growth of water clusters have attracted much interest [4,5] in particular as it has been shown that ion-induced nucleation plays an important role in aerosol formation and hence may influence cloud formation in the Earth's atmosphere [6-7]. Moreover, the geometrical structures of neutral and protonated water clusters were the subject of many studies [8,9] discussing the evolution of the structure when the cluster size increases.

The fragmentation dynamics of water molecules and clusters has been studied with a large variety of electromagnetic radiations such as photons, electrons or ions [10-16] showing that high energy fragments and low energy electrons are produced. These results have direct implications on the understanding of radiation damage in living tissues. As water molecules surround all biological matter, it is important to better understand the response of water to ionizing radiation as well as to identify the influence of a water environment on the damage of the biomolecular system. By so-called indirect effects, radicals which are formed in the water environment may damage biomolecules and DNA. Ion collisions are of special importance as they are applied in cancer treatments by hadrontherapy [17].

In the present work, we report on the fragmentation of pure water clusters as well as adenine clusters embedded in a water environment, induced in gas phase collisions by multiply charged ions (Xe^{20+}).

2. Experimental approach

Ion-cluster collisions are studied in a crossed beams device, the set-up of which is shown in the figure 1. It is described in more detail in a previous communication [18]. In brief, multiply charged ions in different charge states (1-25) are produced in an electron cyclotron resonance ion source (ECRIS) of the installation ARIBE at the GANIL facility [19]. At acceleration voltages between 10 and 20 kV, the ions are mass-selected and guided to the interaction region. They cross a beam of neutral hydrogen-bonded or van der Waals bonded clusters, which have been produced within a cluster aggregation source [20]. The cluster beam is characterized by a wide mass distribution (containing up to ~65 molecules) at a cluster temperature of ~70 K. After the pulsed ion beam (typical pulse width 500 ns, frequency of the order of 2 kHz) has passed the interaction zone, cationic interaction products are extracted and mass-over-charge analyzed by means of a time-of-flight mass spectrometer, from which secondary ion mass spectra are obtained. The spectrometer acceptance allows for 100% detection of fragments with kinetic energies of up to 10 eV; higher energies will provoke forward- and backward-structures in the peak forms. Due to the strong post-acceleration (up to 25 kV) of the ions towards an electron emitting conversion plate, the detection efficiency is constant also for large fragment masses. Time-of-flights can be recorded in an event-by-event mode allowing to analyze correlations between fragments [21].

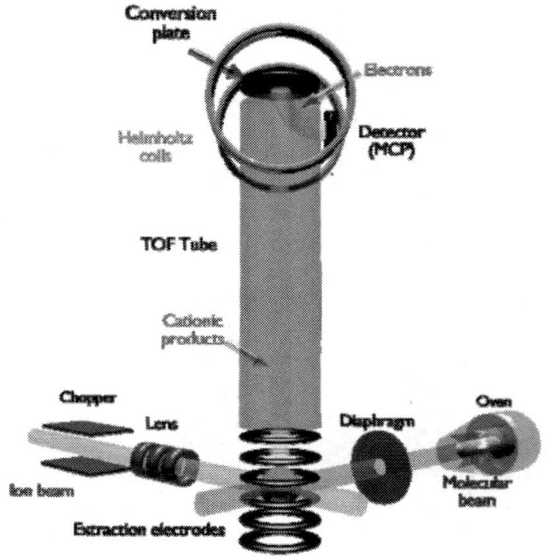

Figure 1: Experimental device with crossed ion and molecular beams and the time-of-flight mass spectrometer.

3. Ion interactions with neutral water clusters

In figure 2 we show an inclusive mass spectrum (containing all events) obtained in collisions of Xe^{20+} projectiles at 300 keV with a wide distribution of neutral water clusters $(H_2O)_n$, with n between 1 and ~ 65. The neutral molecules are ionized by electron capture processes, forming during the collision time (~10 fs) singly and multiply charged water clusters. The observed peak intensities of the protonated water clusters (see discussion below), measured on a time scale of several μs, decrease with increasing cluster size showing the dominance of small size clusters and fragments. The slightly increased relative intensity for n=4 reflects the higher stability of the 'magic number' cluster.

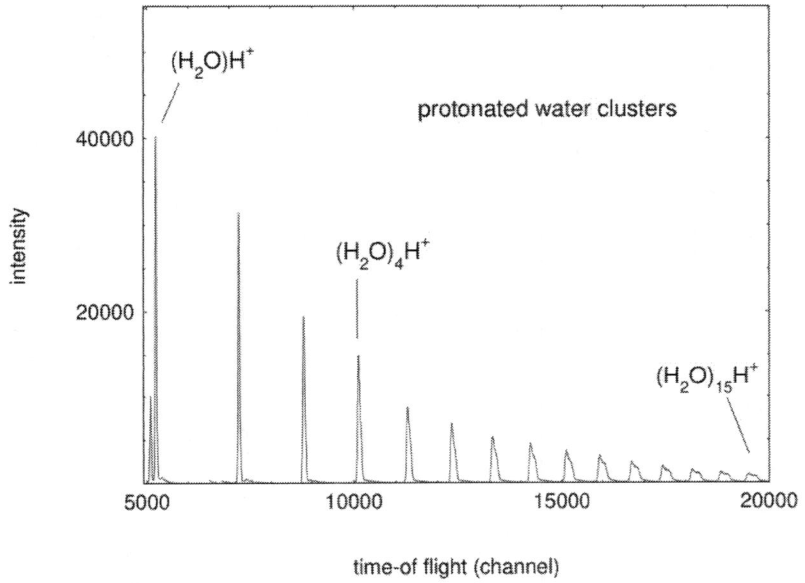

Figure 2. Mass spectrum of protonated water clusters formed in collisions of 300 keV Xe^{20+} ions with neutral water clusters (corrected for N_2 and O_2 background).

Ionized excited clusters may decay by evaporation of a neutral water molecule or via evaporation sequences or by charge-driven fission processes. In the first case, the kinetic energies of the emitted molecules and of the residual ion are rather small in the meV energy range. After evaporation of several molecules, the residual ion may accumulate larger recoil energies in the range of several hundreds of meV. Finally, in charge driven reaction, due to the Coulomb repulsion, kinetic energies in the eV range are typical values which will strongly influence the shape of the measured peaks in particular by increasing their half widths.

In figure 3 more details are given concerning the variation of the peak form with cluster size. In the left upper panel, it is shown that the peak width of small fragments (or cluster residues) increases strongly with size n also showing a tail to higher masses. As shown more clearly in the right upper panel of figure 3 for the cluster with n=10, the widening is due to the kinetic energy release upon fragmentation, which is in the present case of the order of 2 eV as determined by the peak width [22]. The dashed curve is the result of a SIMION trajectory calculation within the time-of-flight mass spectrometer, taking into account a 4π emission angle of the fragment, an energy distribution of 2 eV with a half width of 250 meV, as well as the special detector geometry leading to an asymmetric form. The tail towards larger drift times is due to retarded evaporation processes which occur within the extraction field, i.e. on a μs time scale:

$$(H_2O)_{11}H^+ \rightarrow (H_2O)_{10}H^+ + H_2O.$$

In the intermediate cluster size range (left lower panel of figure 3) we observe the formation of doubly charged water clusters. Whereas in the literature the appearance size for doubly charged species has been reported to be > 35 [23], here we observe these systems in the size range from about n=41 to 61. The upper limit is close to the maximum cluster size in the primary distribution and agrees with the limit for singly charged systems in the spectrum.

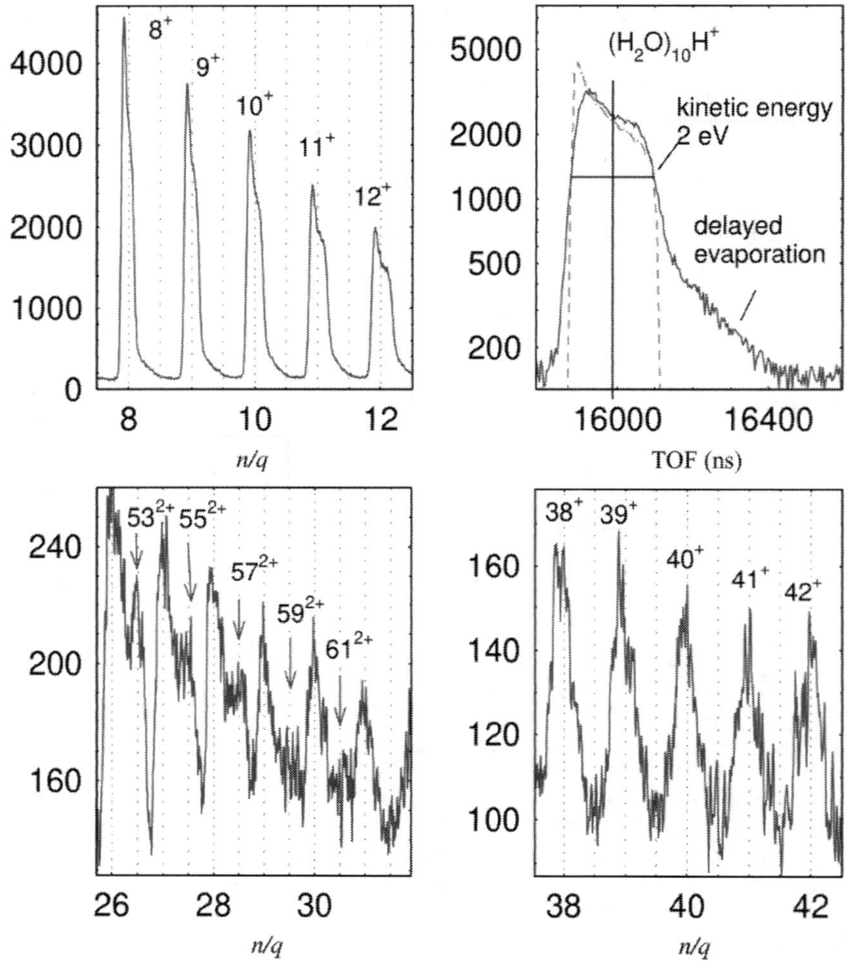

Figure 3. Details of the mass spectrum in regions of different cluster sizes. The dashed line in the upper right panel is the result of trajectory calculations of the $(H_2O)_{10}H^+$ ion with a kinetic energy of 2 eV.

The lower right panel of figure 3 shows symmetric peak structures for larger clusters, corresponding to kinetic fragment energies below 1 eV. An overview of the most likely kinetic energies of the fragments is given in Table 1. The kinetic energy increases with cluster size, passes a maximum and decreases again. It is surprising that the average kinetic energy of the monomer H_3O^+ is that low. In the case of asymmetric fission processes, we would expect a much larger energy release as it is the case for other weakly bound cluster systems like clusters of fullerenes, PAHs or biomolecules,

Journal of Physics: Conference Series **438** (2013) 012007 doi:10.1088/1742-6596/438/1/012007

where in general a wide monomer peak is observed in the fragmentation spectrum [24-26]. Hence, it is likely that the small fragments are formed by evaporation processes. In the intermediate range ($10 < n < 20$) the kinetic energies are of the order of 2 eV indicating the importance of charge driven processes. This finding is confirmed by experiments with He^{2+} ions, which produce water clusters in lower average charge states. Here, the kinetic energy of clusters in the intermediate size range is reduced by a factor of 2. When we analyse spectra where only multiply charged water clusters are formed (so-called multistop spectra) the kinetic energy of the monomer increases to about 1 eV reflecting Coulomb repulsion. Furthermore, high energy H^+ fragments are observed (with average kinetic energies between 15 and ~20 eV) proving the fragmentation of individual water molecules when the system is highly charged.

Table 1. Average kinetic energies of fragments produced in collisions of Xe^{20+} ions with neutral water clusters as a function of their size.

cluster size n	kinetic energy (eV)
1	0.19
3	0.38
5	0.9
10	2.0
20	2.2
41	0.8

As mentioned above, the observed clusters are protonated. Similar results have been obtained in other experiments using 26.5 eV soft X-ray laser [4], fs photoionization [10], chemical ionization [27], electron impact ionization [28] and more recently high energy ion impact [12]. In an experimental study performed with mixed $Ar(H_20)_n$ clusters [29], ionized by photon collisions close to the ionization threshold (photon energies of 11.62 and 11.83 eV), also unprotonated radical cations have been observed for small cluster sizes (n<10). The finding was explained by the emission of the Ar atom after ionization which is supposed to cool the cluster and to stabilize the radical cations.

The observed protonated species are likely to be produced by a fast proton transfer reaction and the loss of OH involving two molecules of the cluster following the ionization process, along the following equation:

$$\mathbf{(H_2O)_n \xrightarrow{Xe^{20+}} (H_2O)_n^+ \rightarrow (H_2O)_{n-1}H^+ + OH} \quad (1)$$

This fast reaction can be followed by the evaporation of further water molecules, depending on the internal energy of the system. Recently it has been shown by TD-DFT Car Parinello calculations that this transfer proceeds on a very short timescale namely within less than 1 ps [12,30]. This is shorter than the average statistical dissociation time.

Although in ionizing collisions with highly charged ions the transferred energy can be as low as 1 eV for single ionization, our results indicate that the potential energy barrier for protonation of water clusters is lower than this value. Thus, we observe radical cations only for n=1 and n=2, whereby for n=2 the intensity is very low (see figure 4). It should be noted that some of the water monomer radical cations observed can come from residual gas in the spectrometer extraction region.

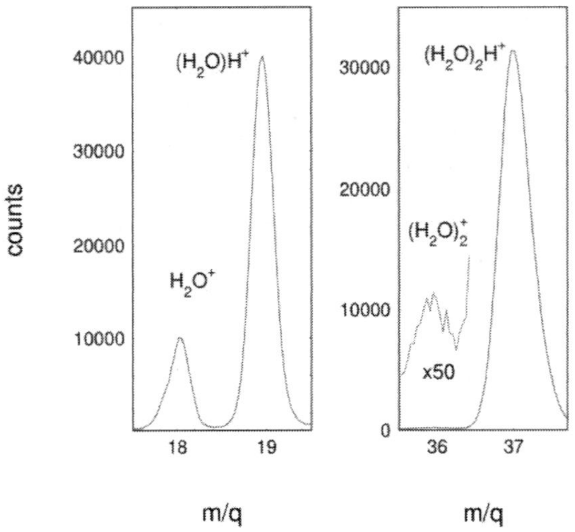

Figure 4: Protonated species and radical cations formed in collisions of Xe^{20+} with $(H_2O)_n$ at a collision energy of 300 keV. Note the low intensity of $(H_2O)_2^+$. For larger *n* values only protonated clusters are observed.

4. Mixed adenine-water clusters

In the following we will describe collision experiments performed with mixed clusters containing water molecules as well as molecules of the nucleobase adenine. These targets are prepared in the cluster aggregation source by introducing water vapor and creating an adenine vapor pressure by heating an oven device. In figure 5 we show a series of mass spectra obtained after collisions of Xe^{20+} projectiles at 300 keV with different mixed cluster distributions (temperatures of the adenine oven are given). At very low temperatures (upper left panel) we find a spectrum of pure water clusters as described in the section above (figure 2, the peak forms are different in both spectra due to a modified time-of-flight system).

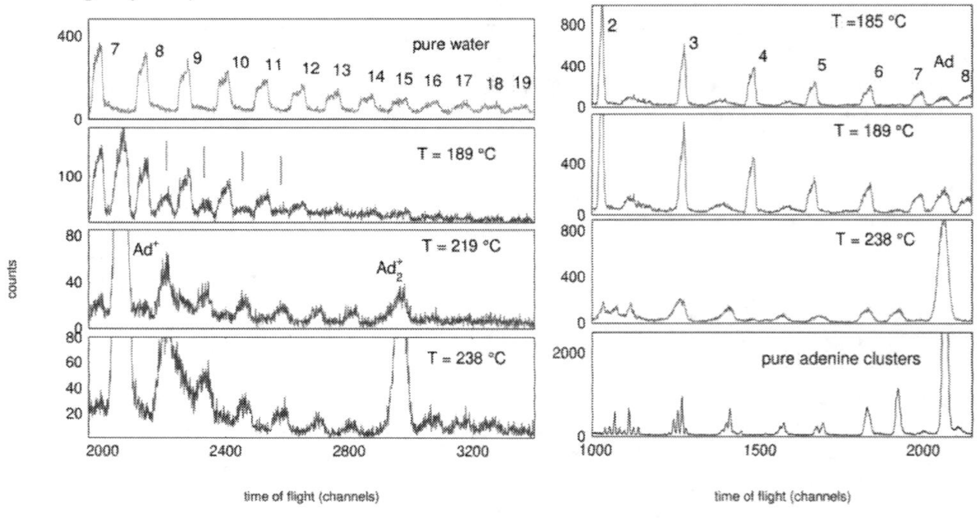

Figure 5. Inclusive mass spectra obtained in Xe^{20+} collisions with water, adenine and mixed water-adenine clusters. The given temperatures are those of the adenine oven and reflect the adenine vapor pressure in the cluster source. Left: mixed cluster distribution; right: fragmentation region of the adenine molecule.

At higher temperatures (T=189°C) we observe in addition to water clusters the adenine monomer peak between $(H_2O)_7H^+$ and $(H_2O)_8H^+$ as well as a series of hydrated adenine clusters. At even higher temperatures no water clusters are observed anymore, the spectrum contains only adenine and hydrated adenine clusters (T=238 °C) with up to 8 water molecules attached. The neutral target might contain even more water molecules as evaporation processes, as discussed above, are expected to occur. The relative integrated intensities (see figure 6) decrease with the number of attached water molecules in the cluster and are more or less independent of the fact whether the adenine monomer or dimer is present.

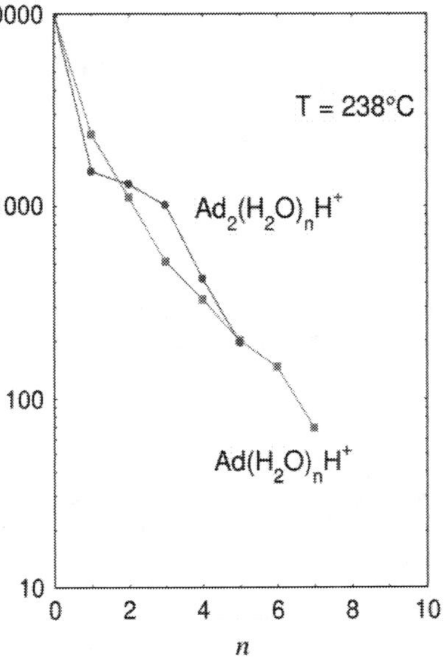

Figure 6: Relative intensities of hydrated adenine clusters.

On the right panel of figure 5, the mass range below the adenine monomer is shown. When comparing the fragmentation of adenine and hydrated adenine, the same types of fragments are observed. In particular, no fragments where water molecules are attached are produced, indicating that all water molecules are lost before the adenine molecule fragments. This is due to the fact that the fragmentation of the adenine molecule requires more energy than the evaporation of a water molecule. This is in agreement with studies of the collision induced dissociation of nanosolvated AMP^- anions, where the loss of water molecules turned out to be the dominant de-excitation mechanism [31].

In experiments with low charged projectiles (O^{3+} at 37.2 keV) [32], we analyzed the peak form of the adenine peak in more detail. In figure 7 we compare the peak shape of the adenine monomer for different targets: In the case of the isolated adenine molecule the peak is rather narrow showing the isotopic distribution as well as the loss of one hydrogen atom. When adenine clusters are irradiated, the peak becomes wider due to evaporation and fragmentation processes. Finally, when using hydrated adenine clusters, the peak becomes even wider and shifts slightly to larger masses. A peak deconvolution shows that the adenine fragment ion is formed partly in its protonated form, which might be explained by a proton transfer from the protonated water to the adenine.

We know, as discussed above in section 3, that when the O^{3+} ion interacts with a water molecule in the hydrated cluster of adenine, a protonated water molecule is fastly formed by the emission of a neutral OH fragment on the picosecond time scale. As the proton affinity of the adenine molecule is higher than that of water [40, 41], a proton transfer to the adenine molecule may occur. Finally, after evaporation of all surrounding water molecules the protonated adenine monomer (AdH^+) is observed.

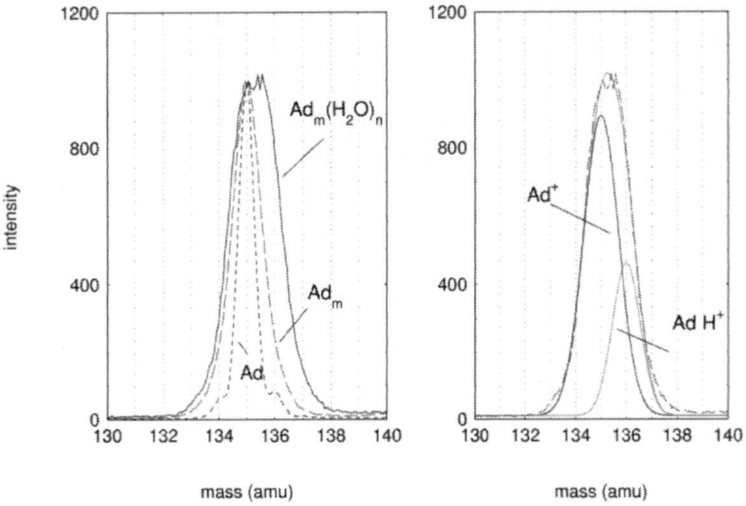

Figure 7. Left: shape of the adenine peak obtained in ion collisions with the isolated adenine molecule (short dashed line); with adenine clusters (long dashed line) and hydrated adenine clusters (full line). Right: deconvolution of the peak obtained with hydrated adenine clusters yielding pure adenine as well as protonated adenine monomers (see also ref. 32).

5. Acknowledgements
The experimental studies have been performed at the ARIBE facility, the low-energy ion facility of GANIL. Financial support received from the ANR Programme Blanc PIBALE/ANR-09-BLAN-013001 and the Conseil Régional de Basse-Normandie is gratefully acknowledged.

References

[1] Zettergren H, Schmidt H T, Jensen J *et al.* 2010 *J. Chem. Phys.* **133**, 104301
[2] Zettergren H, Rousseau P *et al.* 2013 *Phys. Rev. Lett.* submitted
[3] Stillinger F H 1980 *Science* **209**, 4455
[4] Dong F *et al.* 2006 *Chem. Phys.* **124**, 224319
[5] Zamith S, Feiden P, Labastie P *et al.* 2010 *Phys. Rev. Lett.* **104**, 103401
[6] Svensmark H, Bondo T and Svensmark J 2009 *Geophys. Res. Lett.* **36**, L15101
[7] Hvelplund P *et al.* 2010 *Int. J. Mass Spectrom.* **292**, 48
[8] Ludwig R 2004 *ChemPhysChem* **5**, 1495
[9] Zwier T S 1999 *Science* **304**, 1119
[10] Radi P P *et al.* 1999 *J. Chem. Phys.* **111**, 512
[11] Sternovski Z *et al.* 2001 *Phys. Rev.* A **64**, 023203
[12] Adoui L *et al.* 2009 *J. Phys.* B*: At. Mol. Opt. Phys.* **42**, 075101
[13] Alvarado F, Hoekstra R *et al.* 2005 *J. Phys.* B*: At. Mol. Opt. Phys.* **38**, 4085
[14] Legendre S *et al.* 2005 *J. Phys.* B*: At. Mol. Opt. Phys.* **38**, L233
[15] Adoui L *et al.* 2001 *Phys. Scr.* **T92**, 89
[16] Kumarrapan V *et al.* 2003 *Phys. Rev.* A **67**, 063207
[17] Krämer M *et al.* 2000 *Phys. Med. Biol.* **45** 3299
[18] Bergen T, Biquard X, Brenac A *et al.* 1999 *Rev. Sci. Instrum.* **70**, 3244
[19] Bernigaud V, Kamalou O, Ławicki A, *et al.* 2008 *Publ. Astron. Obs. Belgrade* **84**, 83
[20] Chandezon F, Huber B A and Ristori C 1994 *Rev. Sci. Instrum.* **65**, 3344
[21] Capron M *et al.* 2012 *Chem. Eur. J.* **18**, 9321
[22] Tomita S *et al.* 2002 *Phys. Rev.* A **65**, 053201

[23] Shukla A K, Moore C and Stace A J 1984 *Chem. Phys. Lett.* **109**, 324
[24] Manil B, Maunoury L, Huber B A *et al.* 2003 *Phys. Rev. Lett.* **91**, 215504
[25] Holm A I S , Zettergren H, Johansson H A B *et al.* 2010 *Phys Rev Lett.* **105,** 213401
[26] Maclot S *et al.* 2011 *ChemPhysChem* **12**, 930
[27] Shi Z *et al.* 1993 *J. Chem. Phys.* **99**, 8009
[28] Echt O *et al.* 1984 *Chem. Phys. Lett.* **108**, 401
[29] Shinohara H and Nishi N 1986 *J. Chem. Phys.* **84**, 5561
[30] Gaigeot M P *et al.* 2007 *J. Phys.* B*: At. Mol. Opt. Phys.* **40**, 1
[31] Liu B, Nielsen S B, Hvelplund P *et al.* 2006 *Phys. Rev. Lett.* **97**, 133401
[32] Domaracka A, Capron M, Maclot S *et al.* 2012 *J. Phys.: Conf. Series* **373**, 012005

Modelling of water clusters formation processes in the inner atmosphere of comets

N Y Bykov, D V Leshchev

Center for Advanced Studies, St. Petersburg State Polytechnical University, 195251 St. Petersburg, Politechnicheskaja, 29, Russia

E-mail: nbykov2006@yandex.ru

Abstract. The model of water clusters formation has been developed for the direct simulation Monte Carlo method. The model describes hierarchy of reactions which lead to growth and decay of water clusters including initial processes of dimer formation by triple and binary collisions of water molecules. The algorithm takes into account internal state of clusters and energy exchange processes accompanying clusterization kinetics. The model was applied for the research of condensation process in the rarefied flows of water molecules typical for the inner atmospheres of comets.

1. Introduction

The kinetics of molecular clusters formation is of great interest for large number of applications in science and technology. In particular, the formation of clusters in water vapor flows is a very important process in planetary science. Recently European Space Agency (ESA) approved new research mission Rosetta with 67P/Churyumov-Gerasimenko comet (67P/C-G) as a primary target. Rosetta will rendezvous with the Jupiter family comet 67P/C-G in 2014 and provide the probe landing at the comet surface. Preliminary estimations of near-nucleus atmosphere parameters affected strongly by water clusters formation processes [1] are required for landing mission success and correct interpretation of observations results.

The aim of present research is to formulate the model water clusters growth and to research cluster formation phenomenon for the flows in inner atmosphere of 67P/C-G comet. For numerical investigation we employed the direct simulation Monte Carlo method (DSMC) [2] which widely used for modeling of rarefied and near-continuum flows with different physic-chemical processes in volume.

2. Kinetic model of water clusters formation

The model of clusters formation has been elaborated recently for the DSMC method [3,4]. The clusters are characterized by number of monomers k, mass, radii (determined by modified drop spherical model [3]), translation velocity, rotation and vibration degrees of freedom and corresponding rotation and vibration internal energies. For description of vibration state of water cluster we use traditional presentation of vibration energy [2]

$$E_{v,k} = \sum_{i=1}^{3k_a-6}\left(\frac{k_B\theta_i}{\exp[\theta_i/T]-1}\right) \tag{1}$$

where $\theta_i=h\upsilon_i/k_B$ — characteristic temperature of vibration mode i (h — Plank constant, v_i — vibration frequency of mode i), k_a — number of atoms (H and O) in cluster, T — temperature. The known data on frequencies of inter- and intra-molecular vibrations of water clusters are generalized in [3]. The use of (1) gives possibility to calculate effective excited vibration degrees of freedom and to take into account partial excitation of vibration internal modes.

$$
\begin{array}{ll}
i & H_2O+H_2O+H_2O \xrightarrow{p_i} (H_2O)_2 + H_2O^* \\
ii & H_2O+(H_2O)_k \xrightarrow{p_{ii}} (H_2O)_{k+1} \\
iii & (H_2O)_{k_1}+(H_2O)_{k_2} \xrightarrow{p_{iii}} (H_2O)_{k_1+k_2} \\
iv & (H_2O)_{k+1} \xrightarrow{p_{iv}} H_2O+(H_2O)_k \\
v & H_2O+H_2O \xrightarrow{p} (H_2O)_2
\end{array}
$$

where p, $p_{i\text{-}iv}$ — probabilities of reactions.

In general case [3] the model takes into account such kinetic processes as (i) dimer formation under triple collision of monomers, (ii) reaction of the monomer-cluster association under binary collisions of these particles, (iii) reaction of the cluster-cluster association under mutual collisions of clusters, (iv) cluster decay by one monomer evaporation. For the conditions of low temperature flow of water molecules the probabilities of association during particles collisions for reactions (i)-(iii) were supposed to equal 1 [3]. The new feature of the developed model for water clusters formation is in consideration of additional reaction (v) which allow to form dimer directly under binary collision of water molecules. Opposite the case of atomic dimer formation, for our problem this reaction may take place due to enough large capacity of vibration modes of water dimer. For the water vapor flows in inner comet atmospheres the probability of monomers association under binary collisions of water molecules was estimated to be p=0.01 [1]. We use also alternative way for p determination based on approach discussed in [5] for estimation of probability association of atoms under triple collisions. This approach predicts for reaction (v)

$$p=CE_c^x,\ x=b-0.5,\ C=\frac{a\sqrt{\pi m_r}}{k_B^b\sigma_T\sqrt{2}}\frac{\Gamma(\varsigma_c/2)}{\Gamma(b-0.5+\varsigma_c/2)} \tag{2}$$

where E_c, ς_c, m_r — total energy, total number of degrees of freedom and reduced mass of collision pair correspondedly, σ_T — collision cross-section (for hard spheres model), Γ — gamma-function, a and b — constants in Arrenius type equation for the rate of association reactions without activation barrier: $k_{ass}=aT^b$. The data about a and b constants are very poor. The estimations [6] give a in the range 10^{10}-10^{14} cm^3/(mole s), b = 0.

3. Statement of the problem

Under orbital motion of comet 67P/C-G the maximal flux of water molecules through active areas of nucleus Q is observed in perihelion (1.29 AU from the Sun) and is approximately 10^{28}s^{-1} [7]. The usual approach to model flow in near nucleus comet atmosphere is in simplification of the problem toward 1D spherical setting [1]. In this case the nucleus characterized by effective radius $R_{eff}=(\beta S)^{1/2}$, (S — is the total surface area, β=0.05 — the ratio of an active area to the total area of the surface [7]). Below we present simulation results for the case of expansion of water molecules in vacuum evaporated at rate Q from the sphere with R_{eff}=900 m. The surface temperature was T_0=200 K, this value was estimated for equatorial parts on the base of the thermal model of a porous ice nucleus [8]. The considering flow region was $R_{eff}<X<3R_{eff}$. The probability of association of water molecules under binary collisions (reaction (v)) was p=0.01 (case 1) or determined by equation (2) with parameters a = 10^{12} cm^3/(mole s), b=0 (case 2) and a = 10^{10} cm^3/(mole s), b=0 (case 3). The last one may be considered for estimation of minimal degree of condensation process in the comet atmosphere. Other details of the problem setting and parameters of the DSMC method are the same as in [3].

4. Results and discussion

For considering range of initial parameters the characteristic Knudsen number was $Kn=\lambda_w/R_{eff}=0.2*10^{-4}$ (λ_w — free path near the surface). This value corresponds to near continuum flow regime. The data on the degree of supersaturation in such flow predict good conditions for cluster formation process (Figure 1a).

The degree of condensation may be defined as

$$\alpha_s = \sum_{k=2} kN_k / \sum_{k=1} kN_k \quad (3)$$

where N_k — the total number of clusters with size k in the considering flowfield. For the cases 1 and 2 α_s=0.23 and 0.21 accordingly. For case 3 the ratio of association process (v) is less by two orders of magnitude in comparison with case 2. Nevertheless, the value of α_s for this case is surprisingly high (α_s=0.13). The influence of water clusters formation processes on gasdynamic parameters is very high (Figure 1). The density of monomers-clusters mixture drops more quickly in comparison with the flow without condensation. In most area of the flow values of temperature for the case 1 differ from the case without clusterization approximately in five times. The reason of this fact is in release of latent heat of condensation during clusters growth process. The velocity of the flow is also affected by clusterization and sufficiently higher in comparison with flow without cluster formation process.

The clusters formation process starts in the vicinity of the nucleus surface. The role of triple collisions of water molecules (reaction (i)) on the dimer formation process is negligible. The most probable way of dimers appearance in the inner atmosphere of 67P/C-G comet is under binary collisions of water molecules. The maximal value of clusters density is near the nucleus surface, with increase of the distance from the nucleus density drops (Figure 1a). The employed model of cluster formation predicts close velocities of small clusters, monomers and the gas mixture in the considering near-surface region (Figure 1b).

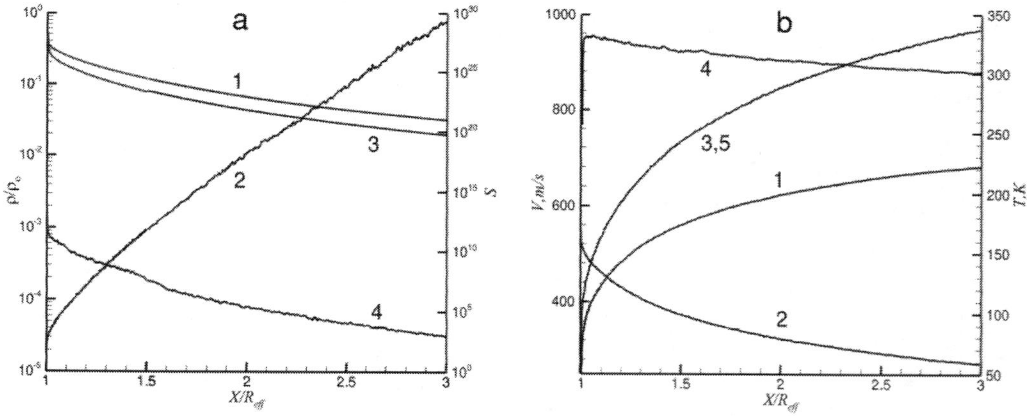

Figure 1 a: radial distributions of density (1) and degree of supersaturation S (2) of gas for case without condensation, density of gas mixture (3) and clusters with size k=10 (4) for case 1. b: radial distributions of velocity (1) and temperature (2) of gas for the case without condensation, velocity(3) and temperature(4) of gas mixture, velocity of clusters with size k=10 (5) for case 1. ρ_0=9,9·10^{-7} kg/m^3

The size distributions of forming in whole flowfield water clusters are presented in Figure 2. The majority of water clusters is dimers for cases 1 and 2. For low rates of association process (case 3) the small clusters in the range $2<k<10$ have got approximately the same representation. The difference between considered cases with intensive condensation (cases 1 and 2) and week rates is very large for clusters with sizes $2<k<20$. For all cases the maximum size of observed clusters exceeds 150 molecules per cluster.

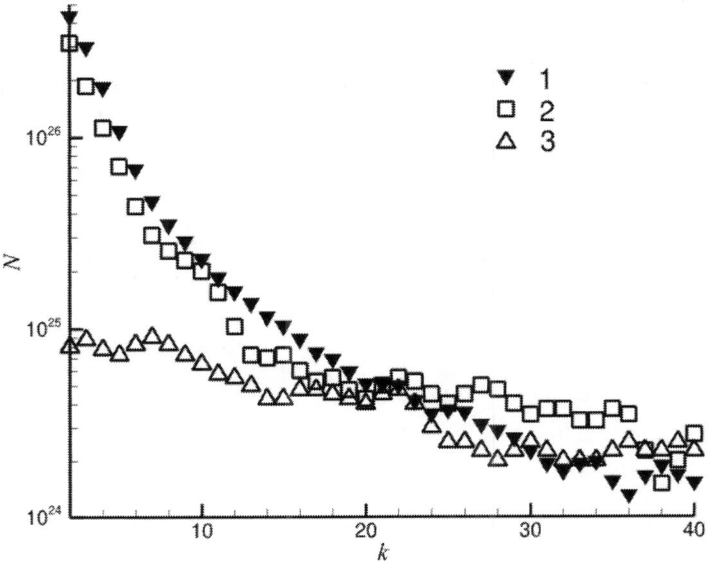

Figure 2. Total over flowfield size distributions of clusters. 1 — case 1, 2 — case 2, 3 — case 3

References

[1] Crifo J F 1990 Water clusters in the coma of comet Halley and their effect on the gas density, temperature, and velocity *Icarus* **84** 414

[2] Bird G A 1994 *Molecular gas dynamics and the direct simulation of gas flows* (Oxford: Clarenton Press)

[3] Bykov N Y 2009 Modelling of condensation at spherical expansion of water vapor into vacuum *Thermophysics and Aeromechanics* **16** 177

[4] Li Z, Zhong J, Levin D A, Garrison B J 2009 Kinetic nucleation model for free expanding water condensation plume simulations *In Proc. of 26th int. symposium on RGD (AIP Conference Proceedings 1084, American Institute of Physics, Melville, NY)* ed Takashi Abe, pp. 613-618.

[5] Bykov N Y, Leshchev D V, Simakova O I 2008 The direct simulation Monte Carlo of near-continuum flows with condensation process, *In Proc. of XXXVI summer school Advanced Problems in Mechanics (Saint-Petersburg: Comilfo)* pp. 159-166

[6] *Physic-chemical processes in gasdynamics* 1995 Handbook, **I**, ed Chernyi G G, Losev S A, (Moscow: Moscow state university press)

[7] P.L. Lamy, I. Toth, B.J.R. Davidsson, et al. 2007 A portrait of the nucleus of comet 67P/Churyumov-Gerasimenko *Space science reviews* **128** 23

[8] M.C. De Sanctis, M.T. Capria, A. Coradini 2006 67P/Churyumov-Gerasimenko nucleus model: Portrayal of the Rosetta target *Advances in space research* **38** 1906

New applications of the jellium model for the study of atomic clusters

R G Polozkov[1], V K Ivanov[1], A V Verkhovtsev[1,2], A V Korol[2,3] and A V Solov'yov[2,4]

[1] St. Petersburg State Polytechnic University, Politekhnicheskaya ul. 29, 195251 St. Petersburg, Russia
[2] Frankfurt Institute for Advanced Studies, Ruth-Moufang-Str. 1, 60438 Frankfurt am Main, Germany
[3] Department of Physics, St. Petersburg State Maritime Technical University, Leninskii prospekt 101, 198262 St. Petersburg, Russia

E-mail: verkhovtsev@fias.uni-frankfurt.de

Abstract. Application of the jellium model for investigation of the electronic structure and photoionization of metal clusters and fullerenes is discussed. The valence electrons are considered either within the Hartree-Fock and the local density approximations. The random phase approximation is utilized to account for the many-electron correlations in the response of a system to an external field. It is shown that the photodetachment cross section and photoelectron angular distribution in metal cluster anions are described reasonably well within the jellium model. Its application to fullerenes requires the use of corrections for a better description of the ground state electron density.

1. Introduction

Electronic structure, as well as photoexcitation and photoionization processes, of clusters have been studied for more than several decades. During this period a number of important achievements have been made. For instance, the geometrical structure of mass-selected clusters has been studied in detail experimentally by means of the low-temperature photoelectron spectroscopy and theoretically by using the density functional theory (DFT).

Among various types of atomic clusters, metal and carbon clusters have been of the main interest for studying (see, e.g., the reviews [1–4] and references therein). For instance, the geometrical structures of sodium cluster anions consisting of up to 350 atoms were determined with high accuracy by comparing results of the DFT calculations with measured photoelectron spectra [5–7]. It was shown that valence electrons form highly discretized density of states, the so-called electronic shell structure. Another achievement is associated with observation of the plasmon resonances in photoabsorption spectra of various small metal clusters, e.g., Na, K and Mg clusters (see [1, 2, 8] and references therein). It was established that in many cases the pattern structure of plasmon resonances is determined by the deformation parameter of a cluster as well as by the interaction between single-electron and collective excitation modes.

[4] On leave from A.F. Ioffe Physical-Technical Institute, St. Petersburg, Russia

Similar achievements have been made for fullerenes. For instance, it was shown that plasmon excitations can be formed in the processes of photon- or electron impact ionization of fullerenes. Having a profound collective nature, the plasmon excitations influence significantly the formation of broad resonance structures, the so-called giant resonances, in the the cross sections of different excitation processes. The plasmon excitations in the C_{60} molecule have been intensively investigated both experimentally and theoretically (see, e.g. [9–17]). It was found that the photoionization cross section for the neutral fullerene C_{60} possesses a very strong giant resonance centered at the photon energy about 20 eV [15]. Later on, a second collective excitation in C_{60} and its ions, centered around the photon energy of 40 eV, was observed and discussed [17–20].

Recently, the photoabsorption spectrum of the C_{60} fullerene was calculated in a broad photon energy range by means *ab initio* methods based on the time-dependent density functional theory [21, 22]. A detailed analysis of the calculated spectrum revealed the the contributions, coming out from single-particle and collective electron excitations [22].

At present, essential progress in experimental study of photoabsorption processes of clusters has been achieved by using new sources for free neutral [23] and charged [24] size-selected metal clusters. These measurements were followed by new theoretical calculations of the photoionization cross section [25–27] and of the photoelectron angular distribution for neutral and charged metal clusters [28, 29].

From the theoretical point of view, the valence electron-shell structure and optical properties of various atomic clusters were successfully described by means of the jellium model [30, 31]. Within the jellium model a cluster is considered as a object consisting of two systems: a system of valence electrons and an ionic core. The basic idea of the jellium model is to replace the real geometry of the ionic core by a smooth positive background in a finite volume and to treat only delocalized electrons explicitly in the mean-field approximation. The electron system is considered in a single-particle approach which can be the Hartree-Fock or the local density approximations. Since the collective many-electron effects play sometimes a crucial role in ionization processes of clusters, the dynamical many-electron correlations are usually also taken into account within the random phase approximation [32].

This article is devoted to some achievements of the recent applications of jellium model to metal clusters and fullerenes.

The atomic system of units, $m_e = |e| = \hbar = 1$, is used throughout the paper.

2. Application of the jellium model to clusters

The characteristic features of small metal clusters, like the electronic shell structure and plasmon excitations, can be well understood in terms of quantum motion of the delocalized valence electrons moving in the field created by themselves and by the positively charged ionic core [31]. This concept, known as the jellium model for neutral metal clusters, can also be applied to the charged systems.

In this section we consider so-called magic sodium clusters, namely Na_{19}^- and Na_{57}^-, which can be treated as spherically symmetric objects due to the completely filled electronic shells. Within the jellium model the complicated ionic structure of the cluster can be reduced to the uniform spherically symmetric positive charge distribution. The potential $U_{\rm core}(r)$ generated by this distribution is equal to

$$U_{\rm core}(r) = \begin{cases} -\dfrac{Z}{2R_0}\left(3-\left(\dfrac{r}{R_0}\right)^2\right) & , \ r \le R_0 \\ -\dfrac{Z}{r} & , \ r > R_0 \end{cases} , \tag{1}$$

where Z is the charge of the ionic core. For a neutral sodium cluster, Z equals to the number of delocalized valence electrons N, coinciding with the number of atoms for clusters of monovalent elements. Therefore, for the singly charged anions the relation $Z = N - 1$ is fulfilled. The core radius R_0 is related to the charge as $R_0 = r_s Z^{1/3}$, where the Wigner-Seitz radius r_s stands for the average distance between atoms in the bulk material. For the bulk sodium $r_s = 4$ a.u.

In a spherically symmetric field, a single-electron wave function is written as a product of radial, angular and spin wave functions:

$$\varphi_{nlm\sigma} = \frac{1}{r} P_{nl}(r) Y_{lm}(\theta, \phi) X_\sigma \ , \tag{2}$$

where n, l, m and σ are the principal, orbital, magnetic and spin quantum numbers, respectively. Solving self-consistently the system of N single-particle equations (where N is the number of delocalized electrons in a system), one obtains the single-electron wave functions φ_i ($i \equiv nlm\sigma$) and the corresponding energies ε_i. The system of equations is usually solved within the Hartree-Fock (HF) or the local density (LDA) approximations.

The electronic shell structure of neutral and charged metal clusters corresponds to that in a short-range spherical potential. When the spin-orbit interaction is neglected, the shell with quantum numbers $\{n, l\}$ accommodates $2(2l+1)$ electrons. Thus, the ground state electronic configurations of Na_{19}^- and Na_{57}^- clusters are defined as:

$$1s^2 2p^6 3d^{10} 2s^2 \text{ for } Na_{19}^- \ , \qquad 1s^2 2p^6 3d^{10} 2s^2 4f^{14} 3p^6 5g^{18} \text{ for } Na_{57}^-.$$

In the present study, the wave functions of a photoelectron are obtained from the HF equations as the solution corresponding to the energy $\varepsilon = \omega - I_p$ and satisfying certain asymptotic behavior (see, e.g. [32]), where I_p is the ionization potential and ω is the photon energy. The wave functions of the excited states are calculated either in the field of the "frozen" core with the created vacancy or in the field of the rearranged residual electron structure of the cluster. The latter approximation is termed as the static rearrangement. The obtained HF radial wave functions $P_{nl}(r)$ and $P_{\varepsilon l}(r)$ are used further to calculate the dipole-photon amplitudes as well as the Coulomb matrix elements to account for many-electron correlations.

Below we discuss in more detail the photodetachment process from metal cluster anions. In recent works [5–7] the first measurements of the angular resolved photoelectron spectra of negatively charged sodium clusters were reported. The experiments were performed in a broad range of cluster sizes, $3 \le Z \le 147$, and allowed one to probe the angular momenta of single-electron orbitals. It was also demonstrated [7] that simple models based on single-electron treatment of the photoionization process fail to describe the angular anisotropy of photoelectrons. The first calculations performed within the framework of the jellium model for the ionic core and the random phase approximation with exchange (RPAE) for the valence electrons demonstrated a crucial role of many-electron correlations in describing the correct behavior of the photoelectron angular distribution [29]. The essential contribution of many-electron correlations in the photoionization process is not a surprise for neutral metal clusters. The correlations become even more pronounced for negative ions because of weaker binding of the valence electrons. As a result, the role of the continuous spectrum of excitations becomes even more important in the description of photoionization of negative ions, which is very interesting from the theoretical point of view.

The partial photodetachment cross section $\sigma_{nl}(\omega)$ of the nl shell is given by the expression:

$$\sigma_{nl}(\omega) = \frac{4\pi^2 \alpha N_{nl}}{3(2l+1)\omega} \left(|d_{l+1}|^2 + |d_{l-1}|^2 \right) \ , \tag{3}$$

where α is the fine-structure constant, ω is the photon energy, N_{nl} is the number of electrons in the nl shell. The reduced HF dipole matrix elements $d_{l\pm1}$ (in the length form) are defined as follows:

$$d_{l\pm1} \equiv \langle \varepsilon l \pm 1 |\, d \,| nl \rangle = (-1)^l \sqrt{l_>} \int_0^\infty P_{nl}(r) P_{\varepsilon l \pm 1}(r)\, r\, dr \,, \tag{4}$$

where $l_> = l+1$ for the $l \to l+1$ transition and $l_> = l$ for the $l \to l-1$ transition. The total photodetachment cross section is obtained by the sum over all partial cross sections.

To account for many-electron correlations we use the RPAE scheme which describes the dynamic collective response of an electron system to an external electromagnetic field [32]. Within the RPAE, the dipole-photon amplitudes $\langle \varepsilon l \pm 1 | D(\omega) | nl \rangle$ are obtained by solving the following integral equation:

$$\langle \varepsilon l \pm 1 | D(\omega) | nl \rangle = \langle \varepsilon l \pm 1 |\, d \,| nl \rangle + \left(\sum_{\substack{\nu_2 > F \\ \nu_1 \le F}} - \sum_{\substack{\nu_1 > F \\ \nu_2 \le F}} \right) \frac{\langle \nu_2 | D(\omega) | \nu_1 \rangle \langle \nu_1, \varepsilon l \pm 1 | U | \nu_2, nl \rangle}{\omega - \varepsilon_2 + \varepsilon_1 + \imath\delta} \,. \tag{5}$$

Here the indices ν_1 and ν_2 denote the quantum numbers $\{n(\varepsilon), l\}$ of the virtual electron-hole states, F is the Fermi energy, and the matrix element $\langle \ldots | U | \ldots \rangle$ stands for the sum of the direct and exchange Coulomb matrix elements [32].

The angular distribution of photoelectrons detached from the nl shell by an unpolarized photon is determined by the differential cross section of the electron emission into the solid angle $d\Omega = \sin\theta d\theta d\phi$ [33]:

$$\frac{\mathrm{d}\sigma_{\mathrm{nl}}(\omega)}{\mathrm{d}\Omega} = \frac{\sigma_{nl}(\omega)}{4\pi} \left[1 - \frac{1}{2} \beta_{nl}(\varepsilon) P_2(\cos\theta) \right] \,, \tag{6}$$

where $P_2(\cos\theta)$ is the Legendre polynomial, $\beta_{nl}(\varepsilon)$ is the angular distribution anisotropy parameter, and ε is the photoelectron energy.

The main feature of the total photoabsorption cross section in metal clusters is the giant plasmon resonance at about 2-3 eV which appears due to the collective response of the systems [8]. For neutral and positively charged clusters plasmon resonance lies in the discrete spectrum of electronic excitations. The simple HF scheme produces the set of single-electron excitations below the ionization potential and the corresponding distribution of the oscillator strengths. When the correlations are accounted for, the most of the oscillator strength becomes concentrated at the energy close to 2-3 eV, thus describing properly the plasmon excitation.

The distinguishing feature of negatively charged clusters is that they either do not have at all or have just 1-2 discrete dipole excitations. Therefore, the main part of the oscillator strength of partial transitions is distributed in the continuous spectrum of photodetachment cross section. This feature manifests itself both within the HF approximation and the RPAE. Thus, the plasmon resonance lies in continuum, and the account for many-electron correlations leads to significant change in the resonance position, its maximum value and the shape of the resonance curve [27].

The typical photodetachment cross section behavior for metal cluster anions, namely Na_{19}^- and Na_{57}^-, is shown in figure 1 and reveals the powerful maxima due to the collective electron excitation. For the Na_{19}^- anion (left panel), six dipole transitions were taken into account in the RPAE calculations. The dipole transitions from the outer $3d$ and $2s$ shells give the main contribution to the total cross section in the vicinity of the plasmon resonance which is located at about $\omega = 2.4$ eV. However, it should be noted that it is not enough to account only for the transitions from the outer shells in order to obtain the final profile of the plasmon excitation. Instead, one should include the interactions between all valence electrons.

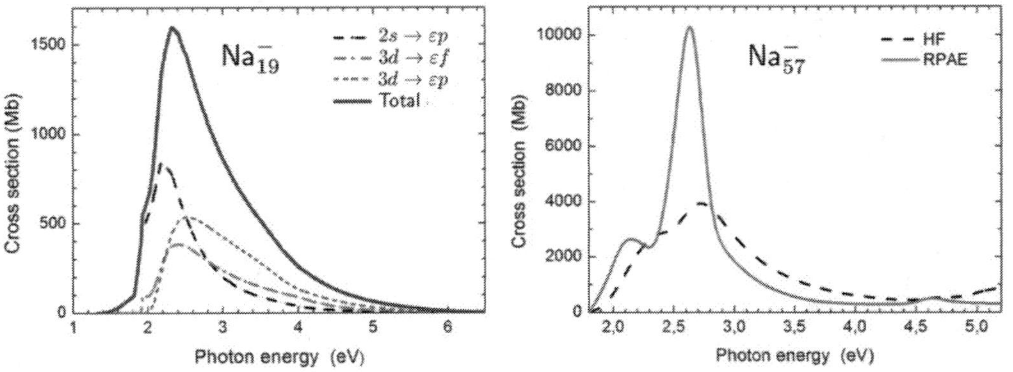

Figure 1. Left panel: the RPAE partial (dashed and dash-dotted lines) and total (solid line) photodetachment cross sections for the Na_{19}^- anion. Right panel: the total cross section for Na_{57}^- within the HF (dashed line) and RPAE (solid line).

For larger cluster anions the number of transitions, which must be accounted for, increases significantly. The total photodetachment cross sections of Na_{57}^- calculated within the HF approximation and the RPAE are shown in the right panel of figure 1. The main contribution to the total cross section comes from the outer $5g$ shell. However, similar to the case of the Na_{19}^- anion, to form the powerful resonance in the total cross section it is essential to take into account the contributions of several shells, namely of the $2s$, $3d$ and $3p$ shells. For Na_{57}^- the position of the plasmon resonance is about $2.6 - 2.7$ eV, and, thus, it does not change significantly with the size of the cluster and lies very close to the classical Mie value.

It should be noted that the total photodetachment cross section is not a very sensitive indicator of the applicability of the jellium model to metal clusters. The photoelectron angular distribution represents a much better test for the conventional jellium model, since the behavior of the anisotropy parameter $\beta(\varepsilon)$ can show whether the shell structure is real and the orbital momentum l is a good quantum number in the cluster system.

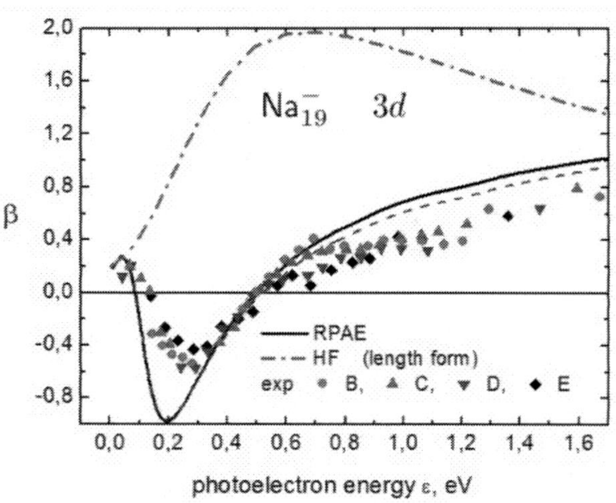

Figure 2. The angular anisotropy parameter $\beta(\varepsilon)$ for the $3d$ shell of the Na_{19}^- anion as a function of photoelectron energy. Comparison between the HF (dash-dotted line), RPAE (solid line) and the experiment [6] (symbols). Experimental data marked as B, C, D and E correspond to the photoionization from the sublevels of the $3d$ orbital split by the crystalline field.

The comparison between the HF, the RPAE and the experimental data [6] for the outer $3d$ shell in the Na_{19}^- anion is presented in figure 2. The experimental points marked as B, C, D and E correspond to the photoionization from the sublevels of the $3d$ orbital which is split by the crystalline field of the cluster in the range $1.75 - 2.15$ eV. It is interesting to note that the general behavior of photoelectron angular distribution is approximately the same for all sublevels. This indicates that all electrons in the sublevels can be characterized by the same orbital quantum number l. The $3d$ binding energy, which is equal to $I_{1d} = 1.92$ eV within the HF approximation, is very close to the center of the multiplet. The calculations within the HF framework (dash-dotted line) fail to explain the dependence of the angular anisotropy parameter on the photoelectron energy. On the contrary, the RPAE results (solid line) are in a good agreement with the experimental data. However, the HF and RPAE schemes based on the jellium model ignore the splitting and, thus, provide the average dependence of the angular anisotropy parameter.

As for larger clusters, we performed the calculations of the angular anisotropy parameter $\beta(\varepsilon)$ for several outer shells of the Na_{57}^- anion. The experimental measurements of the photoelectron spectra were performed not for the closed-shell Na_{57}^- anion, but for Na_{55}^-, which has the unfilled $5g^{16}$ shell [6, 7]. Note that despite of different cluster anions, considered theoretically and experimentally, as well as of the differences in ionization potentials, the general behavior of the angular anisotropy parameter for inner shells (for instance, for $4f$ shell, see the left panel of figure 3) is reproduced quite well by the performed many-body calculations.

However, for the outer $5g$ shell we can only state a qualitative agreement between the calculated results for Na_{57}^- and the experimental data for Na_{55}^- (the right panel of figure 3). Several reasons can be indicated which may explain the quantitative deviation. First, in our calculations we neglected the polarization potential which is rather large for neutral metal clusters and, thus, modifies the dynamics of the outgoing photoelectron. Second, the electrons in the open $5g^{16}$ shell in Na_{55}^- are much stronger influenced by the crystalline field than the electrons in the filled $5g^{18}$ shell of Na_{57}^-.

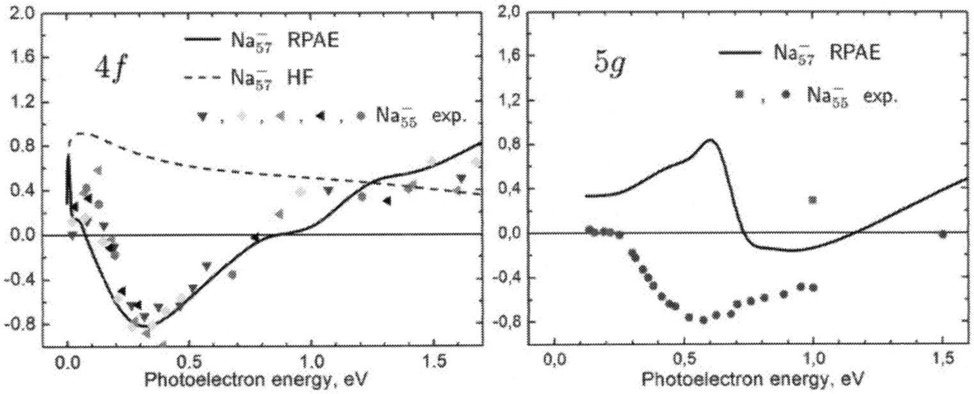

Figure 3. The angular anisotropy parameter for the $4f$ and $5g$ shells of Na_{57}^- anion as a function of the photoelectron energy. Comparison between the HF (dashed line), the RPAE (solid line) results and the experiment data (symbols) in Na_{55}^- taken from [6]. Different experimental points correspond to the photoionization from the sublevels of the orbital split by the crystalline field of the cluster.

The comparison of the existing experimental and theoretical data indicates that further investigation of the process is needed. The theoretical consideration of many-electron system was performed using the simple jellium model for the ionic core. Of course, a more detailed analysis of the problem should go beyond this model and account for the realistic geometrical structure of the core. However, the agreement between the experimental data on the angular distribution and the results of the current theory indicates that the definite shell structure of valence electrons is reproduced quite well even neglecting the interaction with real crystalline field.

3. Application of the jellium model to fullerenes

Ab initio quantum-chemical calculations may provide an accurate quantitative description of the ground state of many-particle systems (atomic clusters and fullerenes, in particular) and allow one to obtain detailed information on geometrical and chemical properties of the system. Meanwhile, the description of dynamic properties, which play an important role in the processes of photoabsorption, elastic and inelastic scattering, electron attachment and photon emission, is a challenging task for the most of contemporary computer packages for *ab initio* based calculations. Dynamic properties (for example, dynamic polarizability) are closely related to the response of a many-electron system to an external electromagnetic field. In case of metal cluster anions and, especially, fullerenes, the properties are governed by collective electron excitations and formation of the plasmon resonances in the continuum [8]. Only recently it became possible to calculate the photoabsorption spectrum of many-electron systems in a broad photon energy range by means of the *ab initio* approach. For instance, it was done for the C_{60} molecule [21,22].

The description of the collective excitations in the continuous spectrum can be performed comparatively easy by means of simplified model approximations. The advantage of such approaches is that they allow one to overcome significant computational difficulties but at the same time take into account the essential features of the processes providing clear physical insight into the phenomena. During the last decades the jellium model have been applied frequently to the description of ground-state properties of fullerenes [34], as well as to the investigation of photoexcitation processes arising in these systems [12,35–38].

In this section we consider the widely investigated C_{60} molecule as a case study. It is assumed that the valence $2s^22p^2$ electrons in each carbon atom form a cloud of delocalized electrons, while the inner-shell $1s^2$ electrons are treated as frozen and are not taken into consideration. Thus, within the jellium model the fullerene core of the 60 charged carbon ions, C^{4+}, is replaced by an uniform distribution of positive charge $Z = 240$ over a spherical layer of a finite thickness, $\Delta R = R_2 - R_1$. The thickness ΔR is chosen to be equal to 1.5 Å which corresponds to a typical diameter of a carbon atom [39] and refers to experimental data from [37]. The potential of the core may be written as

$$U_{\rm core}(r) = \begin{cases} -\dfrac{3Z(R_2^2 - R_1^2)}{2(R_2^3 - R_1^3)} & , \ r < R_1 \\ -\dfrac{Z}{2\left(R_2^3 - R_1^3\right)}\left(3R_2^2 - r^2\left(1 + \dfrac{2R_1^3}{r^3}\right)\right) & , \ R_1 \le r \le R_2 \\ -\dfrac{Z}{r} & , \ r > R_2 \end{cases} , \qquad (7)$$

where $R_1 = R - \Delta R/2$ and $R_2 = R + \Delta R/2$ with R standing for a fullerene radius, $R = 3.54$ Å for C_{60}. Supposing $R_1 \to 0$, one approaches the metal cluster limit, and equation (7) transforms into (1).

Since it is commonly acknowledged [40] that C_{60}, as well as other fullerenes, is formed from fragments of planar graphite sheets, it is natural to match the σ- and π-orbitals of graphite to the nodeless and the single-node wavefunctions of a fullerene, respectively [41]. Carbon atoms within a graphite sheet are connected by σ-bonds, whereas different sheets are connected by π-bonds. In the fullerene, the nodeless σ-orbitals are localized at the radius of the ionic core

Journal of Physics: Conference Series **438** (2013) 012009 doi:10.1088/1742-6596/438/1/012009

while the single-node π-orbitals are oriented perpendicularly to the fullerene surface. The ratio of σ- to π-orbitals in C_{60} should be equal to 3 : 1 due to the sp^2-hybridization of carbon orbitals [42]. Thereby, the electronic configuration of the delocalized electrons in C_{60} is written in the form [34]:

$$1s^2 2p^6 3d^{10} 4f^{14} 5g^{18} 6h^{22} 7i^{26} 8k^{30} 9l^{34} 10m^{18} 2s^2 3p^6 4d^{10} 5f^{14} 6g^{18} 7h^{10} \ .$$

Radial wavefunctions of the $1s \ldots 10m$ shells are nodeless, while the wavefunctions of the $2s \ldots 7h$ shells have one radial node each.

In the present study, the electronic subsystem of C_{60} is treated within the LDA approach. In this case, the single-electron wavefunctions $\varphi_{nlm\sigma}(r)$ (see equation (2)) and the corresponding energies ε_{nl} are determined from a system of self-consistent Kohn-Sham equations.

It is known [43, 44] that the results obtained within the conventional jellium model for fullerenes, while being qualitatively useful, can sometimes lead to unphysical conclusions. For instance, they can lead to unreliable values for the total energy of the ground state [43]. To avoid this, a phenomenological square-well (SW) pseudopotential (see, e.g., [35–37]) has been commonly added to jellium potential (7):

$$U_{\rm core}(r) \to \begin{cases} U_{\rm core}(r) + U_{\rm SW} & , \ R_1 \le r \le R_2 \\ U_{\rm core}(r) & , \ \text{otherwise} \end{cases} \ , \tag{8}$$

where $U_{\rm SW}$ is an adjustable parameter to correct the obtained electronic structure.

It was claimed that accounting for such a pseudopotential increases the accuracy of the jellium-based description and, for instance, allows one to reproduce the experimental value of the first ionization potential of C_{60} [35]. Nonetheless, the applicability of the jellium model for fullerenes and the choice of parameters of the used SW pseudopotential have not been clearly justified so far from a physical viewpoint. Besides, this simple approach cannot describe properly the valence electron density distribution (see the solid red line in the left panel of figure 4) what is critical for the calculation of dynamical polarizability. In the recent work [45], the structured pseudopotential correction ΔU, originated from the comparison of an accurate *ab initio* calculation with the jellium-based one, was proposed:

$$\Delta U(r) = \bar{U}^{\rm QC}_{\rm core}(r) - U^{\rm jel}_{\rm core}(r) \ , \tag{9}$$

where $\bar{U}^{\rm QC}_{\rm core}(r)$ is the potential of the fullerene core obtained from quantum-chemical calculations and averaged by spherical angles variables, and $U^{\rm jel}_{\rm core}(r)$ is the jellium core potential (7). It was shown [45] that such correction to the conventional jellium model allows one to account, at least partly, for the sp^2-hybridization of carbon atomic orbitals, and can predict the shape of the electron density in a more realistic way (see the dash-dotted blue line in the right panel of figure 4).

The photoionization process is studied in a way similar to that described in section 2 for the metal cluster anions. The partial photoionization cross section $\sigma_{nl}(\omega)$ of the nl-shell of a fullerene is defined by equation (3). Using the single-electron wavefunctions (here we used the LDA approach) one can calculate the one-particle transition amplitudes $d_{l\pm1}$. To take into account the collective electron excitations in fullerenes we solved equation (5). Describing the single-particle interaction within the LDA approach, one should then exclude the exchange interaction in the matrix elements $\langle \ldots | U | \ldots \rangle$, replacing the latter by the ordinary direct Coulomb matrix element in equation (5). Thus, one obtains the reduced amplitudes $D_{l\pm1}(\omega)$, which include the many-electron correlations within the random phase approximation (RPA).

Using the correction introduced, one improves the description of the ground state density distribution within the jellium model and obtains a more realistic electron density as compared to the conventional model. Using the corrected jellium potential, we calculated in the present

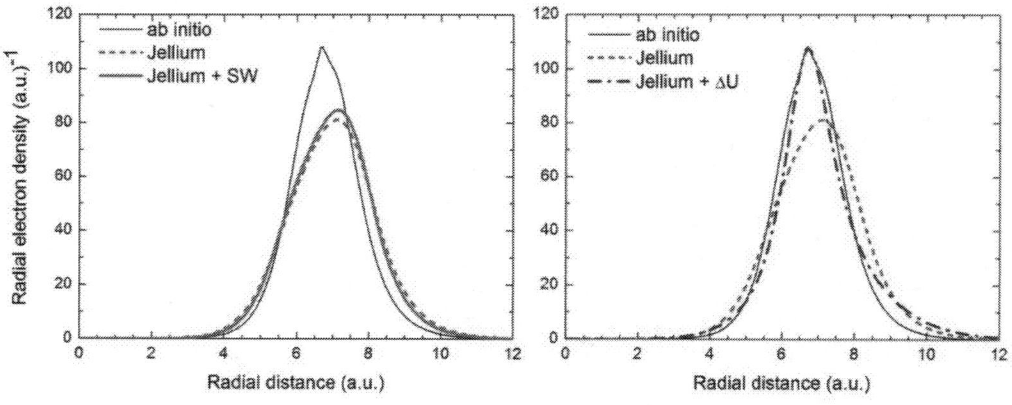

Figure 4. Radial electron density of C_{60} obtained from the *ab initio* calculation (solid black line) and calculated by means of the jellium model: the conventional one (dashed red curve), with the additional square well (SW) pseudopotential (solid red curve) and with the correction ΔU (dash-dotted blue curve).

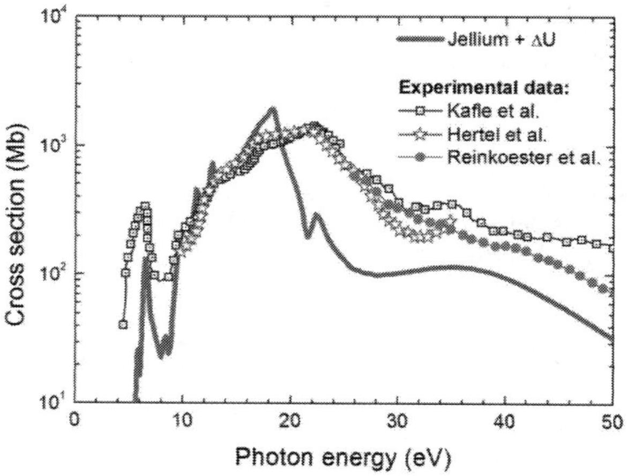

Figure 5. Photoionization cross section of C_{60} obtained from LDA calculations within jellium model with structured correction ΔU (solid green line) and comparison with available experimental data from Hertel *et al.* [15], Reinköster *et al.* [17] and Kafle *et al.* [16]. Many-electron correlations were accounted for in the calculations within the RPA approach.

study the total photoionization cross section of C_{60} within the LDA and RPA approaches. Results of the calculations and the comparison with experimental data [15–17] are presented in figure 5. It is seen that the calculated cross section can describe qualitatively main features of the spectrum in spite of some quantitative discrepancies. It should be noted that the cross section calculated using the corrected jellium model gives a better agreement with experimental data as compared to previous calculations performed within the conventional jellium model (see, e.g. [11, 12]).

4. Conclusion

In this paper we discussed the new applications of the jellium model for the study of photoexcitation processes in various atomic clusters. The consistent many-body theory was applied to calculate the photodetachment cross sections of metal cluster anions as well as the angular distributions of photoelectrons. The case study analysis was performed for sodium cluster anions Na_{19}^- and Na_{57}^- which possess nearly spherical geometry due to the closure of electronic shells of delocalized electrons. The calculations revealed an important role of many-electron correlations in the formation of the giant plasmon resonance and in the angular distributions of photoelectrons, and allowed us to explain the behavior of the angular anisotropy parameter $\beta(\varepsilon)$ versus photoelectron energy in the recent experiments with sodium clusters [6,7].

The theoretical consideration of many-electron system was performed using the jellium model for the ionic core. Of course, a more detailed analysis of the problem should go beyond this model and account for the realistic geometrical structure of the core. However, the agreement between the experimental data on the angular distribution and the results of the current theory indicates that the definite shell structure of valence electrons is reproduced quite well even neglecting the interaction with a real crystalline field.

We discussed also a new type of correction for the description of fullerenes within the spherically symmetric jellium model. The correction represents an additional pseudopotential which originates from the difference between the precise *ab initio* calculation and the one within the jellium model. This potential allows one to mimic partially the sp^2-hybridization, which occurs in formation of fullerenes, and, thus, to import the hybridization effects into the conventional jellium model. It was shown that the correction used improves significantly the electron density distribution as compared to the standard jellium model and the one with an additional square-well pseudopotential. Like the other previously used corrections, it does not allow one to obtain a quantitative agreement with an *ab initio* calculation for the single-electron energy spectrum but reproduces the sequence of energy levels corresponding to the one following from the more precise quantum-chemical calculation [45]. Using this correction to the jellium model we calculated the dynamic response of fullerenes in the photoionization process. We showed that improving the description of the fullerene ground state, one can get a reasonable qualitative agreement with the experimentally measured photoionization cross sections [15–17].

Acknowledgments

R.G.P. and V.K.I. thank the Frankfurt Institute for Advanced Studies for financial support and hospitality.

References

[1] Brack M 1993 *Rev. Mod. Phys.* **65** 677
[2] de Heer W A 1993 *Rev. Mod. Phys.* **65** 611
[3] Connerade J-P and Solov'yov A V 2004 *Latest Advances in Atomic Clusters Collisions: Fission, Fusion, Electron, Ion and Photon Impact* (London: Imperial College Press)
[4] Connerade J-P and Solov'yov A V 2008 *Latest Advances in Atomic Clusters Collisions: Structure and Dynamics from the Nuclear to the Biological Scale* (London: Imperial College Press)
[5] Kostko O, Huber B, Moseler M and von Issendorff B 2007 *Phys. Rev. Lett.* **98** 043401
[6] Bartels C, Hock C, Huwer J, Kuhnen R, Schwöbel J and von Issendorff B 2009 *Science* **323** 1323
[7] Bartels C 2008 PhD thesis (Faculty of Mathematics and Physics, University of Freiburg)
[8] Solov'yov A V 2005 *Int. J. Mod. Phys.* B **19** 4143
[9] Bertsch G F, Bulgac A, Tomanek D and Wang Y 1991 *Phys. Rev. Lett.* **67** 2690
[10] Alasia F, Broglia R A, Roman H E, Serra L, Colo G and Pacheco G M 1994 *J. Phys. B: At. Mol. Opt. Phys.* **27** L643
[11] Ivanov V K, Kashenock G Yu, Polozkov R G and Solov'yov A V 2001 *J. Phys. B: At. Mol. Opt. Phys.* **34** L669
[12] Polozkov R G, Ivanov V K and Solov'yov A V 2005 *J. Phys. B: At. Mol. Opt. Phys.* **38** 4341
[13] Verkhovtsev A V, Korol A V, Solov'yov A V, Bolognesi P, Ruocco A and Avaldi L 2012 *J. Phys. B: At. Mol. Opt. Phys.* **45** 141002
[14] Bolognesi P, Ruocco A, Avaldi L, Verkhovtsev A V, Korol A V and Solov'yov A V 2012 *Eur. Phys. J.* D **66** 254

[15] Hertel I V, Steger H, de Vries J, Weisser B, Menzel C, Kamke B and Kamke W 1992 *Phys. Rev. Lett.* **68** 784
[16] Kafle B P, Katayanagi H, Prodhan M, Yagi H, Huang C and Mitsuke K 2008 *J. Phys. Soc. Jpn.* **77** 014302
[17] Reinköster A, Korica S, Prümper G, Viefhaus J, Godehusen K, Schwarzkopf O, Mast M and Becker U 2004 *J. Phys. B: At. Mol. Opt. Phys.* **37** 2135
[18] Scully S W J *et al* 2005 *Phys. Rev. Lett.* **94** 065503
[19] Korol A V and Solov'yov A V 2007 *Phys. Rev. Lett.* **98** 179601
[20] Scully S W J *et al* 2007 *Phys. Rev. Lett.* **98** 179602
[21] Chen Z and Msezane A Z 2012 *Phys. Rev.* A **86** 063405
[22] Verkhovtsev A V, Korol A V and Solov'yov A V 2013 (submitted)
[23] Huttula M, Mikkelä M-H, Tchaplyguine M and Bjorneholm O 2010 *J. Electron Spectrosc. Relat. Phenom.* **181** 145
[24] von Issendorff B and Cheshnovsky O 2005 *Annu. Rev. Phys. Chem.* **56** 549
[25] Jänkälä K, Mikkelä M-H and Huttula M 2011 *J. Phys. B: At. Mol. Opt. Phys.* **44** 105101
[26] Jänkälä K, Tchaplyguine M, Mikkelä M-H, Bjorneholm O and Huttula M 2011 *Phys. Rev. Lett.* **101** 183401
[27] Polozkov R G, Ivanov V K, Korol A V and Solov'yov A V 2012 *Eur. Phys. J.* D **66** 287
[28] Wopperer P, Faber B, Dinh P M, Reinhard P-G and Suraud E 2010 *Phys. Rev.* A **82** 063416
[29] Solov'yov A V, Polozkov R G and Ivanov V K 2010 *Phys. Rev.* A **81** 021202(R)
[30] Knight W D, Clemenger K, de Heer W A, Saunders W A, Chou M Y and Cohen M L 1984 *Phys. Rev. Lett.* **52** 2141
[31] Ekardt W 1999 *Metal clusters* (New York: John Wiley)
[32] Amusia M Ya 1990 *Atomic Photoeffect* (New York-London: Plenum Press)
[33] Cooper J and Zare R N 1968 *J. Chem. Phys.* **48** 942
Cooper J and Zare R N 1968 *J. Chem. Phys.* **49** 4252
[34] Yabana K and Bertsch G F 1993 *Phys. Scr.* **48** 633
[35] Puska M J and Nieminen R M 1993 *Phys. Rev.* A **47** 1181
Puska M J and Nieminen R M 1994 *Phys. Rev.* A **49** 629 (erratum)
[36] Wendin G and Wästberg B 1993 *Phys. Rev.* B **48** 14764
[37] Rüdel A, Hentges R, Becker U, Chakraborty H S, Madjet M E and Rost J-M 2002 *Phys. Rev. Lett.* **89** 125503
[38] Belyaev A K, Tiukanov A S, Toropkin A I, Ivanov V K, Polozkov R G and Solov'yov A V 2009 *Phys. Scr.* **80** 048121
[39] Östling D, Apell P and Rosen A 1993 *Europhys. Lett.* **21** 539
[40] Zhang Q L, O'Brien S C, Heath J R, Liu Y, Curl R F, Kroto H W and Smalley R E 1986 *J. Phys. Chem.* **90** 525
[41] Martins J L, Troullier N and Weaver J H 1991 *Chem. Phys. Lett.* **180** 457
[42] Haddon R C, Brus L E and Raghavachari K 1986 *Chem. Phys. Lett.* **125** 459
[43] Yannouleas C and Landman U 1994 *Chem. Phys. Lett.* **217** 175
[44] Pavlyukh Y and Berakdar J 2010 *Phys. Rev.* A **81** 042515
[45] Verkhovtsev A V, Polozkov R G, Ivanov V K, Korol A V and Solov'yov A V 2012 *J. Phys. B: At. Mol. Opt. Phys.* **45** 215101

On the role of resonances in photoionization of metal clusters

P. Wopperer[1,2], **P. M. Dinh**[1)], **P. G. Reinhard**[2)], **E. Suraud**[1)]

1) Laboratoire de Physique Théorique, Université P. Sabatier, 118 Rte de Narbonne, F-31062 Toulouse cedex, France

2) Institut für Theoretische Physik II, Universität Erlangen-Nürnberg, Staudstr. 7, D-91058 Erlangen, Germany

E-mail: suraud@irsamc.ups-tlse.fr

Abstract. We analyze electron emission from irradiated clusters by means of time-dependent density-functional theory (TDDFT) in real time. We focus on photo-electron spectra (PES) which deliver an invaluable tool to explore static and dynamical properties of irradiated species. We discuss, in particular, the role of resonances in the PES once the laser frequency is below the emission threshold which implies multiphoton processes. We show that the resonances in the electronic spectrum lead to the occurrence of several peaks in the PES and also strongly affect the standard scaling relations between ionization and the number of required photons for electronic emission.

1. Introduction

The analysis of electron emission from irradiated atoms, molecules and clusters is a key tool to explore both static and dynamical properties of these systems. This brings an invaluable complement to optical absorption measurements, for an overview on clusters see [1, 2, 3, 4]. Short laser pulses offer here a most versatile tool to consider various irradiation scenarios. A first useful indicator is the mere net ionization yield while photo-electron spectroscopy (PES) which measures the distribution of the kinetic energy of emitted electrons, appears as an especially valuable method for delivering information on the energies of occupied single-electron states [5, 6, 7, 8, 9]. Photo-electron angular distribution (PAD) offer an even more involved signal often combined with PES, then providing double-differential cross-sections (energy- and angular-resolved) [10, 11, 12]. The PES nevertheless remains a quite valuable technique by itself. In the one-photon regime, PES provides an image of the sequence of occupied single-particle levels, an aspect which was explored in the early days of cluster physics in cluster anions [8]. Resolving deeper levels requires higher frequencies than visible ones, and hence one finds also UV [7] and X-ray PES [13]. The great variety of modern coherent light sources in large ranges of frequency, intensity and pulse length, nowadays even allow for time-resolved studies in molecules [14, 15, 16] or for deeply lying core states of atoms embedded in a material [17]. Studies of PES are thus now found in all areas of molecular physics, from atoms over simple molecules [18] to complex systems such as clusters [19] and organic molecules [20] and this allows the investigation of both structural and dynamical properties of these various atomic systems.

The general need for a robust theoretical tool of analysis of PES in various dynamical regimes is thus obvious. The traditional approaches relying on (multi-)photon perturbation theory

[21] are naturally limited to low laser intensities. Moreover, they require a good knowledge of the continuum states for the outgoing electrons thus becoming extremely inconvenient for systems with low or without any symmetry. In clusters, time-dependent density-functional theory (TDDFT) at the level of the local-density approximation (TDLDA) directly propagated in the time domain has proven to be a robust, reliable, and efficient method, see e.g. [22, 23, 24]. When using absorbing boundary conditions [25, 26] and when augmented by a self-interaction correction (SIC) [27], TDLDA provides a convenient framework for analyzing electronic emission properties. A simple but efficient procedure for calculating PES based on TDLDA, was thus proposed a few years ago in [28, 29] and has been taken up recently in [30] for a combined analysis of PES and PAD. The technique of [28, 29] has been applied to a variety of systems with possibly large electron emission induced by still moderate laser intensities such as free metal clusters [31, 32, 33], dimer molecules [30], and deposited clusters [34]. It has been recently extended to larger laser intensities [33].

The aim of this paper is to discuss some aspects of PES dynamics in the simple case of small sodium clusters but exploring in detail the dynamical mechanisms at work, and how they affect the actual PES. The paper is organized as follows. Section 2 provides a short summary of the formal framework discussing, in particular, the evaluation of PES. Some illustrative results are discussed in section 3 focusing on the role of resonances in PES.

2. Theoretical framework

Cluster dynamics is described by real-time TDDFT at the level of the time-dependent local-density approximation (TDLDA) using the exchange-correlation functional of [35]. In order to properly analyze electron emission, especially at low energy (namely close to threshold emission), one has to augment TDLDA with a self-interaction correction (SIC) [27] which is chosen here at the level of average-density self-interaction correction (ADSIC), following [36]. The coupling to the ions is mediated by soft local pseudo-potentials for Na [37]. The ionic background is kept frozen which is a legitimate approximation as we consider short laser pulses and direct electron emission proceeding rather fast. ¿From a practical point of view, real-time TDLDA is realized with standard coordinate-space techniques. Emission is basically recorded by means of absorbing boundary conditions. Laser excitation is described by an external classical dipole laser field (as the used wavelengths are much larger than the system size) with a finite temporal envelope $\propto \sin(\pi t/T_{\rm pulse})^2$. For formal and technical details, we refer the reader to [26, 38].

Observables from electron emission deserve somewhat more detail. As already pointed out, the description of electron emission requires absorbing boundary conditions. The absorption is performed after each time step according to the following scheme. We first perform a standard time step from t to $t+\delta t$, $\phi(\mathbf{r},t) \longrightarrow \tilde{\phi}(\mathbf{r})$. We then apply a mask function to the preliminary wavefunction $\tilde{\phi}$ to obtain $\phi(\mathbf{r},t+\delta t) = \mathcal{M}(\mathbf{r})\tilde{\phi}(\mathbf{r})$. This allows to gradually remove any flow towards the bounds. In the following, we have used a spherically symmetric mask profile

$$\mathcal{M} = \cos\left(\frac{|\mathbf{r}| - R_{\rm in}}{R_{\rm out} - R_{\rm in}}\,\frac{\pi}{2}\right)^{1/8} \tag{1}$$

which is active in an absorbing margin $R_{\rm in} < |\mathbf{r}| < R_{\rm out}$. The spherical profile turns out to minimize gridding artifacts when computing angular distributions [39]. For a detailed discussion of absorbing bounds and optimal choice of absorption profiles, see [40].

The absorbing bounds practically reduce in a gradual manner the norm of the wavefunctions, which mimics the dynamical ionization of the system. The net ionization, namely the number of escaping electrons is directly accessible from the single-particle norms as

$$N_{\rm esc} = N_{\rm el}(t=0) - \sum_\alpha (\phi_\alpha|\phi_\alpha). \tag{2}$$

The PES require a bit more effort. They are evaluated for each state separately from the phase oscillations of the single-electron wavefunction at an analyzing point $\mathbf{r}_{\rm meas}$ located near the absorbing bounds, just a bit before the absorbing layer [28]. The single-electron wavefunctions $\phi_\alpha(\mathbf{r}_{\rm meas}, t)$ are thus recorded in time at measuring point(s), and then time-frequency Fourier transformed to $\tilde{\phi}_\alpha(\mathbf{r}_{\rm meas}, \omega)$. This delivers the total PES spectrum as

$$\mathcal{P}(\omega) \propto \sum_\alpha |\tilde{\phi}_\alpha(\mathbf{r}_{\rm meas}, \omega)|^2 \tag{3}$$

which is translated to a kinetic-energy spectrum of emitted electrons by identifying $\hbar\omega = \epsilon_{\rm kin}$. This simple identification is possible thanks to the absorbing bounds which leave only outgoing waves in its vicinity (thus establishing a unique relation between frequency and momentum). High laser intensities require slight modifications of this recipe [33]. An energy- (or state-)resolved PES is, of course, immediately attained by skipping the summation on states in Eq. (3).

3. Results

3.1. A simple example in the one-photon regime

Basic computed features of PES for clusters have been collected in a series of papers [28, 29, 31, 32] . We briefly summarize here the major results. PES for one-photon processes provide a direct map of the ground-state single-particle (s.p.) spectra as used since long in experiments [8, 9]. This regime is easily attained once the laser frequency is larger than the span of the s.p. spectra to be measured and once the excitation is sufficiently gentle to avoid perturbation of the ground-state of the system. Larger intensities blur the signal from PES and we shall not consider them here. Lower frequencies quickly lead into the regime of multiphoton ionization (MPI). In that case, one can spot multiple copies of PES peaks, each of them standing for a certain photon number ν. The envelope of this multiple peak structure exponentially decreases with kinetic energy $\epsilon_{\rm kin}$ of the outgoing electrons. Finally for even higher intensities in the MPI regime, the signal turns to pure exponential decrease which can easily be misinterpreted as thermal emission [32].

We illustrate the simple one-photon case (laser frequency above ionization threshold) in Figure 1. The system Na_8 is treated in the soft jellium approximation [41, 42] which is fully justified for such a simple metal. The jellium parameters have been chosen as: Wigner-Seitz radius $r_s = 3.65\ a_0$, jellium surface width $\sigma = 1\ a_0$. The laser frequency has been chosen as $\omega_{\rm las} = 0.9\,{\rm Ry}$ namely well above the ionization potential (IP) of the system (0.3 Ry), thus safely into the one-photon ionization regime. The figure nicely shows how the PES directly maps the single-electron spectrum. It provides a straightforward proof of the capability of the PES to access structural properties of a system. The case is nevertheless especially forgiving. Indeed, an anomalous behaviour has been noticed for frequencies below the IP and we shall now discuss that point in more detail. The following considerations are restricted to the Na_8 cluster (with explicit ionic structure). However, the effect was also observed for other Na clusters. For the sake of simplicity, the discussion is focused on the $(0, 0, 1)$-orientation of the cluster, where the ionic rings are located in the (x, y)-plane (see Figure 2 for details). The laser will be always polarized along the z-axis.

3.2. Ionization dynamics

The cluster is excited by a laser pulse of $\omega_{\rm las} = 0.08\,{\rm Ry}$, $T_{\rm pulse} = 120\,{\rm fs}$, and $I = 3.1\times10^{11}\ {\rm W/cm^2}$. We recall that with explicit ionic background the Na_8 clusters displays three occupied levels (each furthermore doubly degenerate because of spin). The associated energies are $\epsilon_{1s} = -0.42$, $\epsilon_{1p_{x,y}} = -0.33$ and $\epsilon_{1p_z} = -0.31\,{\rm Ry}$. For $\omega_{\rm las} = 0.08\,{\rm Ry}$ (see right panel of Figure 1). The

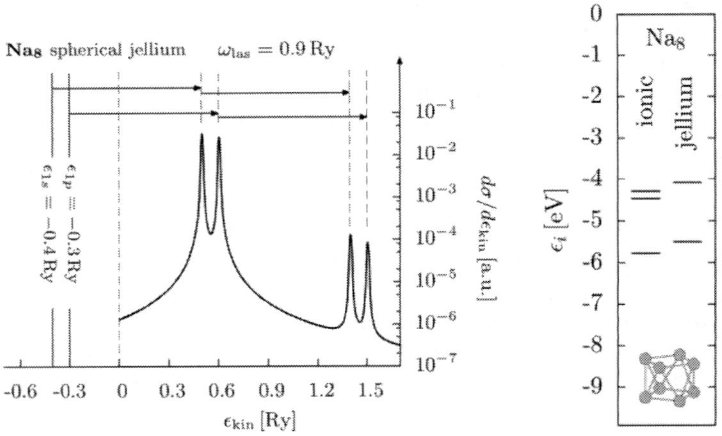

Figure 1: PES (*left panel*) computed for the Na_8 cluster in jellium approximation and in the one-photon regime (ω_{las} > IP). The laser frequency is $\omega_{las} = 0.9\,\mathrm{Ry}$, the intensity $I = 6.9 \times 10^{13}\,\mathrm{W/cm^2}$ and the pulse duration $T_{pulse} = 60\,\mathrm{fs}$. The net ionization is $N_{esc} = 0.003$. The right panel recalls the ionic and electronic spectrum of Na_8 as computed in jellium approximation as well as with full ionic structure.

$1s$ state needs six, the $1p_{x,y}$ states five, and the $1p_z$ state four (which is very close to the threshold) photons to be ionized. The total number of emitted electrons is plotted in Figure 2 as a function of time. Although the laser stops at $t = 120\,\mathrm{fs}$, the cluster surprisingly continues

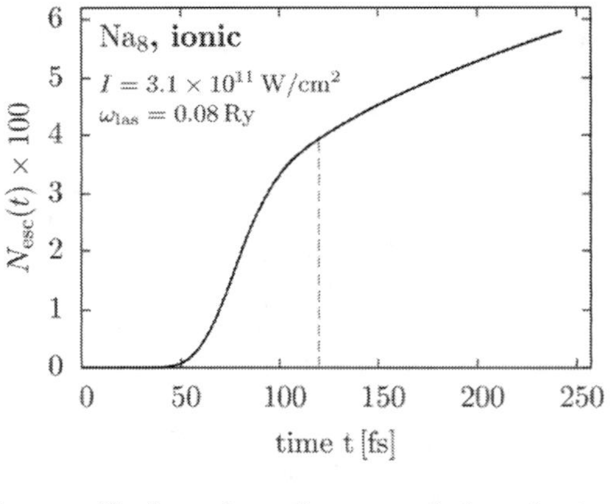

Figure 2: Typical emission profile for a laser frequency below the ionization threshold. The laser pulse begins at $t = 0$ and ends at $t = 120\,\mathrm{fs}$. Significant electron emission starts not until $t \approx 50\,\mathrm{fs}$.

to emit electrons. This is a clear non-perturbative behaviour.

A closer look at the dipole signal $D_z(t)$ in the direction of the laser polarization, see the left panel in Figure 3, reveals what happens. For $t < 120\,\mathrm{fs}$, the dipole follows the electric field as expected. For $t > 120\,\mathrm{fs}$, however, the electronic cloud continues to oscillate, but faster than before. Furthermore, the post-pulse oscillations seem to be very "clean", i.e., with one single

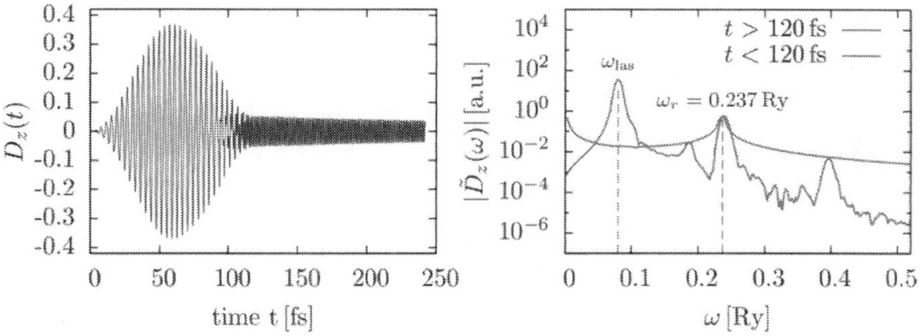

Figure 3: *Left:* Dipole signal D_z in the direction of the laser polarization as a function of time. *Right:* Fourier transformation $|\tilde{D}_z(\omega)|$ of the dipole during ($t < 120$ fs, thin curve) and after the pulse ($t > 120$ fs, thick line).

frequency only. The spectral analysis of the signal from $t = 120 - 250$ fs in the right panel confirms that. The electronic cloud oscillates with $\omega_r = 0.237$ Ry after the pulse is over. During the pulse, the spectrum is mixed: the main frequency is the laser frequency, but ω_r also appears.

This analysis provides the following possible explanation to the phenomenon: the laser pulse has excited a dipole resonance of the cluster. The resonance is triggered by absorption of three photons, since $\omega_r \approx 3\,\omega_{\rm las}$. Figure 4 shows the dipole strength $\mathcal{S}_D(\omega) \sim \Im\{\tilde{D}_z(\omega)\}$ of the dipole signal. The dipole response has been obtained by an initial boost of all s.p. wavefunctions [24, 26],

$$\varphi_\alpha(\mathbf{r},0) \longrightarrow e^{i\vec{p}_{\rm boost}\cdot\mathbf{r}}\,\varphi_\alpha(\mathbf{r},0)\ .$$

The strength function is then directly related to the photoabsorption cross-section [26]. The predominating excitation mode is, of course, the plasmon at $\omega_{\rm pl} = 0.187$ Ry. However, besides the plasmon, one can observe several $1ph$ excitations and resonances, all of them below the IP. Nevertheless, close to the laser frequency, there is only one stronger mode at 0.106 Ry related to $1ph$ excitations from the $1s$ state to the $1p$ states. This mode is too far away for the laser

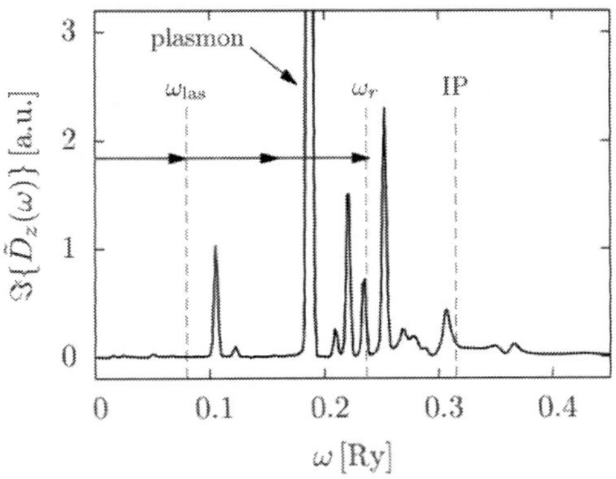

Figure 4: Dipole strength function $\mathcal{S}_D \sim \Im\{\tilde{D}_z(\omega)\}$ of Na_8 with explicit ionic background, in the direction of the laser polarization (denoted here by z). The plasmon frequency is $\omega_{\rm pl} = 0.187$ Ry, the ionization potential at IP $= 0.315$ Ry.

frequency. But the system can still be excited by multiphoton absorption. For laser frequencies below the IP, the laser can therefore very likely hit a resonance. However, it has to be noted that a dipole resonance can only be induced by an odd number of photons. The present case corresponds to three photons. Note that it is also sufficient that the energy $3\,\omega_{\rm las}$ lies just in the close vicinity of the resonance energy ω_r, an exact matching is not compulsory.

3.3. Photo-electron spectra

It is now interesting to analyze the photoemission observables in more detail. Figure 5 shows the PES (in logarithmic scale) for different time spans: during and after the pulse, and for the full calculated time. The *full* PES (line at the top of the panel) exhibits pronounced peaks

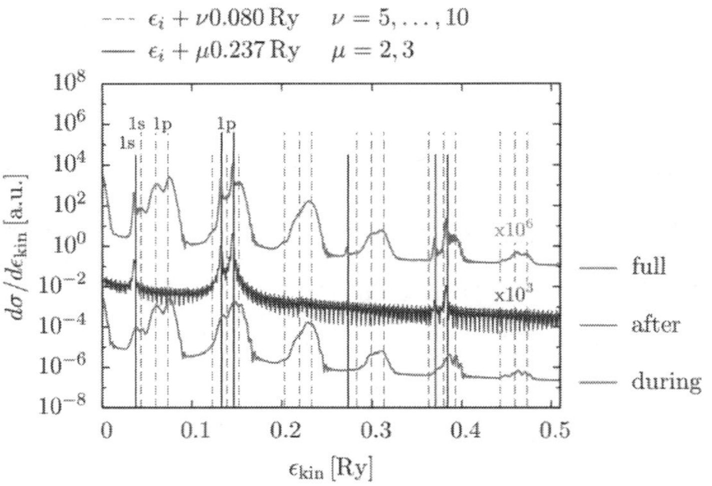

Figure 5: PES of Na_8 after excitation with a 120 fs-pulse with $\omega_{\rm las} = 0.08$ Ry and $I = 3.1 \times 10^{11}\,{\rm W/cm^2}$. The PES is evaluated for different time spans: during ($t = 0 - 120$ fs, green) and after ($t = 120 - 250$ fs, blue) the pulse, and for the full time span (red). The vertical full and dashed lines indicate the kinetic energy deduced from single-particle energies plus multiples of $\omega_{\rm las}$ or ω_r, as indicated.

with certain distance to each other. At first glance, these peaks just resemble MPI peaks. The four-photon peak of the $1p_z$ state should appear closely above the IP, but has not fully passed the threshold. This is due to the positive charge of the residual cluster which shifts the whole spectrum.

More interesting is the shape of the PES *after* the pulse (line in the middle of the panel). Because of numerical artifacts of the Fourier transformation, it shows more noise. Again, the PES exhibits sharp peaks, but many less. These peaks could finally be identified as *multi-resonance peaks*. Their energy exactly corresponds to the s.p. energy plus multiples of the resonance frequency, here $\epsilon_i + \mu\omega_r$, with $\mu = 2, 3$. The last relations have been marked with solid vertical lines. Due to the residual cluster charge, the s.p. spectrum has been shifted by about $\delta\epsilon = -0.012$ Ry in order to achieve better match. The first visible peak at $\epsilon_{\rm kin} = 0.035$ Ry corresponds to the doubly excited ($\mu = 2$) $1s$ state, the following peaks to the $1p_{x,y}$ and $1p_z$ states. The triply excited $1s$ peak disappears behind the noise, but the $1p$ states are still present. No MPI peaks appear in the spectrum *after* the pulse. Hence, the important energy for the post-pulse emission is the resonance energy $\omega_r = 0.237$ Ry and not $\omega_{\rm las} = 0.08$ Ry. The *full* PES, finally, is a superposition of MPI peaks, $\epsilon_i + \nu\omega_{\rm las}$ (also shifted by $\delta\epsilon$ and marked with black dashed lines), and the multi-resonance peaks. The resonance-enhanced emission already starts *during* the pulse (line at the bottom of the panel) and continues after the laser excitation has stopped.

3.4. Impact of laser intensity

We now turn to the impact of laser intensity on actual electron emission in Na_8. Figure 6 shows the total number of emitted electrons $N_{\rm esc}$ as a function of the laser intensity in the range $I \sim 10^{10} - 10^{12}\,{\rm W/cm^2}$. The frequency $\omega_{\rm las} = 0.08\,{\rm Ry}$ and the pulse length $T_{\rm pulse} = 120\,{\rm fs}$ have been kept constant. All curves are normalized to unity. The results indicate that the post-pulse

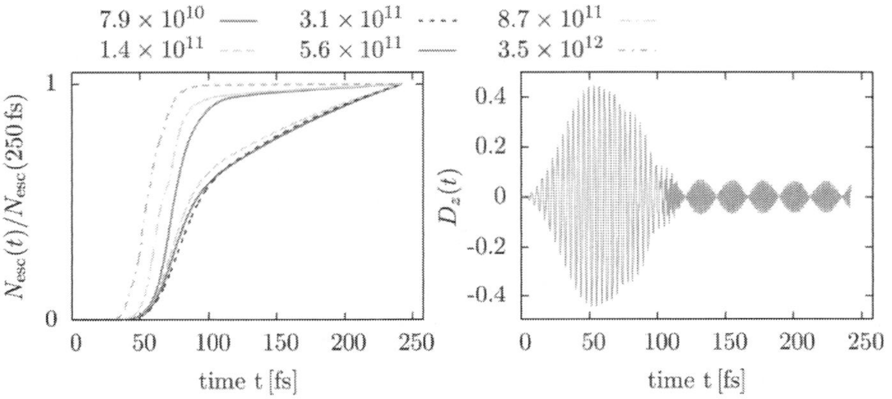

Figure 6: *Left panel:* Emission profiles $N_{\rm esc}(t)$ in Na_8 for different intensities. The lines have been normalized. The final $N_{\rm esc}(250\,{\rm fs})$ ranges from 7×10^{-4} to 1.4 electrons. *Right panel:* Dipole signal $D_z(t)$ for $I = 5.6 \times 10^{11}\,{\rm W/cm^2}$. Two close frequencies are interfering after the pulse.

emission profile sensitively depends on the intensity: for intensities below $3.1 \times 10^{11}\,{\rm W/cm^2}$, all profiles look very similar, typical post-pulse emission occurs. In contrast, for intensities $I \gtrsim 5 \times 10^{11}\,{\rm W/cm^2}$ the post-pulse emission becomes weaker and finally disappears. It is worth noting that the ionization is very large for such intensities, $N_{\rm esc} = 0.2 - 1.4$ for $I \geq 5.6 \times 10^{11}\,{\rm W/cm^2}$. The dipole spectrum of Figure 4 is then blue-shifted during the emission process and resonances are broadened. Consequently, several resonances can be excited at once, as can be seen from the right panel of Figure 6 for $I = 5.6 \times 10^{11}\,{\rm W/cm^2}$. In this case, two close frequencies are interfering. However, at large intensities one could also imagine that the blue shift of the spectrum has the consequence that no resonance is induced. All in all, the conclusion stemming from these various calculations is that the emission behaviour is hard to predict in this intensity range.

The number of emitted electrons $N_{\rm esc}$ for the different intensities is shown in Figure 7 in double-logarithmic scale, during (left panel) and after the pulse (right). For a pure multiphoton process, one expects that $N_{\rm esc} \sim I^\nu$, where ν is at least the minimum number of photons required for ionization, derived from perturbation theory. Actually, the $1p_z$ state needs four photons to be ionized. However, according to the PES in Figure 5 the four-photon peak has not fully passed the threshold. Moreover, the $1s$ state needs six photons while for the $1p_{x,y}$ states already five photons are sufficient. A value between $6 > \nu > 4$ and close to five should thus be reasonable if emission is dominated by multiphoton processes. According to Figure 5, this is not the case.

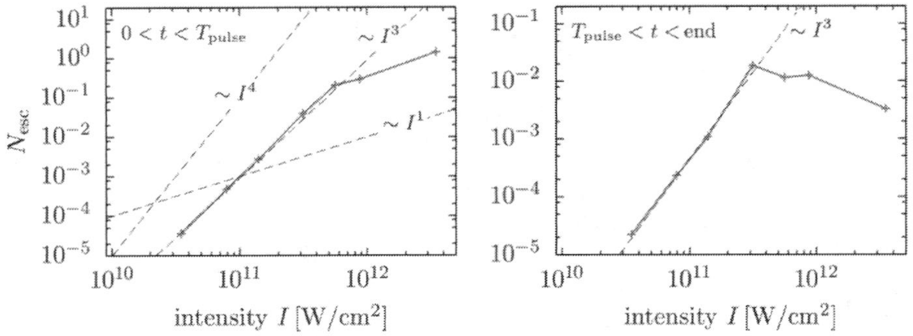

Figure 7: Number of emitted electrons as a function of laser intensity during (*left panel*) and after (*right*) the pulse, in double-logarithmic scale. Different power laws are indicated with dashed lines.

Instead, the number of emitted electrons is proportional $\sim I^p$, with $p \approx 3$. This applies for the emission during and after the pulse, as long as the laser intensity is weak. For higher intensities, the post-pulse emission disappears and $p \approx 1$.

The consequence is that multiphoton processes play only a minor role at weak intensities and the emission is dominated by the resonance at ω_r. Otherwise, one would have expected a much higher power p. Instead, the power corresponds to the number of photons required to induce the resonance. It is interesting to note that the number μ of resonance energies ω_r needed for ionization does not seem to be important, either.

As already mentioned, the presented phenomenon is very sensitive to intensity variation, since for too large intensities, the emission could be on- or off-resonant, or several resonances could be hit due to dynamical shifts of the dipole spectrum. Finally, it should be noted, however, that multiple resonances can also be induced by a variation of the laser frequency. The laser frequency could lie next to two or more resonances which may be excited by one photon. One could also imagine a situation where $\omega_{\rm las}$ is close to an excitation mode and a multiple of the photon energy as well. In all these cases, several excitation modes interfere, the dipole signal appears "dirtier" and the PES should be more puzzling. Moreover, it becomes difficult to distinguish between multiphoton and multi-resonance ionization. In this respect, the chosen frequency $\omega_{\rm las} = 0.08\,{\rm Ry}$ is a kind of exceptional case. The first photon energy is below any excitation, the second photon is in the gap between the first $1ph$ excitation and the plasmon, and the third photon hits almost exactly the resonance whose energy is, in turn, different from the photon energy.

4. Conclusions and perspectives

We have briefly reviewed the theoretical description of photo-electron spectra (PES) focusing on the special case of simple metal clusters. The numerical evaluation relies on a Cartesian coordinate-space grid with absorbing boundary conditions cutting a dynamically active sphere into the 3D grid. The PES are obtained from recording the time evolution of each single-electron wavefunction at a few selected measuring point near the absorbing bounds. This technique adds only a small overhead to the underlying TDDFT propagation thus allowing large scale studies.

We have discussed the basic features of electron emission in the simple case of Na_8. The PES outcome depends very much on the dynamical conditions mainly set by laser frequency and intensity. The one-photon regime of ionization (low intensity, frequency larger than the lowest bound state) is particularly clean allowing a detailed analysis of the single-particle states in the cluster ground-state. It thus provides a direct and detailed structural information on the system as known since long. Higher intensities and/or lower frequencies lead into the regime of multiphoton ionization which turns out to be much more involved. We have focused here on the lower end of intensities, where one still stays in the frequency dominated regime and where one can count photons and sees them acting. This, in particular, shows up in the PES as repeated peaks separated by the photon frequency. A detailed discussion of that case further shows that resonances which can be directly identified from the PES, are also coming into play. Nevertheless, the resulting picture becomes very involved, basically blurring the original simple idea of the single-photon regime. No clear systematics emerges and each case is likely to be considered separately and in detail.

Acknowledgments

This work was supported by the Deutsche Forschungsgemeinschaft, project RE 322/10-1, Institut Universitaire de France, and Agence Nationale de la Recherche.

References

[1] W. A. de Heer, Rev. Mod. Phys. **65**, 611 (1993).
[2] M. Brack, Rev. Mod. Phys. **65**, 677 (1993).
[3] U. Kreibig, M. Vollmer, *Optical Properties of Metal Clusters*, Vol. 25 (Springer Series in Materials Science, 1993).
[4] H. Haberland, ed., *Clusters of Atoms and Molecules 1- Theory, Experiment, and Clusters of Atoms*, Vol. 52 (Springer Series in Chemical Physics, Berlin, 1994).
[5] D. W. Turner and M. I. Al Jobory, J. Chem. Phys. **37**, 3007 (1962).
[6] D.W. Turner *Molecular Photoelectron Spectroscopy* (Wiley, New York 1970).
[7] J.W. Rabalais, *Principles of Ultraviolet Photoelectron Spectroscopy*, (Wiley, New York 1977).
[8] K. M. McHugh, J. G. Eaton, G. H. Lee, H. W. Sarkas, L. H. Kidder, J. T. Snodgrass, M. R. Manaa, and K. H. Bowen, J. Chem. Phys. **91**, 3792 (1989).
[9] D. L. Lichtenberger, K. W. Nebesny, C. D. Ray, D. R. Huffman, and L. D. Lamb, Chem. Phys. Lett. **176**, 203 (1991).
[10] O. Kostko, C. Bartels, J. Schwobel, C. Hock, and B. v Issendorff, J. Phys. : Conf. Ser. **88**, 012034 (2007).
[11] C. Bartels, C. Hok, J. Huwer, R. Kuhnen, J. Schwöbel, and B. von Issendorff, Science **323**, 132 (2009).
[12] M. Kjellberg, O. Johansson, F. Jonsson, A. V. Bulgakov, C. Bordas, E. E. B. Campbell, and K. Hansen. Phys. Rev. A **81**, 023202 (2010).
[13] J.F. Moulder, W.F. Stickle, P.E. Sobol, and K.D. Bomben, *Handbook of X-ray Photoelectron Spectroscopy*, (Perkin-Elmer Corp., Eden Prairie, MN, USA 1962).
[14] T. Seideman, J. Chem. Phys. **107**, 7859 (1997).
[15] T. Seideman, Ann. Rev. Phys. Chem. **53**, 41 (2002).
[16] Y. Suzuki and T. Seideman, J. Chem. Phys. **122**, 234302 (2005).
[17] A. Pietzsch, A. Föhlisch, M. Beye, M. Deppe, F. Hennies, M. Nagasono, E. Suljoti, W. Wurth, C. Gahl, K. Döbrich, and A. Melnikov, New J. Phys. **10**, 033004 (2008).
[18] L. H. Tjeng, A. R. Vos, and G. A. Sawatzky, Surf. Sci. **235**, 269 (1990).
[19] M. A. Hoffmann, G. Wrigge, B. v. Issendorff, J. Muller, G. Gantefor, and H. Haberland, Euro. Phys. J. D **16**, 9 (2001).
[20] H. Liu and R. J. Hamers, Surf. Sci, **416**, 354 (1998).
[21] F. H. M. Faisal, *Theory of Multiphoton Processes* (Plenum Press, New York, 1987).
[22] F. Calvayrac, P.-G. Reinhard, and E. Suraud, Phys. Rev. B **52**, R17056 (1995).
[23] K. Yabana and G. F. Bertsch, Phys. Rev. B **54**, 4484 (1996).
[24] F. Calvayrac, P.G. Reinhard, E. Suraud, Ann. Phys. (NY) **255**, 125 (1997)
[25] C. A. Ullrich, J. Mol. Struct. (THEOCHEM) **501-502**, 315 (2000).
[26] F. Calvayrac, P.G. Reinhard, E. Suraud, C.A. Ullrich, Phys. Rep. **337**, 493 (2000)
[27] J. P. Perdew and A. Zunger, Phys. Rev. B **23**, 5048 (1981).
[28] A. Pohl, P.-G. Reinhard, and E. Suraud, Phys. Rev. Lett. **84**, 5090 (2000).
[29] A. Pohl, P.-G. Reinhard, and E. Suraud, J. Phys. B **34**, 4969 (2001).
[30] U. De Giovannini, D. Varsano, M. A. L. Marques, H. Appel, E. K. U. Gross, and A. Rubio, Phys. Rev. A 85, 062515, (2012).

[31] A. Pohl, P.–G. Reinhard, and E. Suraud, Phys. Rev. A **68**, 053202 (2003).
[32] A. Pohl, P.–G. Reinhard, and E. Suraud, J. Phys. B **37**, 3301 (2004)
[33] P.M. Dinh, P. Romaniello, P.-G. Reinhard, and E. Suraud, subm. Phys. Rev. A (2012); arXiv:1206.3918.
[34] M. Bär, P. M. Dinh, L. V. Moskaleva, P.-G. Reinhard, N. Rösch, and E. Suraud, Physica Status Solidi B **247** 1122 (2010).
[35] J. P. Perdew and Y. Wang, Phys. Rev. B **45**, 13244 (1992).
[36] C. Legrand, P. G. Reinhard, E. Suraud, J. Phys. B **35**, 1115 (2002).
[37] S. Kümmel, M. Brack, P.G. Reinhard, Euro. Phys. J. D **9**, 149 (1999)
[38] P.G. Reinhard, E. Suraud, *Introduction to Cluster Dynamics* (Wiley, New York, 2003)
[39] A. Pohl, P.–G. Reinhard, and E. Suraud, Phys. Rev. A **70**, 023202 (2004).
[40] P.–G. Reinhard, P. D. Stevenson, D. Almehed, J. A. Maruhn, and M. R. Strayer, Phys. Rev. E **73**, 036709 (2006).
[41] A. Rubio, L. C. Balbas, J. A. Alonso, Z. Phys. D **19**,93 (1991).
[42] B. Montag, P. G. Reinhard, Z. Phys. D **32**, 125 (1994).

Plasmon excitations in photo- and electron impact ionization of fullerenes

A V Verkhovtsev[1,2], A V Korol[1,3] and A V Solov'yov[1,4]
[1] Frankfurt Institute for Advanced Studies, Ruth-Moufang-Str. 1, 60438 Frankfurt am Main, Germany
[2] St. Petersburg State Polytechnic University, Politekhnicheskaya ul. 29, 195251 St. Petersburg, Russia
[3] Department of Physics, St. Petersburg State Maritime Technical University, Leninskii prospekt 101, 198262 St. Petersburg, Russia
E-mail: verkhovtsev@fias.uni-frankfurt.de

Abstract. We study collective electron excitations (also referred to as plasmons) in the C_{60} fullerene in the processes of photoionization and electron inelastic scattering. To reveal the contribution of collective electron excitations, we utilize the plasmon resonance approximation. It is shown that within this framework the photoionization cross section is described as a sum of two contributions, which represent two coupled modes of the surface plasmon. The electron energy loss spectra of C_{60} are described by three contributions, namely by the two modes of the surface plasmon and the volume plasmon. The results of calculations are in good agreement with experimental data on photoionization and electron inelastic scattering of C_{60}. We show that the collective excitations play a significant role in the ionization process and provide a dominant contribution to the spectra.

1. Introduction

Since the discovery of fullerenes [1], the peculiar highly symmetric geometry and amazing properties [2] of these molecules have attracted a lot of interest and stimulated intensive experimental and theoretical investigations (see, e.g. [3]). At present, the investigation of fullerenes is active since they are proposed to be used in various fields of science and technology. For instance, excitation of fullerenes, placed in a biological medium, by an external radiation or incident heavy ions may lead to an enhancement in generation of secondary electrons or reactive oxygen species [4]. This allows fullerenes to be potentially used as sensitizers in photodynamic therapy or in ion-beam cancer therapy. A very important fundamental problem closely related to these application is an adequate description of dynamic response of fullerenes to external fields or to the interaction with projectiles.

A characteristic feature of a fullerene molecule is a partial delocalization of its electrons and hybridization of atomic orbitals. The valence $2s^2 2p^2$ electrons in each carbon atom form a cloud of delocalized electrons, which can be excited collectively by an external electric field. This excitation is represented as collective oscillation of delocalized electrons against the positively charged ions. By analogy with the collective electron excitations in condensed media, the collective excitations in fullerenes are referred to as plasmons.

[4] On leave from A.F. Ioffe Physical-Technical Institute, St. Petersburg, Russia

It was predicted theoretically [5] and soon after observed experimentally [6] that the photoionization spectrum of the gas phase C_{60} molecules is characterized by a broad prominent peak located at about 20 eV, the so-called giant resonance, which is formed due to collective electron excitations. Similar to photoionization, a broad resonant-like structure between 20 and 30 eV was observed in the process of inelastic scattering of electrons on C_{60} [7–10]. Recent experiments on photoionization of neutral [11] and charged [12] C_{60} molecules revealed the existence of the second collective resonance at about 40 eV and opened up the discussion [13, 14] on the nature of this resonance.

In the present paper we study the formation of plasmon excitations in the C_{60} fullerene in the processes of photoionization and electron inelastic scattering. To evaluate the contribution of collective electron excitations to the cross sections, we utilize the plasmon resonance approximation [15–17]. It allows one to apply the classical electrodynamics and hydrodynamics to study a collective motion of delocalized electrons of the system. We show that, similar to metal clusters [18], there are two different types of collective electron excitations, namely the surface and the volume plasmons, which can be formed in fullerenes. The dipole surface plasmon is responsible for the formation of the giant resonance in photoionization spectrum of fullerenes, while the volume plasmon modes, which have higher resonance frequencies, provide an essential contribution to the formation of the electron impact ionization cross section. Within the plasmon resonance approximation, the photoionization cross section of C_{60} is described as a sum of two contributions, which represent two coupled modes of the surface plasmon. The electron energy loss spectra of fullerenes are described by three contributions, namely by the two modes of the surface plasmon and the volume plasmon as well. We show that the collective excitations play a significant role in the excitation process and provide a dominant contribution to the spectra.

The atomic system of units, $m = |e| = \hbar = 1$, is used throughout the paper.

2. Methods of investigation

To study the processes of photoionization and electron inelastic scattering from C_{60}, we utilize the following model. We consider the fullerene as a spherically symmetric system with a homogeneous charge distribution over the shell of a finite width, $\Delta R = R_2 - R_1$, where $R_{1,2}$ are the inner and the outer radii of the molecule, respectively [9, 19–21]. The chosen value of the shell's width, $\Delta R = 1.5$ Å, corresponds to the typical size of the carbon atom [20].

The equilibrium electron density distribution, $\rho_0(r)$, is expressed via the number N of delocalized electrons (four valence $2s^2 2p^2$ electrons from each carbon atom, i.e. $N = 240$ in case of C_{60}) and the fullerene volume V:

$$\rho_0 = \begin{cases} N/V & \text{for } R_1 \le r \le R_2 \\ 0 & \text{if otherwise .} \end{cases} \tag{1}$$

The volume of the fullerene shell can be expressed as

$$V = \frac{4\pi}{3}\left(R_2^3 - R_1^3\right) = \frac{4\pi}{3} R_2^3 \left(1 - \xi^3\right) , \tag{2}$$

where $\xi = R_1/R_2 \le 1$ is the ratio of the inner to the outer radii. It should be mentioned that the spherical-shell model defined by equations (1) and (2) is applicable for any spherically symmetric system with an arbitrary value of the ratio ξ. Supposing $\xi = 0$, one obtains a model of a metal cluster [15, 16, 18], while the case $\xi = 1$ represents a fullerene modeled as an infinitely thin sphere [8, 15, 17].

Action of the external electric field causes variation of the electron density, $\delta\rho(r,t)$, in the system. Thus, the total electron density is introduced as

$$\rho(r,t) = \rho_0(r) + \delta\rho(r,t) , \tag{3}$$

where $\rho_0(r)$, defined by equation (1), denotes the stationary distribution of the negative charge.

Treating a collective motion of delocalized electrons of the system as a charged liquid, one can apply the classical electrodynamics and hydrodynamics to describe this motion. The detailed formalism of this approach is presented in [22–24]. Therefore, here we skip the details of the derivation and introduce the general equation for the variation of electron density in an arbitrary spherically symmetric system [23, 24]:

$$\left(\omega^2 - 4\pi\rho_0(r)\right)\delta\rho_l(r) + 4\pi\frac{\rho_0'(r)}{2l+1}\int\limits_0^\infty g_l(r,r')\delta\rho_l(r')\mathrm{d}r' = q^2\rho_0(r)\phi_l(r) - \rho_0'(r)\phi_l'(r) \,, \quad (4)$$

where $\delta\rho_l(r)$ is multipole variation of electron density in a spherically symmetric system under the action of the multipole component $\phi_l(r)$ of the external field, ω is the frequency of the external field, and the function $g_l(r,r')$ is defined as:

$$g_l(r,r') = l\left(\frac{r}{r'}\right)^{l-1}\Theta(r'-r) - (l+1)\left(\frac{r'}{r}\right)^{l+2}\Theta(r-r') \,, \quad (5)$$

where $\Theta(x)$ is the Heaviside step function.

In case of the fullerene model defined above, the equilibrium electron density distribution, $\rho_0(r)$ (see equation (1)), and the corresponding derivative $\rho_0'(r)$ can be written as:

$$\rho_0(r) = \rho_0\,\Theta(r-R_1)\Theta(R_2-r) \,, \qquad \rho_0'(r) = \rho_0\left(\delta(r-R_1) - \delta(r-R_2)\right) \,, \quad (6)$$

where $\delta(x)$ is the delta function.

The solution of equation (4) for a fullerene is then sought in the following form:

$$\delta\rho_l(r) = \delta\varrho_l(r)\Theta(r-R_1)\Theta(R_2-r) + \sigma_l^{(1)}\delta(r-R_1) + \sigma_l^{(2)}\delta(r-R_2) \,, \quad (7)$$

where $\delta\varrho_l(r)$ describes the volume density variation arising inside the fullerene shell, and $\sigma_l^{(1,2)}$ are variations of the surface charge density at the inner and outer surfaces of the shell, respectively. The first term leads to the formation of the volume plasmon, while the two latter terms produce two coupled modes of the surface plasmon, the symmetric and antisymmetric ones [19, 20] (see figure 1).

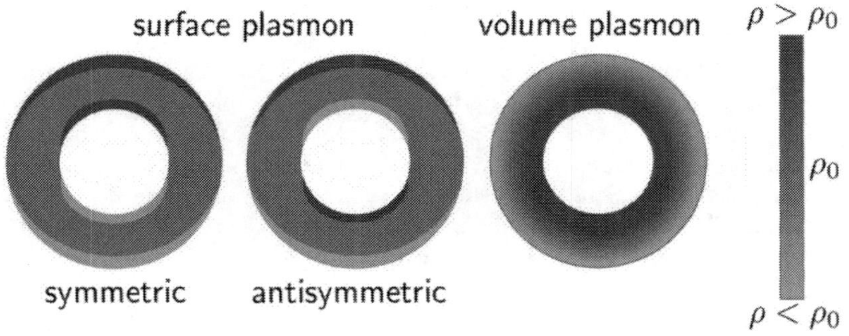

Figure 1. Representation of the two modes of the surface plasmon (left and middle panels) and of the volume plasmon (right panel). Light-blue and dark-blue regions represent the additional positive and negative charge, respectively.

The photoionization and inelastic scattering cross sections are treated within the plasmon resonance approximation [15–17]. It relies on the fact that collective plasmon excitations give the main contribution to the cross section in the vicinity of the giant resonance. A single-particle contribution, which comes from quantum phenomena, is not accounted for in the approximation, since the single-particle effects give a small contribution as compared to the collective modes [16, 18]. The detailed formalism for the description of plasmon excitations in electron- and photon-fullerene collisions is presented in [23, 24].

Within the plasmon resonance approximation, the differential inelastic scattering cross section of fast electrons in collision with fullerenes can be defined as a sum of three contributions:

$$\frac{d^3\sigma}{d\varepsilon_2 d\Omega_{\mathbf{p}_2}} = \frac{d^3\sigma^{(v)}}{d\varepsilon_2 d\Omega_{\mathbf{p}_2}} + \frac{d^3\sigma^{(s_1)}}{d\varepsilon_2 d\Omega_{\mathbf{p}_2}} + \frac{d^3\sigma^{(s_2)}}{d\varepsilon_2 d\Omega_{\mathbf{p}_2}}\,, \tag{8}$$

where

$$\begin{aligned}
\frac{d^3\sigma^{(v)}}{d\varepsilon_2 d\Omega_{\mathbf{p}_2}} &= \frac{2R_2 p_2}{\pi q^4 p_1}\,\omega \sum_l \frac{\omega_p^2\,\Gamma_l^{(v)}\,V_l(q)}{(\omega^2-\omega_p^2)^2+\omega^2\Gamma_l^{(v)2}} \\
\frac{d^3\sigma^{(s_1)}}{d\varepsilon_2 d\Omega_{\mathbf{p}_2}} &= \frac{2R_2 p_2}{\pi q^4 p_1}\,\omega \sum_l \frac{\omega_{1l}^2\,\Gamma_{1l}^{(s)}\,S_{1l}(q)}{(\omega^2-\omega_{1l}^2)^2+\omega^2\Gamma_{1l}^{(s)2}} \\
\frac{d^3\sigma^{(s_2)}}{d\varepsilon_2 d\Omega_{\mathbf{p}_2}} &= \frac{2R_2 p_2}{\pi q^4 p_1}\,\omega \sum_l \frac{\omega_{2l}^2\,\Gamma_{2l}^{(s)}\,S_{2l}(q)}{(\omega^2-\omega_{2l}^2)^2+\omega^2\Gamma_{2l}^{(s)2}}
\end{aligned} \tag{9}$$

are obtained within the plane-wave first Born approximation. Here $\varepsilon_2 = \mathbf{p}_2^2/2$ is the kinetic energy of the scattered electron, $\Omega_{\mathbf{p}_2}$ its solid angle, $\mathbf{p}_1$ and $\mathbf{p}_2$ the initial and the final momenta of the projectile electron, $\mathbf{q} = \mathbf{p}_1 - \mathbf{p}_2$ the transferred momentum, $\omega = \varepsilon_1 - \varepsilon_2$ the energy loss, and ε_1 the kinetic energy of the incident electron. The volume plasmon frequency, $\omega_p = \sqrt{3N/(R_2^3 - R_1^3)}$ (where N is the number of delocalized electrons), is related to the frequencies of the symmetric and antisymmetric surface plasmon modes of the multipolarity l by the following expression [19, 20, 23]:

$$\omega_{(1,2)l}^2 = \frac{\omega_p^2}{2}\left(1 \mp \frac{1}{2l+1}\sqrt{1+4l(l+1)\xi^{2l+1}}\right)\,, \tag{10}$$

where the signs '−' and '+' correspond to the symmetric (ω_{1l}) and the antisymmetric (ω_{2l}) modes, respectively, and $\xi = R_1/R_2$. The quantities $\Gamma_l^{(v)}$ and $\Gamma_{jl}^{(s)}$ ($j = 1, 2$) are the widths of the plasmon resonances. Functions $V_l(q)$, $S_{1l}(q)$ and $S_{2l}(q)$ are the diffraction factors depending on the transferred momentum q. They determine the relative contribution of the multipole plasmon modes in various ranges of electron scattering angles and, thus, the resulting shape of the differential energy loss spectrum. Explicit expressions for these functions are given in [23].

It was shown [15] that the excitations with large angular momenta l have a single-particle rather than a collective nature. It follows from the fact that with the increase of l the wavelength of the plasmon excitation becomes smaller than the characteristic wavelength of the delocalized electrons in the fullerene. In the case of C_{60}, the estimates show [15] that the excitations with $l > 3$ are formed by single-electron transitions rather than by the collective excitations. Hence, only terms corresponding to the dipole ($l = 1$), quadrupole ($l = 2$) and octupole ($l = 3$) plasmon excitations should be accounted for in the sum over l in equations (8) and (9).

The theory presented relies on the number of multipole terms to be accounted for ($l_{\max} = 3$) and the widths of the plasmon resonances, $\Gamma_{jl}^{(s)}$ and $\Gamma_l^{(v)}$, which are not just the fitting parameters of the model but the real physical quantities. Their calculation can be performed considering another physical process, namely, the decay of the collective excitation mode into the incoherent sum of single-electron excitations. Such calculations are beyond the scope of the present paper. Nevertheless, one can estimate the widths of the plasmon excitations in the fullerene using the relation similar to the Landau damping of plasmon oscillations. Such an estimate results in $\Gamma_{jl}^{(s)} \sim l_{v_F}/\Delta R$, where v_F is the velocity of the fullerene electrons on the Fermi surface. A similar estimate was successfully applied to the investigation of collective excitations in metal clusters [18].

In Ref. [13, 23, 24], it was shown that in case of photoionization, when the system interacts with a uniform external electric field, only the surface plasmon can occur. Indeed, the wave length of electromagnetic radiation is assumed to be much larger than the characteristic size of the system, i.e. the condition $\omega R \ll 1$ is fulfilled. This condition implies the validity of the dipole approximation, i.e. only dipole excitations ($l = 1$) may arise in the system due to interaction with the external electromagnetic field. In this limit, equation (4), describing interaction with an external field, is simplified due to $q = 0$. Thus, the photoionization cross section is defined as a sum of two contributions which correspond to the two modes of the surface plasmon [21, 23]:

$$\sigma(\omega) = \frac{4\pi\omega^2}{c} \left(\frac{N_1 \Gamma_1}{(\omega^2 - \omega_1^2)^2 + \omega^2\Gamma_1^2} + \frac{N_2 \Gamma_2}{(\omega^2 - \omega_2^2)^2 + \omega^2\Gamma_2^2} \right) . \tag{11}$$

Here ω is the photon energy, $\omega_1 \equiv \omega_{11}$ and $\omega_2 \equiv \omega_{21}$ are the frequencies of the symmetric and antisymmetric surface plasmon modes, respectively, $\Gamma_1 \equiv \Gamma_{11}^{(s)}$ and $\Gamma_2 \equiv \Gamma_{21}^{(s)}$ are the corresponding widths. N_1 and N_2 are the number of delocalized electrons, involved in each collective excitation, which should obey the sum rule $N_1 + N_2 = N$. The frequencies of the collective excitations are defined as [19–21]

$$\omega_{1,2}^2 = \omega_0^2 + \frac{\omega_p^2}{6}\left(3 \mp \sqrt{1 + 8\xi^3}\right) , \tag{12}$$

where the signs '$-$' and '$+$' correspond to the symmetric and antisymmetric modes, respectively, ω_p is the volume plasmon frequency defined above, and $\xi = R_1/R_2$. The term ω_0^2 defines the free-electron picture threshold [20]. Following Ref. [20], we use the threshold value $\omega_0 = 14$ eV in the calculations. Below ω_0, some of the electrons are treated as bound ones and, therefore, are not involved in the formation of the plasmon excitations.

In the calculations, we used the ratio $\gamma_1 = \Gamma_1/\omega_1 = 0.6$ for the symmetric mode, which corresponds to the experimental values obtained from the photoionization and energy loss experiments on neutral C_{60} [6, 8]. For the antisymmetric mode, we used the value $\gamma_2 = \Gamma_2/\omega_2 = 1.0$ which corresponds to the widths of the second plasmon resonance obtained in the study of photoionization of C_{60}^{q+} ($q = 1 - 3$) ions [12].

3. Results and discussion

In this section, we present the results of calculations of photo- and electron impact ionization spectra of the C_{60} fullerene based on the formalism described above. Figure 2 represents the photoionization spectrum of C_{60} calculated using the plasmon resonance approximation (see equation (11)). The left panel shows the spectrum obtained in the broad range of photon energies, namely up to 100 eV, while the right panel shows in more detail the region of the plasmon resonances, up to 50 eV. The symmetric and antisymmetric modes of the surface plasmon are shown by the dashed red and dash-dotted blue curves, respectively. The widths Γ_1 and Γ_2 of the symmetric and antisymmetric modes of the surface plasmon are equal to 11.4 and 33.2 eV, respectively, and correspond to the results of experimental measurements [6, 12]. The sum of the two modes is shown by the thick solid green line. Theoretical curves are compared to the results of experimental measurements of photoabsorption of C_{60} [6, 25, 26]. Open squares represent the recent compilation of experimental data made by Kafle *et al.* [25] in a broad energy range up to 100 eV (see the left panel). Open circles and stars represent, respectively, the compilation of experimental data made by Berkowitz [26] and the results of experimental measurements by Hertel *et al.* [6] (see the right panel).

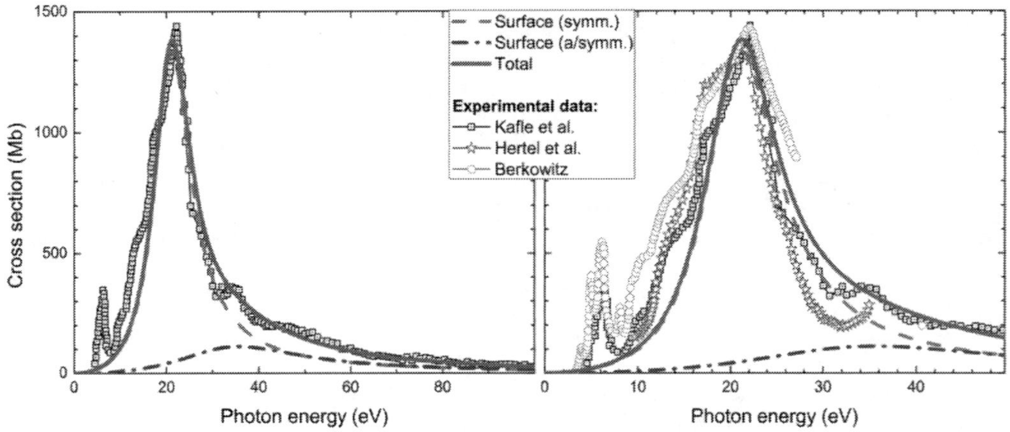

Figure 2. The photoabsorption cross section of C_{60} calculated within the plasmon resonance approximation (thick green line). Contribution of the symmetric and antisymmetric modes of the surface plasmon are shown by the dashed red and dash-dotted blue lines, respectively. Theoretical curves are compared to the experimental data of Kafle *et al.* [25] (open squares), Berkowitz [26] (open circles) and Hertel *et al.* [6] (open stars). The left panel shows the spectrum obtained in the broad range of photon energies, up to 100 eV, and the right panel shows in more detail the region of the plasmon resonances, up to 50 eV.

As it is seen from figure 2, the plasmon resonance approximation gives a good agreement with the experimental data. The oscillator strength, calculated in the energy range up to 100 eV, is equal to 216.3, and corresponds to the experimentally measured value of 230.5 [25]. The plasmon resonance approximation describes quite well the main features of the giant resonance, such as height, width and position of the peak. Therefore, such a model can be considered as a useful tool for interpretation of experimental results. It should be also noted that the approach is focused on the study of collective electron excitations above the ionization threshold, which is about 7.6 eV for C_{60} [6]. Therefore, it does not account for the series of discrete excitations, which manifest themselves in the low-energy region of the spectrum below 10 eV [27].

Figure 3 represents the electron energy loss spectra of C_{60} calculated within the plasmon resonance approximation (see equations (8) and (9)). The calculations were performed for the scattering angles $\theta = 3°, 5°, 7°$ and $9°$, and the curves obtained were compared to the results of recent experimental measurements of the inelastic scattering of fast (1 keV) electrons on C_{60} [9, 10]. The results of the comparison are presented in figure 3. For the sake of convenience, both the experimental and the theoretical curves are normalized to unity at the point of maximum.

Similar to the study of the photoionization process, we used the ratios $\gamma_{1l}^{(s)} = 0.6$ and $\gamma_{2l}^{(s)} = 1.0$ for the symmetric and antisymmetric modes, respectively. The widths of the volume

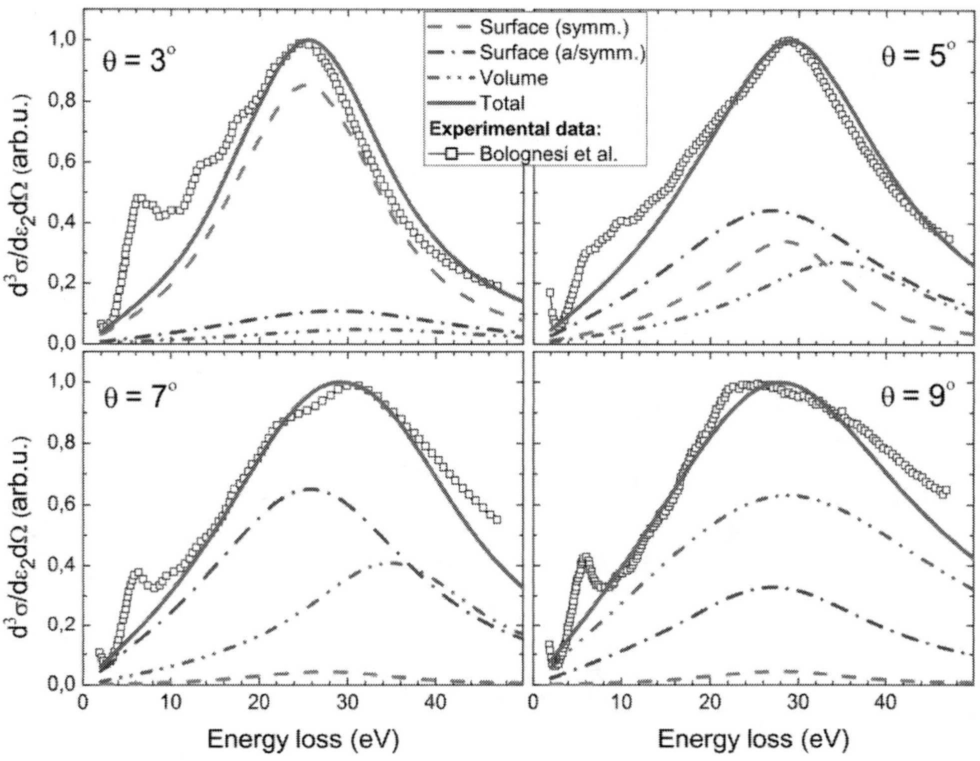

Figure 3. Comparison of the electron energy loss spectra, calculated within the plasmon resonance approximation, with the experimental spectra measured for the incident energy range 1002-1050 eV and for the scattering angles $\theta = 3° \ldots 9°$. The symmetric and antisymmetric modes of the surface plasmon are shown by dashed red and the dash-dotted blue lines, respectively; the volume plasmon contribution is shown by the double-dotted purple line. The total cross section is shown by the thick green line. Open squares represent the experimental data [9, 10]. For the sake of convenience, both the experimental and the theoretical curves are normalized to 1 at the point of maximum. The energy scale is the same for all panels.

plasmon were varied to obtain a better agreement with the experimental data. In the present calculations, the ratios $\gamma_l^{(v)} = \Gamma_l^{(v)}/\omega_p$ were considered within the range $0.7 \ldots 1.3$. The values of the plasmon frequencies and the corresponding widths for all three collective excitations are summarized in table 1.

In figure 3, the dashed red and dash-dotted blue curves represent, respectively, the symmetric and antisymmetric modes of the surface plasmon, the double-dotted purple line shows the contribution of the volume plasmon. The sum of the three collective excitations is shown by the thick green line. Open squares represent the experimental data [9, 10].

At the small scattering angle, $\theta = 3°$, the symmetric mode of the surface plasmon (dashed red line) dominates the cross section. A similar behavior is observed in the photoionization process (see figure 2). In fact, in the case of the uniform external field ($q \to 0$), there is no volume plasmon excitation in the system and the symmetric plasmon mode exceeds significantly the antisymmetric mode. Non-uniformity of the external field causes the formation of the volume

Table 1. Peak positions and the widths of the two surface plasmon modes and of the volume plasmon used in the present calculations. All values are given in eV.

	$l=0$	$l=1$	$l=2$	$l=3$
ω_{1l}	0	19.0	25.5	30.5
$\Gamma_{1l}^{(s)}$	0	11.4	15.3	18.3
ω_{2l}	37.1	33.2	31.0	29.5
$\Gamma_{2l}^{(s)}$	37.1	33.2	31.0	29.5
ω_p	37.1			
$\Gamma_l^{(v)}$	26.0 - 48.3			

plasmon whose contribution to the cross section is insignificant when the scattering angle is small. With increasing the scattering angle ($\theta = 5°$ and $7°$), the symmetric mode of the surface plasmon becomes less relevant and the antisymmetric mode (dash-dotted blue line) more prominent. At the larger angle ($\theta = 9°$), the symmetric surface plasmon almost does not contribute to the cross section while the volume plasmon (double-dotted purple line) becomes dominant. The contribution of the antisymmetric surface and the volume plasmons can explain the origin of the two peaks in the energy loss range from 20 to 30 eV at the scattering angle $\theta = 7°$.

The peak position of each plasmon resonance (9) as well as of the resulting cross section $d^3\sigma/d\varepsilon_2 d\Omega_{\mathbf{p}_2}$ is influenced by the manifestation of the diffraction effects. In Ref. [8], it was shown that plasmon modes with different angular momenta provide dominating contributions to the differential cross section at different scattering angles, which leads to the significant angular dependence of the energy loss spectrum. This phenomenon was described in terms of the electron diffraction at the fullerene edge [8]. As it is seen from equation (9), the resonance peak of each plasmon is defined not only by the plasmon frequencies ω_p, ω_{1l} and ω_{2l} but also by the multipolar diffraction factors $V_l(q)$, $S_{1l}(q)$ and $S_{2l}(q)$ which depend on the transferred momentum q. In the limiting case of an infinitely thin layer, this dependence is described by the spherical Bessel functions $j_l^2(qR)$ which oscillate with q and, thus, give suppression and enhancement of the partial plasmon modes at certain angles [8]. The incident energy of the projectile does not influence on this behavior and defines only the absolute value of the cross section.

4. Conclusion

To conclude, we studied the formation of plasmon excitations in the C_{60} fullerene in the processes of photoionization and electron inelastic scattering. To reveal the contribution of collective electron excitations, we utilized the plasmon resonance approximation. We showed that the photoionization cross section of C_{60} is described within the utilized approach as a sum of two contributions, which represent two coupled modes of the surface plasmon. In case of electron inelastic scattering, non-uniformity of the external electric field causes also the formation of the volume plasmon, whose contribution to the cross section is insignificant for small scattering angles and becomes dominant, as the angle increases. The results of calculations are in good agreement with experimental data on photoionization and electron inelastic scattering of C_{60}. The plasmon resonance approximation describes quite well the main features of the giant resonance, such as height, width and position of the peak as the model is proved to be a useful tool for interpretation of experimental results.

References

[1] Kroto H W *et al* 1985 *Nature* **318** 162
[2] Hebard A F, Rosseinsky M J, Haddon R C, Murphy D W, Glarum S H, Palstra T T M, Ramirez A P and Kortan A R 1991 *Nature* **350** 600
[3] Sattler K D 2010 *Handbook of Nanophysics: Clusters and Fullerenes* (Boca Raton: CRC Press)
[4] Mroz P *et al* 2008 *Medicinal Chemistry and Pharmacological Potential of Fullerenes and Carbon Nanotubes* vol 1, ed F Cataldo and T Da Ros (New York: Springer Science+Business Media B.V.) p 79
[5] Bertsch G F, Bulgac A, Tomanek D and Wang Y 1991 *Phys. Rev. Lett.* **67** 2690
[6] Hertel I V, Steger H, de Vries J, Weisser B, Menzel C, Kamke B and Kamke W 1992 *Phys. Rev. Lett.* **68** 784
[7] Keller J W and Coplan M A 1992 *Chem. Phys. Lett.* **193** 89
[8] Gerchikov L G, Efimov P V, Mikoushkin V M and Solov'yov A V 1998 *Phys. Rev. Lett.* **81** 2707
[9] Verkhovtsev A V, Korol A V, Solov'yov A V, Bolognesi P, Ruocco A and Avaldi L 2012 *J. Phys. B: At. Mol. Opt. Phys.* **45** 141002
[10] Bolognesi P, Ruocco A, Avaldi L, Verkhovtsev A V, Korol A V and Solov'yov A V 2012 *Eur. Phys. J.* D **66** 254
[11] Reinköster A, Korica S, Prümper G, Viefhaus J, Godehusen K, Schwarzkopf O, Mast M and Becker U 2004 *J. Phys. B: At. Mol. Opt. Phys.* **37** 2135
[12] Scully S W J *et al* 2005 *Phys. Rev. Lett.* **94** 065503
[13] Korol A V and Solov'yov A V 2007 *Phys. Rev. Lett.* **98** 179601
[14] Scully S W J *et al* 2007 *Phys. Rev. Lett.* **98** 179602
[15] Gerchikov L G, Solov'yov A V, Connerade J-P and Greiner W 1997 *J. Phys. B: At. Mol. Opt. Phys.* **30** 4133
[16] Gerchikov L G, Ipatov A N and Solov'yov A V 1997 *J. Phys. B: At. Mol. Opt. Phys.* **30** 5939
[17] Gerchikov L G, Ipatov A N, Solov'yov A V and Greiner W 1998 *J. Phys. B: At. Mol. Opt. Phys.* **31** 3065
[18] Gerchikov L G, Ipatov A N, Polozkov R G and Solov'yov A V 2000 *Phys. Rev.* A **62** 043201
[19] Lambin Ph, Lucas A A and Vigneron J-P 1992 *Phys. Rev.* B **46** 1794
[20] Östling D, Apell P and Rosen A 1993 *Europhys. Lett.* **21** 539
[21] Lo S, Korol A V and Solov'yov A V 2007 *J. Phys. B: At. Mol. Opt. Phys.* **40** 3973
[22] Lundqvist S and March N H 1983 *Theory of the Inhomogeneous Electron Gas* (New York: Plenum Press)
[23] Verkhovtsev A V, Korol A V and Solov'yov A V 2012 *Eur. Phys. J.* D **66** 253
[24] Connerade J-P and Solov'yov A V 2002 *Phys. Rev.* A **66** 013207
[25] Kafle B P, Katayanagi H, Prodhan M, Yagi H, Huang C and Mitsuke K 2008 *J. Phys. Soc. Jpn.* **77** 014302
[26] Berkowitz J 1999 *J. Chem. Phys.* **111** 1446
[27] Saito S and Oshiyama A 1991 *Phys. Rev. Lett.* **66** 2637

Electron interactions with Ar clusters and liquid Ar

Francisco Blanco[1] **and Gustavo García**[2,3]

[1]Departamento de Física Atómica Molecular y Nuclear, Facultad de Ciencias Físicas, Universidad Complutense de Madrid, Ciudad Universitaria, 28040 Madrid, Spain
[2]Instittuto de Física Fundamental, Consejo Superior de Investigaciones Científicas, Serrano 113-bis, 28006 Madrid, Spain
[3]Centre for Medical Radiation Physics, University of Wollongong, NSW 2522, Australia

g.garcia@iff.csic.es

Abstract. Differential and integral electron scattering cross section from some Ar clusters (dimer, trimer and tetramer) are calculated for incident energies ranging from 1 to 500 eV by using a screening corrected additivity rule based on an independent atom representation (IAM-SCAR). The possibility of using this method to derive electron scattering cross section in the liquid phase is discussed and electron scattering cross section data for Ar liquid are provided.

1. Introduction

Secondary electrons have been proved to be responsible for the energy deposition pattern and related damage when irradiating matter with different high energy particles (photons, electrons, positrons or ions). Depending on the type and energy of primary radiation these electrons can be generated over a broad energy range from nearly zero up to the high energy of the primary particle (in the case of photoelectrons). Energy deposition models including secondary electron effects are required by some medical applications of radiations in the fields of radiotherapy, radiodiagnostics and radiation protection. One of the main difficulties to develop these models is that the subject targets are usually in the condensed phase, liquid or solid, while available cross section data are related to single atoms or molecules. The main goal of this paper is to study how target condensation can affect to the calculated electron scattering cross sections. We have chosen Ar for being a single atom with abundant cross section data available in the literature. In addition accurate theoretical and experimental cluster geometrical configurations for its dimmer, trimmer and tetramer have been recently published [1,2]. This allows us to apply our screening corrected additivity rule (SCAR) [3-6] to an independent atom representation (IAM) in order to follow the evolution of the differential (DCS) and integral (ICS) scattering cross section as a function of the number of atoms forming the cluster. This can be regarded as the way that atom condensation affects to the single atom scattering cross sections. Finally we can model the liquid state as a large cluster by assuming a homogeneous atom distribution defined by its density.

In this study we use the above mentioned IAM-SCAR method to calculate differential and integral electron scattering cross section for the Ar atom and its dimer, trimer and tetramer cluster configurations providing a systematic of the cross section value evolution as a function of the number of atoms in the cluster. We also present an approach to model electron scattering processes in liquid argon by using this modified independent atom representation. The applicability of the present electron scattering data to model electron transport in liquids is finally discussed.

2. Calculation procedure

To describe the electron Ar scattering process an optical potential method has been followed. Both the potential and the calculation procedure have been extensively described elsewhere [3-5] and therefore only a brief summary will be reported here.

We represent the atomic target by an interacting complex potential (i.e. the optical potential), whose real part accounts for the elastic scattering of the incident electrons, while the imaginary part represents the inelastic processes that are considered as 'absorptions' from the incident beam. To construct this complex potential for each atom, the real part of the potential is represented by the sum of three terms: (i) a static term derived from a Hartree-Fock calculation of the atomic charge distribution [6], (ii) an exchange term to account for the indistinguishability of the incident and target electrons [7] and (iii) a polarisation term [8] for the long-range interactions which depend on the target dipole polarisability [9]. The imaginary part, following the procedure of Staszewska et al.[10], then treats inelastic scattering as electron–electron collisions. However, we initially found some major discrepancies in the available scattering data, which were subsequently corrected when a physical formulation of the absorption potential [3] was introduced. Further improvements to the original formulation, such as the inclusion of screening effects, local velocity corrections and in the description of the electrons' indistinguishability [4], finally led to a model that provides a good approximation of electron–atom scattering over a broad energy range. An excellent example of this was for elastic electron–atomic iodine (I) scattering [11], where the optical potential results compared very favourably with those from an independent highly sophisticated Dirac-B-spline R-matrix computation.

2.1. Ar dimer, trimer and tetramer electron scattering calculation

To calculate the cross sections for electron scattering from Ar clusters, we follow the IAM, in the same way it is used for molecules [4], by applying what is commonly known as the additivity rule (AR). In this approach, the cluster scattering amplitude is derived from the sum of all the relevant atomic amplitudes, including the phase coefficients, which lead to the cluster DCSs for the atomic structure in question.

Integral cross sections (ICS) can then be determined by integrating those DCSs. The sum of the elastic and the absorption (for all inelastic processes except rotations and vibrations) gives the TCSs. Alternatively, the ICSs for the cluster can also be derived from the relevant atomic ICSs in conjunction with the optical theorem [4]. We found an inherent contradiction in the original formulation of the AR procedure between the ICSs derived from those two approaches, which suggested that the optical theorem was being violated [12]. This problem has been resolved by employing a normalization procedure during the calculation of the DCSs which ensures that the ICSs derived from the two approaches are now entirely consistent [12]. An additional limitation of the AR method is that no target structure is considered, so that it is really only applicable when the incident electrons are so fast that they effectively see the target cluster as a sum of individual atoms (typically above ~100 eV). To reduce this limitation, we [13, 14] introduced the SCAR method, which considers the geometry of a relevant atomic structure (atomic positions and bond lengths) by using some screening coefficients. With this correction the range of validity of the IAM-SCAR approach might be extended to incident electron energies of about 20 eV. Indeed it is the elastic DCS and the elastic and inelastic ICS results from the application of the IAM–SCAR method to Ar dimer, trimer and tetramer that we report on here. Atomic positions and bond lengths for these clusters have been taken from [1].

2.2. Liquid Ar electron scattering calculation

We can consider a liquid as formed by a homogeneous mixture of atoms which interatomic distances are given by its temperature and density conditions (1430 kg/m^3 in the case of Ar). By assuming that the screening corrections are not very dependent on the direction of the incident electrons we can apply the SCAR procedure, in a similar way we accomplished for molecules and clusters, in order to calculate the effective cross section (σ^{eff}) of an Ar atom within the liquid. A sketch of the screening situation for different electron energies is shown in Figure 1. For high energies electrons (a), above 100 eV, where atomic electron scattering cross sections do not overlap, an independent atom representation of the solid is appropriate. However, for intermediate energies (b), between 20 and 100 eV, overlapping between atomic cross sections occur and the effective cross section of the *i* atom (σ_i^{eff}) is less than the atomic cross section (σ_i) by the σ_{ij} screening coefficients provided by the *j* surrounding atoms. At low energies (c), below 20 eV, the effective cross section approaches to the cell size $\sigma^{eff}=\sigma^c$.

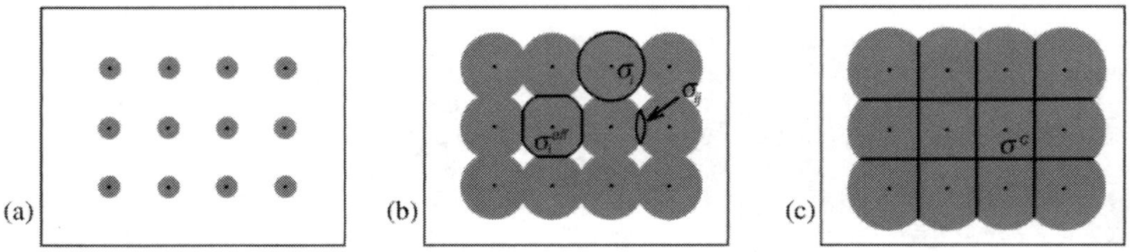

FIGURE 1. Geometric representation of the Ar atomic cross section for (a) high, (b) intermediate and (c) low incident energy electrons

3. Results and discussion

Differential electron elastic scattering cross section for Ar atom, dimer, trimer and tetramer are shown in Table 1 and plotted in Figure 2 for selected energies ranging from 1 to 1000 eV. By integrating over the whole scattering angle range we derived the corresponding integral elastic cross sections which, in conjunction with the optical theorem, provided the total scattering cross sections and therefore, the integral inelastic cross sections. Integral electron scattering cross sections for Ar atoms and Ar clusters are shown in Table 2. Due to the small binding energy of these Ar clusters, a few meV, and their relatively large internuclear distances we can expect no much different scattering behavior from that of the independent atoms for the energy range considered here. However, it has been recently shown [14] that the multicenter structure of small noble gas clusters modifies the electron impact ionization dynamics even for incident energies of 100 eV.

TABLE 1. Differential electron scattering cross sections from Argon atoms (Ar_1) and Argon clusters (Ar_2, Ar_3, Ar_4), in atomic units (a_0^2/Sr), for selected energies.

Ang (deg)	1 eV				10 eV				100 eV			
	Ar_1	**Ar_2**	**Ar_3**	**Ar_4**	**Ar_1**	**Ar_2**	**Ar_3**	**Ar_4**	**Ar_1**	**Ar_2**	**Ar_3**	**Ar_4**
0	0.0917	0.193	0.305	0.432	33.4	86.6	153	229	50	144	278	451
1	0.0906	0.191	0.301	0.427	33.2	86	152	227	48.4	138	267	431
2	0.0874	0.184	0.291	0.412	32.6	84.3	149	222	45.4	127	244	391
3	0.0823	0.173	0.275	0.389	31.9	82.1	145	216	42.7	116	219	347
4	0.0757	0.16	0.253	0.36	31.1	79.8	141	209	40.1	105	193	301
5	0.068	0.144	0.229	0.325	30.3	77.6	136	202	37.5	93.6	166	254

6	0.0597	0.127	0.202	0.287	29.7	75.3	132	195	35.1	82.8	142	210
7	0.0512	0.109	0.174	0.249	29	73	128	188	32.8	72.8	119	170
8	0.0429	0.092	0.148	0.212	28.2	70.7	123	181	30.5	63.9	99.4	136
9	0.0353	0.0763	0.123	0.178	27.5	68.3	118	173	28.4	56.1	83	108
10	0.0286	0.0624	0.102	0.147	26.8	65.9	113	165	26.3	47.6	63.5	73.1
11	0.023	0.0507	0.0835	0.122	26.1	63.5	108	158	24.3	40.4	48.4	47.8
12	0.0186	0.0416	0.0693	0.102	25.5	61.2	104	150	22.4	35.2	38.7	33.1
13	0.0154	0.0351	0.0591	0.0879	24.8	58.8	98.9	142	20.6	31.6	33.5	26.9
14	0.0136	0.0313	0.0532	0.0797	24.1	56.4	94.2	135	18.9	29.2	31.5	26.8
15	0.0131	0.0303	0.0517	0.0775	23.4	54.1	89.5	127	17.3	27.6	31.7	30.6
16	0.014	0.0321	0.0544	0.0813	22.8	51.8	84.9	120	15.8	26.5	32.9	36.1
17	0.0162	0.0366	0.0615	0.0912	22.1	49.6	80.4	113	14.4	25.6	34.4	41.7
18	0.0198	0.0441	0.0731	0.107	21.4	47.4	76.1	106	13.1	24.7	35.4	46.1
19	0.0248	0.0545	0.0892	0.13	20.8	45.2	71.8	99.1	11.8	23.3	34.7	46.3
20	0.0314	0.0679	0.11	0.159	20.2	43.1	67.7	92.6	10.6	21.4	32.5	44.1
21	0.0394	0.0845	0.136	0.194	19.5	41.1	63.8	86.4	9.55	19.6	30.1	41.2
22	0.0489	0.104	0.166	0.237	18.9	39.2	60	80.5	8.55	17,7	27.4	37.7
23	0.0599	0.127	0.201	0.285	18.3	37.3	56.4	74.9	7.63	15.8	24.5	33.7
24	0.0724	0.152	0.241	0.34	17.7	35.5	53	69.6	6.78	14	21.6	29.5
25	0.0862	0.18	0.284	0.401	17.1	33.7	49.7	64.6	6,01	12.3	18.8	25.3
26	0.101	0.211	0.332	0.467	16.5	32.1	46.7	60	5.3	10.7	16.1	21.5
27	0.117	0.245	0.383	0.538	15.9	30.5	43.8	55.6	4.66	9.27	13.8	18
28	0.135	0.28	0.438	0.613	15.3	29	41.1	51.6	4.08	7.99	11.6	14.7
29	0.153	0.317	0.495	0.692	14.8	27.6	38.6	47.9	3.57	6.75	9.48	11.6
30	0.171	0.355	0.554	0.774	14.3	26	35.4	42.3	3.11	5.71	7.78	9.24
31	0.191	0.395	0.615	0.857	13.7	24.3	32.1	37	2.69	4.85	6.47	7.54
32	0.21	0.435	0.677	0.943	13.2	22.7	29	32.2	2.33	4.15	5.48	6.37
33	0.231	0.476	0.74	1.03	12.7	21.3	26.3	28.1	2.01	3.58	4.75	5.59
34	0.251	0.518	0.804	1.12	12.2	20	23.9	24.5	1.73	3.11	4.19	5.07
35	0.272	0.56	0.869	1.21	11.8	18.8	21.9	21.5	1.48	2.71	3.75	4.68
36	0.293	0.602	0.934	1.3	11.3	17.7	20	18.9	1.26	2.38	3.38	4.36
37	0.313	0.645	0.999	1.39	10.9	16.7	18.5	16.8	1.08	2.09	3.05	4.05
38	0.334	0.687	1.06	1.47	10.5	15.9	17.2	15.1	0.922	1.82	2.7	3.61
39	0.355	0.73	1.13	1.56	10.1	15.1	16	13.7	0.785	1.57	2.36	3.17
40	0.376	0.771	1.19	1.65	9.69	14.4	15.1	12.7	0.669	1.35	2.05	2.77
41	0.396	0.813	1.25	1.73	9.32	13.7	14.3	11.9	0.571	1.16	1.77	2.4
42	0.416	0.853	1.32	1.82	8.98	13.2	13.7	11.4	0.488	0.996	1.52	2.07
43	0.436	0.893	1.38	1.9	8.64	12.7	13.3	11.1	0.418	0.855	1.31	1.77
44	0.455	0.931	1.43	1.98	8.33	12.3	12.9	10.9	0.36	0.736	1.12	1.51
45	0.473	0,968	1.49	2.05	8.03	11.9	12.6	11	0.313	0.636	0.966	1.29
46	0.491	1	1.54	2.13	7.75	11.6	12.4	11.1	0.274	0.554	0.836	1.11
47	0.508	1.04	1.6	2.19	7.48	11.3	12.3	11.4	0.242	0.488	0.731	0.964
48	0.524	1.07	1.64	2.26	7.24	11	12.3	11.7	0.217	0.433	0.639	0.825
49	0.54	1.1	1.69	2.32	7	10.8	12.3	12.1	0.198	0.388	0.566	0.721
50	0.555	1.13	1.74	2.39	6.78	10.6	12.3	12.6	0.184	0.356	0.513	0.649
51	0.57	1.16	1.78	2.44	6.58	10.5	12.4	13.1	0.173	0.333	0.476	0.602

52	0.584	1.19	1.82	2.5	6.4	10.4	12.6	13.6	0.167	0.319	0.456	0.579
53	0.597	1.22	1.86	2.55	6.22	10.2	12.7	14.1	0.163	0.312	0.448	0.575
54	0.61	1.24	1.9	2.6	6.05	10.1	12.8	14.7	0.162	0.312	0.452	0.588
55	0.623	1.27	1.94	2.65	5.9	10.1	13	15.2	0.164	0.318	0.466	0.616
56	0.635	1.29	1.97	2.7	5.77	10	13.2	15.8	0.167	0.328	0.488	0.655
57	0.647	1.31	2.01	2.74	5.64	9.96	13.4	16.3	0.172	0.343	0.517	0.704
58	0.658	1.34	2.04	2.78	5.52	9.92	13.6	16.8	0.179	0.361	0.548	0.745
59	0.668	1.36	2.07	2.82	5.41	9.89	13.7	17.3	0.186	0.379	0.578	0.787
60	0.678	1.38	2.1	2.86	5.32	9.87	13.9	17.8	0.195	0.398	0.61	0.833
61	0.688	1.39	2.12	2.89	5.22	9.85	14.1	18	0.205	0.42	0.644	0.879
62	0.696	1.41	2.15	2.92	5.14	9.77	14	18	0.216	0.442	0.679	0.925
63	0.704	1.43	2.17	2.95	5.06	9.69	14	18	0.227	0.465	0.713	0.968
64	0.712	1.44	2.19	2.98	4.99	9.61	13.9	18	0.239	0.488	0.746	1.01
65	0.719	1.45	2.21	3	4.92	9.52	13.9	18	0.251	0.512	0.778	1.05
66	0.725	1.47	2.23	3.02	4.85	9.44	13.8	18	0.264	0.535	0.809	1.08
67	0.731	1.48	2.24	3.04	4.79	9.37	13.7	18	0.277	0.557	0.838	1.11
68	0.736	1.49	2.25	3.05	4.73	9.29	13.7	17.9	0.29	0.58	0.866	1.13
69	0.74	1.49	2.27	3.07	4.67	9.21	13.6	17.9	0.303	0.601	0.886	1.14
70	0.744	1.5	2.28	3.08	4.61	9.12	13.5	17.8	0.316	0.619	0.901	1.15
71	0.747	1.51	2.28	3.09	4.55	9.04	13.4	17.7	0.329	0.637	0.918	1.16
72	0.75	1.51	2.29	3.09	4.5	8.95	13.3	17.6	0.341	0.655	0.938	1.18
73	0.752	1.51	2.29	3.09	4.43	8.85	13.2	17.4	0.354	0.675	0.962	1.22
74	0.753	1.52	2.29	3.09	4.37	8.74	13.1	17.3	0.366	0.695	0.991	1.26
75	0.754	1.52	2.29	3.09	4.31	8.63	12.9	17.1	0.377	0.717	1.02	1.31
76	0.753	1.52	2.29	3.08	4.24	8.51	12.7	16.8	0.388	0.74	1.06	1.38
77	0.752	1.51	2.28	3.07	4.17	8.37	12.5	16.6	0.398	0.763	1.11	1.45
78	0.75	1.51	2.27	3.06	4.09	8.23	12.3	16.3	0.407	0.787	1.15	1.53
79	0.747	1.5	2.26	3.04	4.02	8.08	12.1	16	0.415	0.81	1.2	1.61
80	0.743	1.49	2.25	3.02	3.93	7.91	11.8	15.6	0.422	0.832	1.24	1.66
81	0.738	1.48	2.23	3	3.84	7.73	11.6	15.3	0.428	0.85	1.27	1.7
82	0.733	1.47	2.22	2.97	3.75	7.54	11.3	14.9	0.433	0.864	1.3	1.74
83	0.727	1.46	2.2	2.94	3.65	7.34	11	14.5	0.437	0.875	1.32	1.77
84	0.721	1.44	2.17	2.91	3.55	7.14	10.7	14	0.44	0.883	1.33	1.79
85	0.713	1.43	2.15	2.88	3.44	6.91	10.3	13.6	0.441	0.887	1.34	1.8
86	0.705	1.41	2.12	2.84	3.33	6.69	9.98	13.1	0.441	0.888	1.34	1.8
87	0.697	1.4	2.1	2.81	3.21	6.45	9.62	12.6	0.44	0.886	1.33	1.78
88	0.688	1.38	2.07	2.76	3.08	6.2	9.24	12.1	0.437	0.88	1.32	1.76
89	0.678	1.36	2.04	2.72	2.96	5.94	8.86	11.6	0.433	0.87	1.31	1.73
90	0.668	1.34	2.01	2.68	2.82	5.68	8.46	11.1	0.428	0.857	1.28	1.69
91	0.658	1.31	1.97	2.63	2.69	5.41	8.06	10.6	0.421	0.841	1.25	1.65
92	0.646	1.29	1.94	2.59	2.55	5.14	7.66	10	0.413	0.822	1.22	1.59
93	0.635	1.27	1.9	2.54	2.41	4.86	7.25	9.51	0.404	0.801	1.18	1.52
94	0.623	1.24	1.87	2.49	2.27	4.58	6.84	8.98	0.394	0.775	1.13	1.45
95	0.611	1.22	1.83	2.44	2.12	4.29	6.42	8.44	0.382	0.747	1.09	1.38
96	0.599	1.2	1.79	2.38	1.98	4.01	6.02	7.92	0.369	0.718	1.04	1.32
97	0.586	1.17	1.75	2.33	1.83	3.73	5.61	7.4	0.356	0.688	0.992	1.26

98	0.573	1.14	1.71	2.28	1.68	3.45	5.21	6.89	0.341	0.657	0.946	1.21
99	0.56	1.12	1.67	2.22	1.54	3.18	4.82	6.34	0.326	0.627	0.903	1.16
100	0.547	1.09	1.63	2.17	1.39	2.9	4.41	5.8	0.309	0.595	0.86	1.11
101	0.534	1.07	1.59	2.12	1.25	2.63	4.02	5.31	0.292	0.564	0.818	1.07
102	0.521	1.04	1.55	2.06	1.11	2.37	3.64	4.83	0.275	0.533	0.777	1.02
103	0.508	1.01	1.51	2.01	0.981	2.11	3.28	4.39	0.257	0.5	0.735	0.976
104	0.496	0.987	1.48	1.96	0.852	1.87	2.95	3.98	0.239	0.468	0.694	0.929
105	0.483	0.961	1.44	1.91	0.729	1.65	2.63	3.61	0.221	0.436	0.651	0.879
106	0.47	0.935	1.4	1.84	0.613	1.43	2.35	3.27	0.203	0.403	0.608	0.824
107	0.457	0.909	1.35	1.75	0.504	1.24	2.08	2.96	0.185	0.371	0.563	0.762
108	0.444	0.881	1.3	1.67	0.404	1.06	1.85	2.68	0.167	0.338	0.514	0.699
109	0.431	0.851	1.25	1.59	0.313	0.899	1.63	2.44	0.149	0.304	0.466	0.637
110	0.418	0.82	1.19	1.52	0.232	0.757	1.45	2.23	0.132	0.272	0.419	0.575
111	0.405	0.791	1.14	1.44	0.161	0.635	1.29	2.05	0.116	0.24	0.373	0.514
112	0.393	0.761	1.1	1.37	0.103	0.535	1.16	1.9	0.101	0.21	0.329	0.455
113	0.38	0.733	1,05	1.3	0.0567	0.456	1.06	1.79	0.086	0.182	0.287	0.399
114	0.368	0.705	1	1.24	0.0239	0.4	0.993	1.71	0.0726	0.155	0.247	0.347
115	0.355	0.678	0.958	1.18	0.00509	0.369	0.953	1.67	0.0604	0.131	0.211	0.299
116	0.343	0.651	0.915	1.12	0.0009	0.361	0.944	1.66	0.0495	0.11	0.179	0.256
117	0.331	0.625	0.873	1,06	0.012	0.38	0.967	1.68	0.0401	0.0912	0.151	0.219
118	0.319	0.6	0.833	1	0.039	0.425	1.02	1.75	0.0323	0.0757	0.128	0.188
119	0.307	0.575	0.794	0.95	0.0825	0.497	1.11	1.84	0.0262	0.0635	0.11	0.164
120	0.296	0.551	0.757	0.899	0.143	0.598	1.24	1.98	0.0219	0.055	0.0976	0.148
121	0.284	0.527	0.721	0.851	0.221	0.726	1.4	2.15	0.0196	0.0504	0.0906	0.138
122	0.273	0.504	0.686	0.804	0.316	0.885	1.59	2.37	0.0193	0.0497	0.0894	0.137
123	0.262	0.481	0.652	0.76	0.43	1.07	1.83	2.63	0.0211	0.0531	0.0943	0.142
124	0.251	0.459	0.619	0.717	0.562	1.29	2.1	2.93	0.0251	0.0608	0.105	0.156
125	0.24	0.438	0.587	0.676	0.713	1.54	2.41	3.27	0.0313	0.073	0.123	0.177
126	0.23	0.417	0.557	0.638	0.882	1.82	2.76	3.66	0.0399	0.0894	0.146	0.206
127	0.22	0.397	0.527	0.6	1.07	2.14	3.15	4.1	0.0509	0.11	0.175	0.243
128	0.209	0.377	0.499	0.565	1.28	2.48	3.59	4.59	0.0643	0.136	0.211	0.288
129	0.2	0.359	0.472	0.532	1.5	2.86	4.06	5.12	0.0801	0.166	0.254	0.341
130	0.19	0.34	0.446	0.5	1.75	3.27	4.58	5.72	0.0983	0.2	0.302	0.402
131	0.181	0.323	0.422	0.47	2.01	3.71	5.13	6.35	0.119	0.239	0.358	0.473
132	0.172	0.306	0.398	0.441	2.29	4.18	5.74	7.05	0.142	0.283	0.42	0.551
133	0.163	0.29	0.376	0.414	2.59	4.69	6.39	7.8	0.168	0.331	0.488	0.639
134	0.155	0.274	0.354	0.389	2.91	5.23	7.08	8.61	0.196	0.384	0.564	0.736
135	0.147	0.259	0.334	0.365	3.24	5.79	7.81	9.46	0.227	0.442	0.646	0.844
136	0.139	0.245	0.315	0.343	3.59	6.39	8.58	10.4	0.259	0.504	0.735	0.961
137	0.132	0.232	0.297	0.322	3.95	7.01	9.4	11.3	0.295	0.571	0.832	1.09
138	0.125	0.219	0.279	0.302	4.34	7.67	10.3	12.4	0.332	0.642	0.936	1.23
139	0.118	0.206	0.263	0.283	4.73	8.35	11.2	13.4	0.372	0.718	1,05	1.37
140	0.112	0.195	0.248	0.266	5.14	9.06	12.1	14.6	0.413	0.799	1.16	1,53
141	0.105	0.184	0.233	0.249	5.56	9.79	13.1	15.7	0.457	0.883	1.29	1.7
142	0.0994	0.173	0.219	0.234	5.99	10.5	14.1	17	0.502	0.972	1.42	1.88
143	0.0938	0.163	0.206	0.219	6.43	11.3	15,1	18.2	0.55	1.06	1.56	2.07

144	0.0885	0.154	0.194	0.206	6.89	12.1	16,2	19.6	0.598	1.16	1.7	2.27
145	0.0834	0.145	0.182	0.194	7.35	12.9	17,3	20.9	0.649	1.26	1.85	2.47
146	0.0787	0.136	0.172	0.182	7.82	13.8	18,4	22.3	0.701	1.36	2.01	2.68
147	0.0743	0.129	0.162	0.171	8.29	14.6	19,5	23.7	0.754	1.47	2.17	2.89
148	0.07	0.121	0.153	0.161	8.77	15.5	20.7	25.2	0.808	1.58	2.34	3.1
149	0.0661	0.114	0.144	0.152	9.25	16.3	21.9	26.7	0.863	1.69	2.5	3.32
150	0.0625	0.108	0.136	0.144	9.74	17.2	23.1	28.2	0.919	1.8	2.67	3.54
151	0.059	0.102	0.128	0.136	10.2	18.1	24.3	29.7	0.975	1.91	2.84	3.77
152	0.0558	0.0966	0.122	0.129	10.7	19	25.5	31.3	1.03	2.03	3	3.99
153	0.0528	0.0914	0.115	0.122	11.2	19.8	26.8	32.8	1.09	2.14	3.18	4.22
154	0.05	0.0866	0.109	0.116	11.7	20.7	28	34.4	1.14	2.25	3.34	4.44
155	0.0473	0.082	0.103	0.11	12.2	21.6	29.2	35.9	1.2	2.37	3.51	4.67
156	0.0448	0.0778	0.0982	0.104	12.6	22.5	30.4	37.5	1.26	2.48	3.68	4.89
157	0.0425	0.0738	0.0933	0.0995	13.1	23.3	31.6	39	1.31	2.59	3.84	5.11
158	0.0403	0.0701	0.0888	0.0949	13.6	24.2	32.8	40.5	1.37	2.7	4.01	5.33
159	0.0383	0.0667	0.0846	0.0907	14	25	33.9	42	1.42	2.81	4.17	5.54
160	0.0364	0.0636	0.0808	0.0868	14.4	25.8	35.1	43.5	1.48	2.91	4.33	5.75
161	0.0347	0.0607	0.0773	0.0833	14.9	26.6	36.2	44.9	1.53	3.02	4.49	5.96
162	0.0332	0.058	0.074	0.0801	15.3	27.3	37.3	46.3	1.58	3.12	4.64	6.16
163	0.0317	0.0556	0.0711	0.0771	15.7	28.1	38.3	47.7	1.63	3.22	4.78	6.35
164	0.0304	0.0534	0.0684	0.0745	16	28.8	39.3	49	1.67	3.31	4.92	6.53
165	0.0292	0.0514	0.066	0.0721	16.4	29.5	40.3	50.2	1.72	3.4	5.05	6.7
166	0.0281	0.0496	0.0638	0.0699	16.8	30.1	41.2	51.4	1.76	3.48	5.18	6.86
167	0.0272	0.0479	0.0619	0.068	17.1	30.7	42.1	52.6	1.8	3.56	5.3	7.02
168	0.0263	0.0465	0.0601	0.0662	17.4	31.3	42.9	53.6	1.84	3.64	5.41	7.17
169	0.0255	0.0451	0.0585	0.0647	17.7	31.8	43.7	54.6	1.87	3.71	5.52	7.31
170	0.0248	0.0439	0.057	0.0632	17.9	32.3	44.4	55.5	1.91	3.78	5.62	7.44
171	0.0241	0.0428	0.0557	0.062	18.2	32.8	45.1	56.4	1.94	3.84	5.71	7.56
172	0.0235	0.0419	0.0546	0.0609	18.4	33.2	45.6	57.2	1.97	3.89	5.79	7.66
173	0.023	0.041	0.0536	0.0598	18.6	33.6	46.2	57.8	1.99	3.94	5.86	7.76
174	0.0226	0.0403	0.0527	0.059	18.8	33.9	46.6	58.5	2.01	3.98	5.92	7.83
175	0.0222	0.0396	0.0519	0.0582	18.9	34.1	47	59	2.03	4.02	5.98	7.91
176	0.0219	0.0391	0.0513	0.0577	19	34.4	47.3	59.4	2.05	4.05	6.03	7.97
177	0.0216	0.0387	0.0508	0.0572	19.1	34.5	47.6	59.7	2.06	4.08	6.06	8.01
178	0.0214	0.0384	0.0504	0.0568	19.2	34.7	47.8	60	2.07	4.09	6.09	8.05
179	0.0213	0.0382	0.0502	0.0566	19.2	34.7	47.9	60.1	2.07	4.1	6.1	8.07
180	0.0213	0.0381	0.0501	0.0565	19.2	34.8	47.9	60.1	2.08	4.11	6.11	8.08

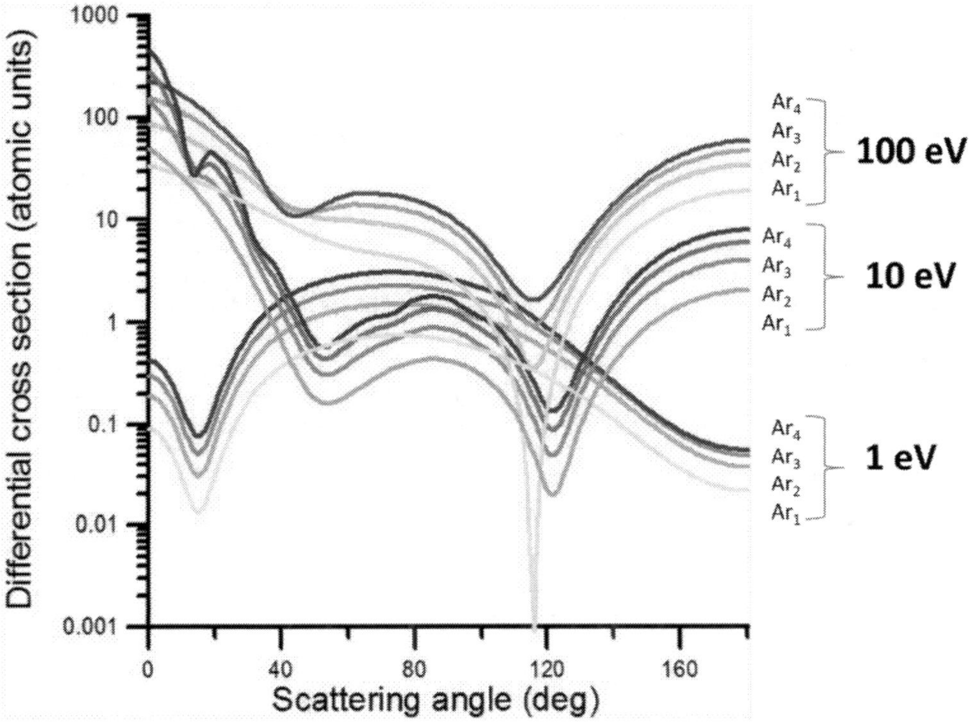

FIGURE 2. Differential electron scattering cross sections from Argon atoms (Ar_1) and Argon clusters (Ar_2, Ar_3, Ar_4), in atomic units (a_0^2/Sr) for selected energies

TABLE 2. Integral elastic (ECS), inelastic (ICS) and total (TCS) electron scattering cross sections from Argon atoms (Ar_1) and Argon clusters (Ar_2, Ar_3, Ar_4), in atomic units (a_0^2) for incident energies ranging from 1 to 500 eV

	Ar_1			Ar_2			Ar_3			Ar_4		
E(eV)	ECS	ICS	TCS	ECS	ICS	TCS	ECS	ICS	TCS	ECS	ICS	TCS
1	5.39	0	5.39	10.7	0	10.7	16	0	16	21.3	0	21.3
1.5	11.3	0	11.3	22.5	0	22.5	33.4	0	33.4	44.2	0	44.2
2	16.9	0	16.9	33.3	0	33.3	49.3	0	49.3	64.9	0	64.9
3	26.2	0	26.2	51.4	0	51.4	75.5	0	75.5	98.8	0	98.8
4	33.8	0	33.8	65.9	0	65.9	96.4	0	96.4	125	0	125
5	40.7	0	40.7	78.8	0	78.8	115	0	115	148	0	148

7	53.5	0	53.5	103	0	103	148	0	148	189	0	189
10	72.2	0	72.1	136	0	136	193	0	193	245	0	245
15	83.2	0.235	83.5	156	0.44	156	219	0.62	220	275	0.778	276
20	71.9	2.47	74.4	136	4.65	140	192	6.59	199	243	8.34	251
30	48.2	6.67	54.9	92.3	12.8	105	133	18.4	151	170	23.6	194
40	35.6	8.4	44	68.9	16.2	85.1	99.9	23.5	123	129	30.4	159
50	28.6	9.11	37.7	55.5	17.7	73.2	80.9	25.8	107	105	33.5	138
70	21.1	9.41	30.5	41.1	18.4	59.5	60.3	26.9	87.2	78.7	35.1	114
100	15.9	9.03	24.9	31.2	17.7	48.9	45.9	26.1	72	60.1	34.1	94.2
150	12.1	8.12	20.2	23.8	16	39.8	35.1	23.6	58.7	46.2	31	77.2
200	10.2	7.32	17.5	20.1	14.4	34.6	29.8	21.4	51.2	39.3	28.2	67.4
300	8.19	6.15	14.3	16.2	12.2	28.4	24	18	42.1	31.7	23.8	55.5
400	7.02	5.33	12.4	13.9	10.6	24.5	20.7	15.7	36.4	27.3	20.7	48
500	6.23	4.72	11	12.4	9.36	21.7	18.4	13.9	32.3	24.3	18.4	42.7

In order to quantify the modification of the single atomic cross sections induced by the surrounding atoms we have evaluated the deviation of the cluster cross sections with those corresponding to the single atom multiplied by the number of atoms in the cluster. For the differential elastic scattering cross sections, these deviations, in percentage, are represented in Figure 3 as *deviation (%)*$=100(Ar_n-nAr_1)/nAr_1$. As can be seen in this figure, for 1 eV incident energy there is a clear enhancement of the differential cross section for low angles, from 0° to about 60° with a clear maximum around 15°, while differential values clearly decrease for high angles above 100°.This behavior shows a clear interference pattern indicating how the cluster structure modifies the angular distribution of scattered electrons with respect to that given by isolated atoms.

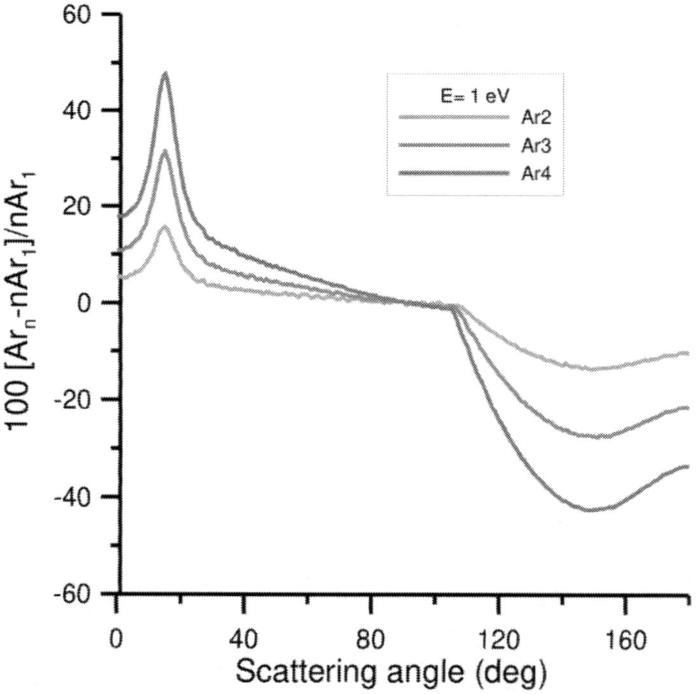

Journal of Physics: Conference Series **438** (2013) 012012 doi:10.1088/1742-6596/438/1/012012

FIGURE 3. Percentage deviation of the differential elastic electron scattering cross section of Ar_n (n=2, 3, 4) clusters with respect to the corresponding to isolated atoms for 1 eV incident energy.

Concerning integral cross sections, similar deviations have been plotted in Figure 4 for the total scattering cross sections of the Ar clusters with respect to those of isolated atoms. As seen from this figure, the total cross sections of clusters tend to be systematically lower than the corresponding to isolated atoms. For energies above 200 eV, these differences are nearly a constant factor. However, below these values, they tend to be quite energy dependent reaching the higher deviations around 15 eV

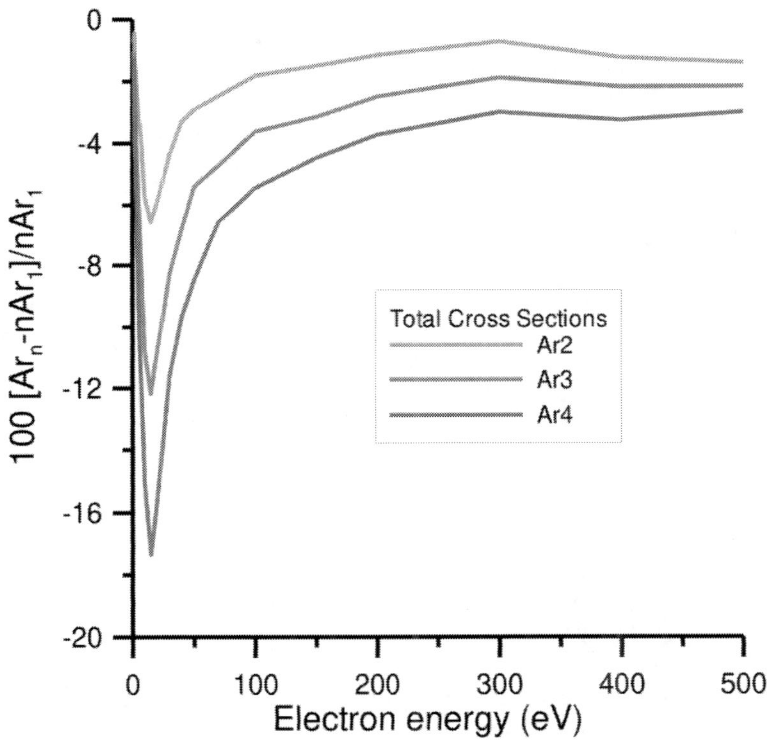

FIGURE 4. Percentage deviation of the total electron scattering cross section of Ar_n (n=2, 3, 4) clusters with respect to the corresponding to isolated atoms

Finally integral electron scattering cross sections for liquid Ar derived with the procedure described above are shown in Figure 5. As expected from the tendency showed by the Ar clusters, the total scattering cross sections of Ar in the liquid state is lower than those for isolated Ar atoms. This decrement is a consequence of the multiple scattering with surrounding atoms and provides an indication of the relevance of condensation effects as a function of the incident energy. Although multiple-scattering effects can be observed even at 100 eV [15], they are dominant at lower energies, between 3 and 40 eV, reaching their maximum contribution to the scattering processes around 15 eV.

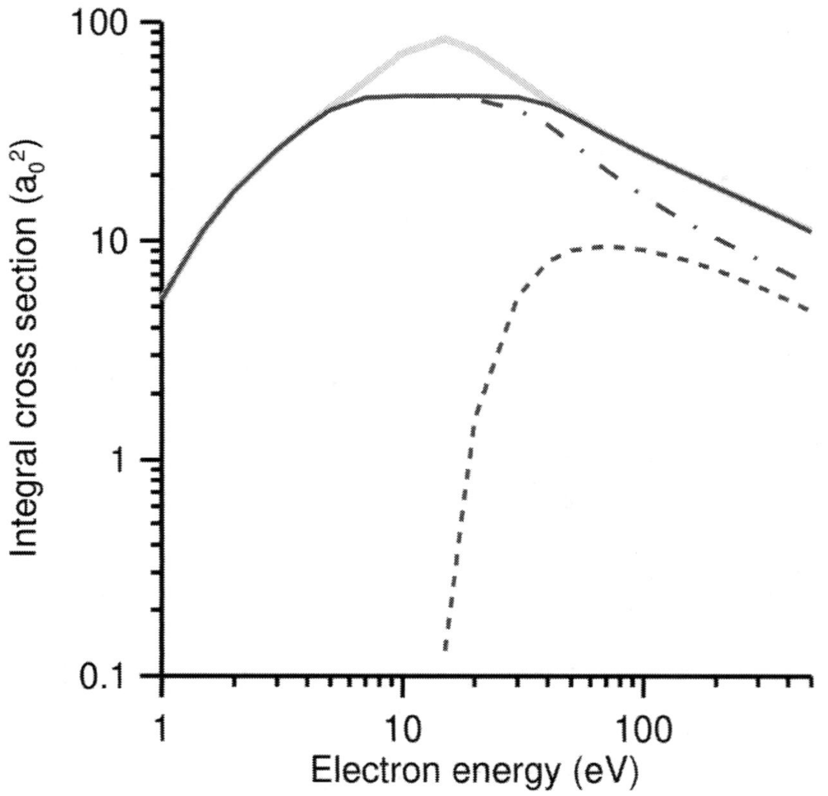

FIGURE 5. Integral cross sections (-.-.-, elastic, ---, inelastic, — total) for liquid Ar. The total cross sections for isolated Ar atoms (—) are also included for comparison

4. Conclusions

The screening corrected additivity rule (SCAR) over an independent atom representation has been applied to calculate differential and integral electron scattering cross sections by Ar clusters (dimer, trimer and tetramer) for incident energies ranging from 1 to 500 eV. This calculation confirmed that, as observed by Plüger et al. [14], multiple scattering in the cluster produces a redistribution of the scattering angles and important changes on the integral cross sections with respect to those for the atomic Ar, even at relatively high energies. These effects are especially pronounced for incident energies ranging from 3 to 40 eV.

Finally the calculation procedure has been applied to obtain electron scattering cross sections in liquid Ar, confirming the tendency showed by the Ar clusters. Electron scattering cross sections in liquids are required to model electron tracks in biologically relevant media for dosimetric purposes [16]. These results open the possibility of using the SCAR method to the scattering of electrons in molecular liquids, as water and other biomolecular systems.

References

[1] Ulrich B, Vredenborg A, Malakzadeh A, Schmidt L Ph H, Havermeier T, Meckel M, Cole K, Smolarski M, Chang Z, Jahnke T, and Dörner R 2011 *J. Phys. Chem.* A **115** 6936
[2] Pérez de Tudela R, Márquez-Mijares M, González-Lezana T, Roncero O, Miret-Artés S, Delgado-Barrio G, and Villarreal P 2010 *J. Chem. Phys.* **132** 244303
[3] Blanco F and García G 2002 *Phys. Lett.* A **295** 178
[4] Blanco F and García G 2003 *Phys. Rev.* A **67** 022701
[5] Blanco F and García G 2003 *Phys. Lett.* A **317** 458
[6] Cowan R D 1981 *The Theory of Atomic Structure and Spectra* (London: University of California Press)
[7] Riley M E and Truhlar D G 1975 *J. Chem. Phys.* **63** 2182
[8] Zhang X Z, Sun J F, and Liu Y F 1992 *J. Phys. B* **25** 1893
[9] Raju G G 2009 *IEEE Trans. Dielectrics and Elect. Insul.* **16** 1199
[10] Staszewska G, Schwenke D W, Thirumalai D, and Truhlar D G 1983 *Phys. Rev.* A **28**,2740
[11] Zatsarinny O, Bartschat K, García G, Blanco F, Hargreaves L R, Jones D B, Murrie R, Brunton J R, Brunger M J, Hoshino M, and Buckman S J 2011 *Phys. Rev.* A **83**, 042702
[12] Maljković J B, MilosavljevićA R, Blanco F, Šević D, García G, and Marinković 2009 *Pys. Rev. A* **79**052706
[13] Blanco F and García G 2004 *Phys. Lett.* A **330** 230
[14] Blanco F and García G 2009 *J. Phys.* B **42** 145203
[15] Pflüger T, Senftleben, Ren X, Dorn A, and Ulrich J 2011 *Phys. Rev. Lett.* **107** 223201
[16] Sanz A G, Fuss M C, Roldán A M, Blanco F, Limão-Vieira P, Brunger M J, Buckman S J and García G 2012 *Int. J. Radiat. Biol.* **88** 71

International Conference on Dynamics of Systems on the Nanoscale (DySoN 2012) IOP Publishing
Journal of Physics: Conference Series **438** (2013) 012013 doi:10.1088/1742-6596/438/1/012013

Vibrations of ordered counterions around left- and right-handed DNA double helixes

S.M. Perepelytsya, S.N. Volkov

Bogolyubov Institute for Theoretical Physics, NAS of Ukraine, 14-b Metrologichna Str., Kiev, 03680, Ukraine

E-mail: perepelytsya@bitp.kiev.ua; snvolkov@bitp.kiev.ua

Abstract. The vibrations of ordered counterions around right- and left-handed DNA double helix are studied. To determine the modes of DNA conformational vibrations the structure of the double helix with counterions is considered as ionic lattice (ion-phosphate lattice). Using the developed approach the frequencies and Raman intensities for right-handed B-form and left-handed Z-form of the double helix with Na^{+}, K^{+}, Rb^{+}, Cs^{+}, and Mg^{2+} counterions are calculated. The obtained frequencies of vibrations of internal structure elements of the double helix ($<$100 cm^{-1}) weakly depend on counterion type. In contrast, the vibrations of the ion-phosphate lattice are determined by counterion mass and charge. The frequencies of ion-phosphate vibrations of alkali metal counterions decrease from 180 to 100 cm^{-1}, while their Raman intensities increase as the counterion mass increases for the both B- and Z-DNA. In the case of Z-DNA new mode of ion-phosphate vibrations near 150 cm^{-1} is found. This mode is characterized by vibrations of Mg^{2+} counterions with respect to the phosphates of different strands of the double helix. Our results explain the experimental Raman spectra of Z-DNA.

1. Introduction

Under the natural conditions DNA forms the double helix consisted of negatively charged phosphate groups outside and nucleic bases inside the macromolecule. For the stability of the double helix the DNA phosphate groups must be neutralized by some positively charged ions of the solution (counterions) [1]. In water solution with monovalent metal counterions DNA double helix is usually in the right-handed B-form [2, 3], but increasing the concentration of counterions it may take the left-handed Z-form [1]. Z-DNA has a significant importance for many processes of DNA biological functioning [4]. The counterion concentration and type determine the structure and dynamics of the double helix in many respects [5, 6, 7, 8], but in the same time the role of counterions in conformational transformations of the double helix is not completely understood.

In solid samples of DNA the counterions occupy defined positions with respect to macromolecule atomic groups [9, 10, 11]. The monovalent counterions are usually localized near the oxygen atoms of phosphate groups from outside of the macromolecule [9, 10], while the counterions of higher charge may be also localized between phosphate groups of different DNA strands or bind to the nucleic bases [9, 11]. In solution the counterions are mobile, and together with water molecules they form the dynamical ion-hydrate shell around DNA macromolecule [1, 2, 8]. In the experiments for DNA solutions the counterions of the ion-hydrate shell are observed as a cloud around the double helix [12, 13, 14, 15]. The existence of coutnerion cloud around DNA has been predicted within the framework of counterion condensation theory and numerical calculations of PoissonBoltzmann equation [16, 17, 18, 19]. Increasing the concentration of counterions this cloud shrinks and counterions become closer to the surface of the double helix [12, 13, 14, 15]. Under such conditions the phosphate groups of DNA with counterions should form the regular structure along the double helix that may be considered as the lattice of ionic type (ion-phosphate lattice).

The molecular dynamics simulations of DNA with counterions show that the lifetime of counterion-phosphate complex is about 1 ns [20, 21, 22, 23, 24]. This time is rather long comparing to the period of molecular vibrations, therefore the dynamics of DNA ion-phosphate lattice should be characterized by counterion vibrations with respect to the phosphate groups (ion-phosphate vibrations). In the same time, the dynamics of the ion-phosphate lattice should be the part of conformational dynamics of DNA double helix, and its vibrations should be coupled with the internal dynamics of the double helix. Thus the determination of vibrational modes of the ion-phosphate lattice is of paramount importance for the understanding of counterion role for DNA structure and dynamics.

The modes of DNA ion-phosphate vibrations should be localized in the low-frequency spectra range (<200 cm^{-1}) the same as modes of ion vibrations in case of ionic crystals and electrolyte solutions [25, 26]. This spectra range is known to be characterized by the modes of DNA conformational vibrations describing the vibrations of atomic groups of the double helix (phosphates, nucleosides and nucleic bases) [27, 28, 29, 30, 31, 32, 33]. In our previous works [34, 35, 36, 37, 38, 39] the model of conformational vibrations of the right-handed DNA double helix with counterions has been developed basing on the approach [40, 41, 42, 43]. As the result the frequencies of ion-phosphate vibrations for DNA with Na^+, K^+, Rb^+, and Cs^+ counterions are determined. The obtained frequency values of DNA ion-phosphate vibrations decrease from 180 to 100 cm^{-1} as counterion mass increases that agree with the experimental data [28, 32]. The vibrations of heavy counterions (Cs^+) influence the internal dynamics of the double helix, while light counterions (Na^+) play the role of counteractive charges for the phosphate groups and do not disturbs the motions in nucleotide pairs. In contrast, the Raman intensities of the ion-phosphate modes increase as the counterion mass increases [36, 37] that agree with the experimental data [44].

The experimental data for the left-handed DNA double helix show that in the low-frequency spectrum new mode near 150 cm^{-1} is observed [33]. This mode may characterize the specific vibrations of counterions and atomic groups of DNA. To determine the origin of this mode the approach [34, 35, 36, 37, 38, 39] should be extended for the case of the ion-phosphate lattice of DNA in the left-handed form of the double helix.

The goal of the present work is to find the modes of ion-phosphate vibrations of Z-DNA and compare them with the ion-phosphate modes of right-handed DNA double helix. To solve this problem in section 2 the model of conformational vibrations of Z-DNA with counterions is introduced. In section 3 frequencies of Z-DNA are calculated and the Raman spectra are for the right- and left-handed DNA with counterions are analyzed. As the result the specific modes of Z-DNA ion-phosphate vibrations are determined. The calculated Raman spectrum of Z-DNA agree with the experimental data.

2. Models of conformational vibrations of DNA ion-phosphate lattices

The structure of DNA ion-phosphate lattice depends on the double helix form and counterion type. According to the experimental data [9] and molecular dynamics simulations [21, 22, 20, 23] in case of the right-handed double helix the monovalent metal ions usually neutralize the negatively charged phosphate groups, localizing from the outside of the double helix (single-stranded position of counterion). The left-handed double helix has significantly different structure and for its stabilization the both monovalent and bivalent counterions are necessary

[1]. The structure of Z-DNA is favorable for the localization of counterions outside of the double helix and inside of its minor grove. The counterions inside the minor grove of Z-DNA neutralize the phosphate groups of different strands along the line orthogonal to the helical axis (cross-stranded neutralization). The models of DNA ion-phosphate lattice in case of the right- and left-handed DNA double helix are shown in the figure 1a.

The dynamics of DNA ion-phosphate lattice we describe in framework of phenomenological approach for conformational vibrations of the double helix [40, 41, 42], which in the previous works has been extended for consideration of counterion vibrations in the right-handed DNA double helix [34, 35, 38, 39]. In the present work the developed model of the ion-phosphate lattice dynamics is improved for the description of the left-handed double helix.

The model of the ion-phosphate lattice dynamics presents the DNA macromolecule as the chain of monomer links. In case of the right-handed double helix the monomer link consists of nucleotide pairs, while in case of the left-handed double helix it consists of two nucleotide pairs (Figure 1b). The nucleotides are modelled as masses of phosphate groups m_0 (PO_4+$C_{5'}$) and masses of nucleosides m. The nucleosides rotate as the physical pendulums with respect to the phosphate groups in plane of nucleotide pair. The physical pendulums are characterized by reduced length l. The nucleosides of different chains are paired by H-bonds (Figure 2). The counterions are modelled as charges tethered to the phosphate groups. In case of single-strand neutralization one counterion with the mass m_a is tethered to one phosphate group from the

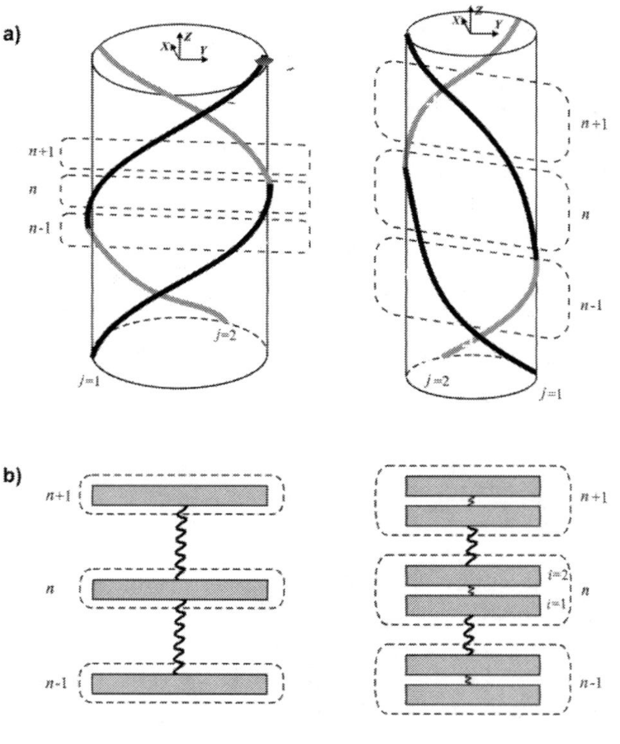

Figure 1. DNA with counterions. (a) Ion-phosphate lattices of the right- and left-handed DNA double helix. (b) Chain of monomer links in case of right- and left-handed double helix. The rectangles symbolize the nucleotide pairs.

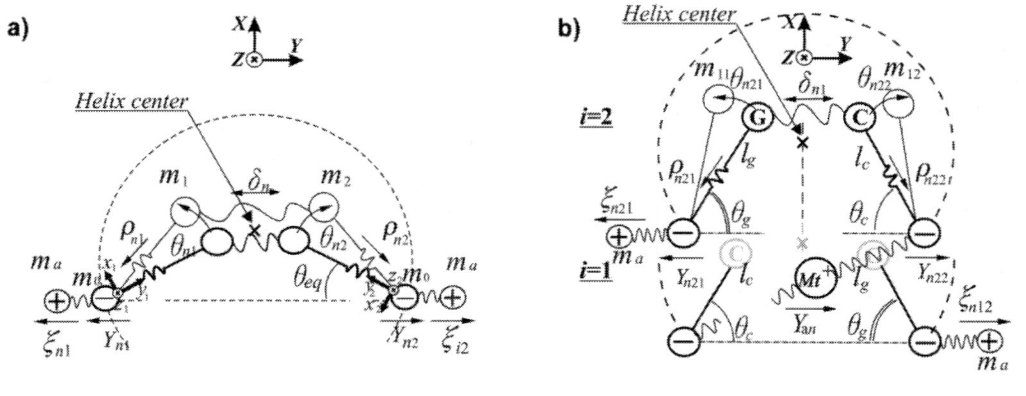

Figure 2. Monomer links of the right-handed (a) and left-handed (b) double helixes. l is reduced length of pendulum-nucleoside; θ_0 is equilibrium angle; m, m_0, and m_a are masses of nucleosides, phosphate groups, and counterions, respectively; X, Y, θ, ρ, and ξ are vibrational coordinates of the model (see text). The arrows indicate positive directions of displacements; XYZ and xyz are the reference frames connected with the monomer link and nucleoside, respectively.

outside of macromolecule. The counterions in cross-stranded positions with the mass M_a are localized between phosphate groups in the minor groove of the double helix (Figure 2). The motions of structural elements of the monomer link are considered in the plane orthogonal to the helical axis (transverse vibrations). The longitudinal vibrations of the macromolecule atomic groups have much higher frequencies [40, 41, 42] and are beyond the scope of this work.

The displacements of nucleosides and phosphate groups in DNA monomer link are described by coordinates X and Y. The coordinates θ describe the deviations of pendulum-nucleosides from their equilibrium position in the plane of complementary DNA pair (angle θ_0). The vibrations of deoxyribose and base with respect to each other, inside the nucleoside (intranucleoside vibrations), are described by changes of pendulum lengths ρ. The vibrations of counterions in single-stranded positions are described by coordinates ξ. For description of vibrations of a counterion between phosphate groups the coordinate Y_a is used. The vibrational coordinates of the model and the positive directions of displacements are showed on Figure 2.

Within the framework of introduced model of the dynamics of DNA ion-phosphate lattice the energy of vibrations of double helix structural elements may be written as follows:

$$E = \sum_n (K_n + U_n + U_{n,n-1}), \tag{1}$$

where K_n and U_n are the kinetic and potential energies of the monomer link, $U_{n,n-1}$ is the potential energy of interaction along the chain.

Let us consider the kinetic and potential energies of the monomer link as a sum of vibrational energy of DNA atomic groups and energy of counterion vibrations: $K_n + U_n = K_{0n} + U_{0n} + K_{an} + U_{an}$. In such a way the energy of vibrations of structure elements in monomer link n of right-handed double helix may be written as follows:

$$K_{0n}^R = \frac{1}{2}\sum_j [M_j \dot{Y}_{nj}^2 + m_j(\dot{\rho}_{nj}^2 + l_j^2\dot{\theta}_{nj} + 2l_j^s\dot{\theta}_{nj}\dot{Y}_{nj} + 2b_j\dot{\rho}_{nj}\dot{Y}_{nj})], \tag{2}$$

$$U_{0n}^R = \frac{1}{2}\alpha\delta_n^2 + \frac{1}{2}\sum_j \left[\sigma\rho_{nj}^2 + \beta\theta_{nj}^2\right], \tag{3}$$

where $l_j^s = l_j a_j$; $a_j = \sin\theta_{0j}$; $b_j = \cos\theta_{0j}$; the index $j = 1, 2$ enumerates the chain of the double helix; the force constants α, σ, and β describe H-bond stretching in base pairs, intranucleoside mobility, and rotation of nucleosides with respect to the backbone chain in base-pair plane, respectively. The variable δ_n describes stretching of H-bonds in the base pairs (Fig. 1b): $\delta_n \approx l_1^s\theta_{n1} + l_2^s\theta_{n2} + Y_{n1} + Y_{n2} + b_1\rho_{n1} + b_2\rho_{n2}$.

Analogically, for the left-handed double helix:

$$K_{0n}^L = \frac{1}{2}\sum_i^2\sum_j^2[M_{ij}\dot{Y}_{nij}^2 + m_{ij}(\dot{\rho}_{nij}^2 + l_{ij}^2\dot{\theta}_{nij} + 2l_{sij}\dot{\theta}_{nij}\dot{Y}_{nij} + 2b_{ij}\dot{\rho}_{nij}\dot{Y}_{nij})], \tag{4}$$

$$U_{0n}^L = \frac{1}{2}\sum_i^2[\alpha\delta_{ni}^2 + \sum_j^2(\sigma_{ij}\rho_{nij}^2 + \beta_{ij}\theta_{nij}{}^2)] + \frac{1}{2}\sum_j^2\{g_1\left[(\theta_{n1j} - \theta_{n2j})^2 + (\rho_{n1j} - \rho_{n2j})^2\right] + g_2(Y_{n1j} - Y_{n2j})^2\}. \tag{5}$$

In the formulae (4) and (5) the index n enumerates dimers of macromolecule and the index i enumerates the nucleotide pairs in the dimers. The force constants g_1 and g_2 describe the interaction of nucleic bases and the interaction between phosphate groups in dimers.

The energy of counterion vibrations in case of the ion-phosphate lattice of the right-handed double helix may be written as:

$$K_{an}^R = \frac{m_a}{2}\sum_j\left(\dot{\xi}_{nj}^2 + \dot{Y}_{nj}^2\right), \tag{6}$$

$$U_{an}^R = \frac{\gamma_1}{2}\sum_j\xi_{nj}^2. \tag{7}$$

Analogically, for the left-handed double helix:

$$K_{an}^L = \frac{M_a}{2}\dot{Y}_{an}^2 + \frac{m_a}{2}[(\dot{Y}_{n11} + \dot{\xi}_{n11})^2 + (\dot{Y}_{n22} + \dot{\xi}_{n22})^2], \tag{8}$$

$$U_{an}^L = \frac{\gamma_1}{2}(\xi_{n11}^2 + \xi_{n22}^2) + \frac{\gamma_2}{2}\left[(Y_{an} - Y_{n22})^2 + (Y_{an} + Y_{n11})^2\right]. \tag{9}$$

In the formulae (7) and (9) the parameters γ_1 and γ_2 are the force constants for the counterions in single-stranded and cross-stranded positions, respectively. Note, the energy of vibrations for the ion-phosphate lattice of DNA in the right-handed form of the double helix may be obtained from (8) and (9) by putting parameters γ_2, g_1, and g_2 to zero and omitting the index of summation i.

Accordingly to our approach [34, 35] we will consider the limited long-wave vibrational modes of the ion-phosphate lattice that are sufficient for interpretation of the experimental vibrational spectra. As known only long-range lattice vibrations interact with the electro-magnetic fields and manifest themselves in vibrational spectra [25]. In long-wave limit, when the wave vector leads to zero ($\bar{k} \to 0$), the frequencies of optical types modes weakly depend on the $\bar{k}$ value. From the point of the theory of the lattice vibrations such approximation is the same as the neglecting interaction along the chain. So, in the following consideration we will neglect by the interaction term: $U_{n,n-1} \approx 0$. Within the framework of this approximation the equations of motions may be written as follows:

$$\frac{d}{dt}\frac{\partial K_{0n}}{\partial \dot{q}_n} + \frac{d}{dt}\frac{\partial K_{an}}{\partial \dot{q}_n} - \frac{\partial U_{0n}}{\partial q_n} - \frac{\partial U_{an}}{\partial q_n} = 0, \tag{10}$$

where q_n denotes some vibrational coordinate in the monomer link of model (Figure 2). The equations of motions (10) may be solved using the substitution: $q_n = \tilde{q_n}\exp(i\omega t)$, where $\tilde{q_n}$ and

ω are amplitude and frequency for some coordinate of vibrations, respectively. The equations of motion (10) in explicit form are shown in Appendix for right- and left-handed double helix of DNA.

The Raman intensities for the modes of DNA conformational vibrations are calculated within the framework of phenomenological approach developed in our previous work [36, 37]. It is based on the valence-optic theory [45] and our model for the conformational vibrations of DNA with counterions [34, 35]. The analytical expression for some mode of DNA conformational vibrations is shown in Appendix.

To estimate the frequencies of conformational vibrations of B-DNA the force constants $\alpha = 85$ kcal/mole Å^2, $\beta = 40$ kcal/mole, and $\sigma = 43$ kcal/mole Å^2 are taken from [40, 41, 42]. In case of Z-DNA the constant α is the same as in B-DNA, while the constants β and σ for guanosine (G) and cytosine (C) nucleosides are taken the same as in A- and B-forms, respectivel. The difference of these constants for G and C appear due to the dimeric structure of the left-handed double helix. In case of A-DNA the constants β and σ have the following values 46 kcal/mole and 22 kcal/mole Å^2, respectively [40, 41, 42]. The constant of ion-phosphate vibrations γ_1 has been determined by us earlier for Na^+, K^+, Rb^+, and Cs^+ counterions, and it vary from 42 to 52 kcal/mole Å^2 depending on counterion type. The constant $\gamma_2 = 62$ kcal/mole Å^2 is taken for the case of Mg^{2+} counterions in cross-stranded position [39]. The magnesium counterion are considered with the hydration shell because the size of hydrated Mg^{2+} ion corresponds to the distances between phosphate groups of the left-handed double helix. The hydration shell of magnesium ion consists of 4 water molecules strongly bond to the ion [46]. The structure parameters of the model for Z-DNA is determined using the X-ray data (pdb code: 1dcg) [47]. Using such parameters the frequencies and amplitudes of Z-DNA conformational vibrations are estimated by the formulae (1) – (10).

3. Frequencies of DNA conformational vibrations

Using the developed approach the frequencies of right- and left-handed double helix with different alkali metal counterions are calculated. The right-handed double helix is considered in natural B-form of DNA with the alkali metal counterions in single-stranded position outside macromolecule. In case of left-handed double helix bivalent magnesium counterions are localized between phosphate groups of the DNA strands, and the monovalent alkali metal counterions are localized outside the double helix. The obtained frequency values for B- and Z-DNA conformational vibrations are shown in the Table 1. According to the character of motions of structural elements in DNA nucleotide pairs the obtained modes may be classified as the modes of ion-phosphate vibrations (**Ion**), H-bond stretching modes (**H**), modes of intranucleoside vibrations (**S**), and modes of backbone vibrations (**B**). The low-frequency spectra of B- and Z-DNA have different number of modes with different frequency values. Such difference appear because in Z-DNA the stacking interaction between nucleic bases differ from B-DNA, and the additional interaction between phosphate groups via cross-linked Mg^{2+} counterion makes the structure of nucleotide dimer more rigid.

The obtained results show that the low-frequency spectra of DNA may be parted on three ranges: the lowest range (from 10 to 30 cm^{-1}), the middle range (from 30 to 110 cm^{-1}), and the high range (from 110 to 180 cm^{-1}). At the lowest frequency range the modes of backbone vibrations are observed. In case of B-DNA there are two **B** modes with close frequency values ($\omega_{B1} \approx \omega_{B2}$). In case of Z-DNA the number of **B** modes is twice higher and the frequencies ω_{B1} and ω_{B2} are different, while the frequencies $\omega_{B3} \approx \omega_{B4}$. The additional modes in the spectra of Z-DNA appear due to the dimeric structure of the left-handed double helix. Increasing the counterion mass the frequency values ω_{B1} and ω_{B2} decrease, while $\omega_{B3} \approx \omega_{B4}$ remain practically the same.

The middle frequency range is characterized by the modes of intranucleoside vibrations and

H-bond stretching. In case of *B*-DNA there are two **H** and one **S** modes, while in case of *Z*-DNA there are four **H** and three **S** modes. The **S** modes of *Z*-DNA have close frequency values that are essentially lower than in *B*-DNA. The modes of H-bond stretching have two characteristic frequency values near 110 cm^{-1} and 60 cm^{-1}. The modes of the middle frequency range gradually decrease as counterion mass increases.

At the high frequency range the modes of ion-phosphate vibrations are observed. The frequencies of ion-phosphate vibrations depend on counterion position. In case of counterions in single-stranded position there are two degenerated modes (ω_{Ion1} and ω_{Ion2}) with close frequencies. These modes are characteristic for the both right- and left-handed double helix forms. The vibrations of hydrated Mg^{2+} ions in cross-stranded position are characterized by one mode (ω_{Ion3}). The vibrations of Mg^{2+} counterion in cross-stranded position are not sensitive to the vibrations of counterions outside of the double helix. Increasing the counterion mass the frequencies of ion-phosphate vibrations ω_{Ion1} and ω_{Ion2} decrease from 180 to 110 cm^{-1}. The same frequency dependence on counterion type has been observe in infrared spectra of DNA dry films with alkali metal counterions [32].

Table 1. Frequencies of *B*- and *Z*-DNA conformational vibrations with different alkali metal counterions (cm^{-1}). The monomer links of *Z*-DNA contain Mg^{2+} counterion between phosphate groups of different strands.

		Na^+		K^+		Rb^+		Cs^+	
		B-form	*Z*-form	*B*-form	*Z*-form	*B*-form	*Z*-form	*B*-form	*Z*-form
	ω_{Ion1}	182	181	151	146	120	110	118	110
Ion	ω_{Ion2}	182	181	151	146	113	107	108	107
	ω_{Ion3}	–	153	–	152	–	152	–	152
	ω_{H1}	111	112	110	110	99	105	94	100
	ω_{H2}	–	107	–	106	–	103	–	96
H	ω_{H3}	–	99	–	96	–	95	–	92
	ω_{H4}	57	63	54	59	47	59	42	59
	ω_{S1}	79	51	75	44	64	40	58	37
S	ω_{S2}	–	46	–	43	–	42	–	42
	ω_{S3}	–	42	–	39	–	36	–	35
	ω_{B1}	16	29	14	25	13	25	13	24
B	ω_{B2}	15	29	15	20	14	18	12	17
	ω_{B3}	–	13	–	11	–	11	–	11
	ω_{B4}	–	12	–	11	–	11	–	10

To determine the manifestations of counterion effects in the experimental spectra the Raman intensities for the modes of *B*- and *Z*-DNA are calculated using the approach developed in [38, 39]. As the result the low-frequency Raman spectra are built for the frequency range from 90 to 200 cm^{-1}, which is characterized by the ion-depended modes (Figure 3). The spectra intensity is normalized per intensity of the mode ω_{H1}. The halfwidth of spectra lines is 5 cm^{-1}

It is seen that in case of *B*-DNA the intensities of ion-phosphate modes increase as their frequency decrease (Figure 3a). Thus in case of light counterions (Na^+ and K^+) the intensities of these modes are low, while in case of heavy counterions (Rb^+ and Cs^+) they are high due to the coupling with the modes of H-bond stretching. Our calculations for *B*-DNA agree with

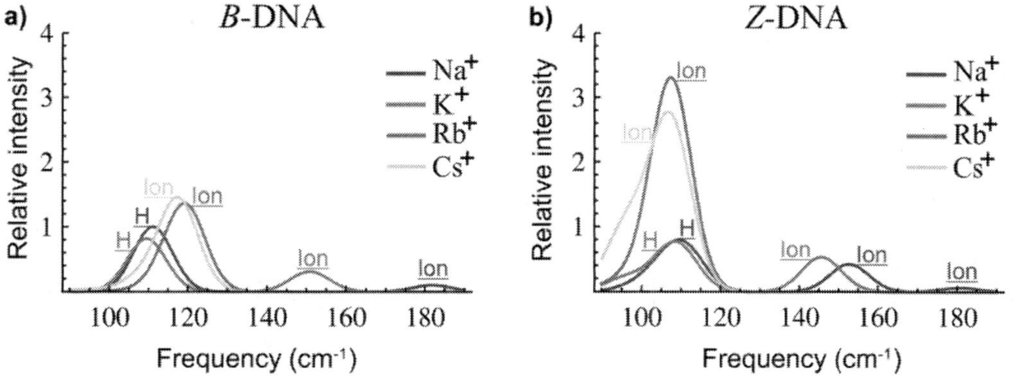

Figure 3. The low-frequency Raman spectra of *B*-DNA and *Z*-DNA. (a) *B*-DNA with Na^+, K^+, Rb^+, and Cs^+ counterions in single-stranded position. (b) *Z*-DNA with Mg^{2+} counterions in cross-stranded position and Na^+, K^+, Rb^+, and Cs^+ counterions in single-stranded position. The halfwidth of spectra lines is 5 cm^{-1}.

the experimental spectra for Cs-DNA water solution [44] where the intensity increase near 100 cm^{-1} has been observed. In the spectra of *Z*-DNA the modes of ion-phosphate vibrations ω_{Ion1} and ω_{Ion2} have about the same intensity ratio as in case of *B*-DNA (Figure 3b). The mode ω_{Ion3} is rather intensive, and in experiment it is observed near 150 cm^{-1} [33]. In case of K^+ counterions the modes ω_{Ion1} and ω_{Ion2} have about the same frequencies as the mode ω_{Ion3} (Table 1), and they form a common band with increased intensity. In case of Rb^+ and Cs^+ counterions the intensities of the ion-phosphate modes are essentially higher than in *B*-DNA. Thus the calculations of the modes of DNA conformational vibrations show that the counterions essentially influence the dynamics of both right- and left-handed double helix. The manifestations of counterion influence are prominent in DNA low-frequency Raman spectra.

4. Conclusions

The vibrations of counterions neutralizing the negatively charged phosphate groups of right- and left-handed double helix of DNA are studied. To find vibrational modes of the double helix the phenomenological approach is used considering the structure of DNA with counterions as the lattice of ionic type (ion-phosphate lattice). In case of right-handed double helix the counterions are localized outside of the double helix near the phosphate groups, while in case of the left-handed DNA bivalent counterions localized between phosphate groups of different strands are also presented. Using the developed model the frequencies and the Raman intensities of vibrational modes are calculated for *B*- and *Z*-DNA with Na^+, K^+, Rb^+, and Cs^+ counterions. In case of *Z*-DNA the Mg^{2+} countrions are also considered. The results show that the modes of internal vibrations of the left-handed double helix (H-bond stretching in nucleotide pairs, intranucleoside vibrations, and backbone vibrations) change the frequency values comparing to the right-handed DNA double helix that is due to the dimeric structure of *Z*-DNA. The frequencies of these modes gradually decrease as counterion mass increases. The modes of ion-phosphate vibrations are at frequency range from 100 to 180 cm^{-1}. The Raman intensities in case of heavy counterions (Rb^+ and Cs^+) are higher than in case of light counterions (Na^+ and K^+). New mode of ion-phosphate vibrations about the frequency 150 cm^{-1} is determined, which characterizes the vibrations of Mg^{2+} counterions with respect to the phosphate groups of different strands of *Z*-DNA. The intensity of this mode is rather large, therefore it is observed in the experimental Raman spectra of Z-DNA. The frequency of vibration of magnesium counterions does not depend on the vibrations of counterions outside the double helix. According to our calculation the modes of ion-phosphate vibrations of Mg^{2+} and K^+ counterions have close frequencies therefore in the low-frequency spectra of Z-DNA these modes form a common band with high intensity. The intensity increase of the band 150 cm^{-1} due to the modes of ion-phosphate vibrations should be observed in the experimental spectra of Z-DNA with K^+ and Mg^{2+} counterions.

Acknowledgments
The present work was partially supported by the Project of the NAS of Ukraine (0110U007540).

Appendix

For a more convenient form of the equations of motion the following variables are used in case of the right-handed double helix: $Y_n = Y_{n1} + Y_{n2}$, $y_n = Y_{n1} - Y_{n2}, \theta_n = \theta_{n1} + \theta_{n2}$, $\eta_n = \theta_{n1} - \theta_{n2}$, $\rho_n = \rho_{n1} + \rho_{n2}$, $r_n = \rho_{n1} - \rho_{n2}$, $\xi_n = \xi_{n1} + \xi_{n2}$, $\varsigma_n = \xi_{n1} - \xi_{n2}$. In these variables the system of equations of motion splits into two subsystems of coupling equations for Y_n, θ_n, ρ_n, ξ_n and y_n, η_n, r_n, ς_n coordinates. In a long-limit approximation the equations of motion (10) for B-DNA may be written as follows:

$$\begin{cases} \ddot{Y}_n + \frac{ml_s}{M}\ddot{\theta}_n + \frac{mb}{M}\ddot{\rho}_n + \frac{m_a}{M}\ddot{\xi}_n = -\alpha_0 \left(l_s\theta_n + Y_n + b\rho_n\right); \\ \ddot{\rho}_n + b\ddot{Y}_n = -\alpha_0 b\frac{M}{m}\left(l_s\theta_n + Y_n + b\rho_n\right) - \sigma_0\rho_n; \\ \ddot{\theta}_n + \frac{a}{l}\ddot{Y}_n = -\alpha_0 l_s\frac{M}{I}\left(l_s\theta_n + Y_n + b\rho_n\right) - \beta_0\theta_n; \\ \ddot{\xi}_n + \ddot{Y}_n = -\gamma_{01}\xi_n. \end{cases}$$

$$\begin{cases} \ddot{y}_n + \frac{ml_s}{M}\ddot{\eta}_n + \frac{mb}{M}\ddot{r}_n + \frac{m_a}{M}\ddot{\varsigma}_n = 0; \\ \ddot{r}_n + b\ddot{y}_n = -\sigma_0 r_n; \\ \ddot{\eta}_n + \frac{a}{l}\ddot{y}_n = -\beta_0\eta_n; \\ \ddot{\varsigma}_n + \ddot{y}_n = -\gamma_{01}\varsigma_n, \end{cases}$$

where $\alpha_0 = 2\alpha/M$; $\sigma_0 = \sigma/m$; $\beta_0 = \beta/ml^2$; $\gamma_0 = \gamma/m_a$. Note, the same as in our previous works [34, 35] for B-DNA we consider the average values of the reduced lengths of pendulum-nucleosides ($l_j \equiv l$), equilibrium angles ($\theta_{0j} \equiv \theta_0$), and masses ($m_j \equiv m$).

Analogically for Z-DNA the following variables are used: $Y_c = Y_{11} + Y_{22}$, $y_c = Y_{11} - Y_{22}$, $\theta_c = \theta_{11} + \theta_{22}$, $\eta_c = \theta_{11} - \theta_{22}$, $\rho_c = \rho_{11} + \rho_{22}$, $r_c = \rho_{11} - \rho_{22}$, $Y_g = Y_{21} + Y_{12}$, $y_g = Y_{21} - Y_{12}$, $\theta_g = \theta_{21} + \theta_{12}$, $\eta_g = \theta_{21} - \theta_{12}$, $\rho_g = \rho_{21} + \rho_{12}$, $r_g = \rho_{21} - \rho_{12}$, $\xi_1 = \xi_{21} + \xi_{12}$, $\xi_2 = \xi_{21} - \xi_{12}$. As the result the equations of motion (10) for Z-DNA may be written as follows:

$$\begin{cases} \ddot{Y}_c + \frac{m_c b_c}{M_c}\rho_c + \frac{m_c l_c^s}{M_c}\ddot{\theta}_c + \frac{M_a}{M_c}\gamma_{02}Y_c - g_{2c}(Y_c - Y_g) + \alpha_c\left(Y_g + Y_c + l_c^s\theta_g + l_c^s\theta_c + b_g\rho_g + b_c\rho_c\right) = 0; \\ \ddot{Y}_g + \frac{m_g b_g}{M_g}\rho_g + \frac{m_g l_g^s}{M_g}\ddot{\theta}_g + \frac{m_a}{M_g}(\ddot{Y}_g + \ddot{\xi}_1) - g_{2g}(Y_c - Y_g) + \alpha_g\left(Y_g + Y_c + l_g^s\theta_g + l_c^s\theta_c + b_g\rho_g + b_c\rho_c\right) = 0; \\ \ddot{\theta}_g + \ddot{Y}_g\frac{a_g}{l_g} + \beta_g\theta_g - g_{1g}(\theta_c - \theta_g) + \alpha_g\frac{M_g a_g}{m_g l_g}\left(Y_g + Y_c + l_c^s\theta_g + l_c^s\theta_c + b_g\rho_g + b_c\rho_c\right) = 0; \\ \ddot{\theta}_c + \ddot{Y}_c\frac{a_c}{l_c} + \beta_c\theta_c - g_{1c}(\theta_c - \theta_g) + \alpha_c\frac{M_c a_c}{m_c l_c}\left(Y_g + Y_c + l_g^s\theta_g + l_c^s\theta_c + b_g\rho_g + b_c\rho_c\right) = 0; \\ \ddot{\rho}_g + b_g\ddot{Y}_g + \sigma_g\rho_g - g_{1g}\left(\rho_c - \rho_g\right) + \alpha_g\frac{M_g b_g}{m_g}\left(Y_g + Y_c + l_g^s\theta_g + l_c^s\theta_c + b_g\rho_g + b_c\rho_c\right) = 0; \\ \ddot{\rho}_c + b_c\ddot{Y}_c + \sigma_c\rho_c - g_{1c}(\rho_c - \rho_g) + \alpha_c\frac{M_c b_c}{m_c}\left(Y_g + Y_c + l_g^s\theta_g + l_c^s\theta_c + b_g\rho_g + b_c\rho_c\right) = 0; \\ \ddot{Y}_g + \ddot{\xi}_1 + \gamma_{01}\xi_1 = 0; \end{cases}$$

$$\begin{cases} \ddot{y}_c + \frac{m_c b_c}{M_c}\ddot{r}_c + \frac{m_c l_c^s}{M_c}\ddot{\eta}_c - \frac{M_a}{M_c}\gamma_{02}\left(2y_a - y_c\right) + g_{2c}\left(y_c + y_g\right) + \alpha_c\left(y_g + y_c + l_g^s\eta_g + l_c^s\eta_c + b_g r_g + b_c r_c\right) = 0\,; \\ \ddot{y}_g + \frac{m_g b_g}{M_g}\ddot{r}_g + \frac{m_g l_g^s}{M_g}\ddot{\eta}_g + \frac{m_a}{M_g}(y_g + \xi_2) + g_{2g}(y_c + y_g) + \alpha_g\left(y_g + y_c + l_g^s\eta_g + l_c^s\eta_c + b_g r_g + b_c r_c\right) = 0\,; \\ \ddot{\eta}_g + \ddot{y}_g\frac{a_g}{l_g} + \beta_g\eta_g + g_{1g}\left(\eta_c + \eta_g\right) + \alpha_g\frac{M_g a_g}{m_g l_g}\left(y_g + y_c + l_g^s\eta_g + l_c^s\eta_c + b_g r_g + b_c r_c\right) = 0\,; \\ \ddot{\eta}_c + \ddot{y}_c\frac{a_c}{l_c} + \beta_c\eta_c + g_{1c}\left(\eta_c + \eta_g\right) + \alpha_c\frac{M_c a_c}{m_c l_c}\left(y_g + y_c + l_g^s\eta_g + l_c^s\eta_c + b_g r_g + b_c r_c\right) = 0\,; \\ \ddot{r}_g + b_g\ddot{y}_g + \sigma_g r_g + g_{1g}\left(r_c + r_g\right) + \alpha_g\frac{M_g b_g}{m_g}\left(y_g + y_c + l_g^s\eta_g + l_c^s\eta_c + b_g r_g + b_c r_c\right) = 0\,; \\ \ddot{r}_c + b_c\ddot{y}_c + \sigma_c r_c + g_{1c}\left(r_c + r_g\right) + \alpha_c\frac{M_c b_c}{m_c}\left(y_g + y_c + l_g^s\eta_g + l_c^s\eta_c + b_g r_g + b_c r_c\right) = 0\,; \\ \ddot{y}_g + \ddot{\xi}_2 + \gamma_{01}\xi_2 = 0\,; \\ 2\ddot{y}_a + 2\gamma_{02}\left(2Y_a - y_c\right) = 0\,, \end{cases}$$

where $\alpha_g = \alpha/M_g$, $\alpha_c = \alpha/M_c$, $\beta_g = \beta/m_g l_g^2$, $\beta_c = \beta/m_c l_c^2$, $\sigma_g = \sigma/m_g$, $\sigma_c = \sigma/m_c$, $\gamma_{01} = \gamma_1/m_a$, $\gamma_{02} = \gamma_2/M_a$, $g_{1g} = g_1/m_g l_g^2$, $g_{1c} = g_1/m_c l_c^2$, $g_{2g} = g_2/m_g$, $g_{2c} = g_2/m_c$, $a_c = \sin\theta_{0c}$, $a_g = \sin\theta_{0g}$, $b_c = \cos\theta_{0c}$, $b_g = \cos\theta_{0g}$, $l_c^s = l_c\sin\theta_{0c}$, $l_g^s = l_g\sin\theta_{0g}$.

According to the developed approach [36, 37] the intensity of some mode of DNA conformational vibrations that is observed in the Stokes part of the Raman spectra at the right angle geometry is determined as follows:

$$J_s \approx 3\kappa J_0(\nu_0 - \nu_s)^4(A^2 + B^2)\left[1 - e^{-\frac{h\nu_s}{k_B T}}\right],$$

where:

$$A = \sum_{ij} A_{ij}^s\left(\widetilde{\theta}_{ij}^s + \frac{\widetilde{\rho}_{ij}^s}{l_{ij}}\right), \quad B = \sum_{ij} B_j^s\left(\widetilde{\theta}_{ij}^s + \frac{\widetilde{\rho}_{ij}^s}{l_{ij}}\right).$$

The parameters A_{ij}^s and B_{ij}^s are the combinations of the components of nucleoside polarizability tensors and structural parameters of the model:

$$A_{ij}^s = [(b_{ij}^{yy} - b_{ij}^{xx})\sin 2\theta_{0ij} + 2(-1)^j b_{ij}^{xy}\cos 2\theta_{0ij}],$$
$$B_{ij}^s = [(b_{ij}^{yy} - b_{ij}^{xx})(-1)^j\cos 2\theta_{0ij} - 2b_{ij}^{xy}\sin 2\theta_{0ij}],$$

where $\kappa = 13\cdot 2^8\pi^5/(9c^4)$; J_0 and ν_0 are intensity and frequency of incident light; ν_s is frequency of a molecular normal vibration that has usual interdependence with cyclic frequency $2\pi\nu_s = \omega$; index s enumerates the mode of normal vibrations; c is velocity of light; h is the Plank constant; k_B is the Boltzmann constant; b_{ij}^{xx}, b_{ij}^{xy}, and b_{ij}^{yy} are components of nucleoside polarizability tensor determined in the nucleoside reference frame xyz (Figure 2). The components of nucleoside polarizability tensor are calculated as a sum of polarizability tensors of nucleoside chemical bonds analogically to our previous works [36, 37].

5. References

[1] Saenger W 1984 *Principles of Nucleic Acid Structure* (New York: Springer)
[2] Blagoi Yu P, Galkin V L, Gladchenko G O *et al.* 1991 *The complexes of nucleic acids and metals in the solutions* (Kiev: Naukova Dumka)
[3] Maleev V Ya, Semenov M A, Gassan M A and Kashpur V I 1993 *Biofizika* **38** 768 [Moscow]
[4] Rich A and Zhang S 2003 Nature Reviews **4** 566-569
[5] Ivanov V I, Minchenkova L E, Schyolkina A K and Poletayev A I 1973 *Biopolymers* **12** 89-110
[6] Williams L D and Maher L J 2000 *Annu. Rev. Biophys. Biomol. Struct.* **29**, 497-521
[7] Levin Y 2002 *Rep. Prog. Phys.* **65** 1577-1632

[8] Kornyshev A A, Lee D J, Leikin S and Wynveen A 2007 *Rev. Mod. Phys.* **79** 943-966
[9] Tereshko V, Wilds C J, Minasov G, Prakash T P, Maier M A, Howard A, Wawrzak Z, Manoharan M and Egli M 2001 *Nucleic Acids Res.* **29** 1208-1215
[10] Tereshko V, Minasov G and Egli M 1999 *J. Am. Chem. Soc.* **121** 3590-3595
[11] Hud N V and Polak M 2001 *Current Opinion Struct. Biol.* **11** 293-301
[12] Das R, Mills T T, Kwok L W, Maskel G S Millet I S, Doniach S, Finkelstein K D, Herschlag D and Pollak L 2003 *Phys. Rev. Lett.* **90** 188103
[13] Andersen K, Das R, Park H Y, Smith H, Kwok L W, Lamb J S, Kirkland E J, Herschlag D, Finkelstein K D and Pollak L 2004 Phys. Rev. Lett. **93** 248103
[14] Andresen K, Qui X, Pabit S A *et al.* 2008 *Biophys. J.* **95** 287
[15] Qiu X, Kwok L W, Park H Y *et al.* 2008 *Phys. Rev. Lett.* **101** 228101
[16] Manning G S 1978 *Q. Rev. Biophys.* **11** 179-246
[17] Soumpasis D-M 1984 *Proc. Natl. Acad. Sci. USA* **81** 5116-5120
[18] Frank-Kamenetskii M D, Anshelevich V V and Lukashin A V 1987 *Sov. Phys. Usp.* **151** 595-618
[19] Klement R, Soumpasis D-M and Jovin T M 1991 *Proc. Natl. Acad. Sci. USA* **88** 4631-4635
[20] Korolev N, Lubartsev A P, Laaksonen A and Nordenskiöld L 2003 *Nucl. Acids Res.* **31** 5971
[21] Varnai P and Zakrzewska K 2004 *Nucleic Acids Res.* **32** 4269-4280
[22] Ponomarev S Y, Thayer K M and Beveridge D L 2004 *Proc. Natl. Acad. Sci. USA* **101** 14771-14775
[23] Cheng Y, Korolev N and Nordenskiöld L 2006 *Nucl. Acids Res.* **34** 686
[24] Sen S, Andreatta D, Ponomarev S Y, Beveridge D L and Berg M A 2009 *J. Am. Chem. Soc.* **131** 1724
[25] Kittel C 1954 *Introduction to Solid State Physics* (New York: John Wiley and Sons)
[26] Heisler I A, Mazur K and Meech S R 2011 *J. Phys. Chem.* B **115** 1863
[27] Tominaga Y, Shida M, Kubota K, Urabe H, Nishimura Y and Tsuboi M 1985 *J. Chem. Phys.* **83** 5972-5972
[28] Powell J W, Edwards G S, Genzel L, Kremer F and Wittlin A 1987 *Phys. Rev. A.* **35** 3929-3939
[29] Lamba Om P, Wang A H-J, Thomas G J, Jr 1989 *Biopolymers* **28** 667-678
[30] Urabe H, Kato M, Tominaga Y, and Kajiwara K 1990 *J. Chem. Phys.* **92** 768-774
[31] Weidlich T, Lindsay S M, Rui Qi, Rupprecht A, Peticolas W L and Thomas G A 1990 *J. Biomolec. Struct. Dyn.* **8** 139-171
[32] Weidlich T, Powell J W, Genzel L and Rupprecht A 1990 *Biopolymers* **30** 477-480
[33] Weidlich T, Lindsay S M, Peticolas W L and Thomas G A 1990 *J. Biomolec. Struct. Dyn.* **7** 849-858
[34] Perepelytsya S M and Volkov S N 2004 *Ukr. J. Phys.* **49** 1074-1080
[35] Perepelytsya S M and Volkov S N 2007 *Eur. Phys. J.* E **24** 261-269
[36] Perepelytsya S M and Volkov S N 2009 *Biophysical Bulletin* **23(2)** 5-19 arXiv:1007.0374v1
[37] Perepelytsya S M and Volkov S N 2010 *Eur. Phys. J.* E **31** 201-205
[38] Perepelytsya S M and Volkov S N 2010 *Ukr. J. Phys.* **55** 1182-1188
[39] Perepelytsya S M and Volkov S N 2011 *J. Molecular Liquids* **5** 1182-1188
[40] Volkov S N and Kosevich A M 1987 *Mol. Biol.* **21**, 797-806 [Moscow]
[41] Volkov S N and Kosevich A M 1991 *J. Biomolec. Struct. Dyn.* **8** 1069-1083
[42] Volkov S N 1991 *Biopolimery i Kletka* **7** 40-49 [Kiev]
[43] Kosevich A M and Volkov S N 1995 in *Nonlinear excitations in biomolecules* Edited by Peyrard M (New-York, Springer)
[44] Bulavin L A, Volkov S N, Kutovy S Yu and Perepelytsya S M 2007 *Reports of the NAS of Ukraine* **10** 69; arXiv: q-bio.BM/0805.0696v1.
[45] Volkenshtein M V, Eliashevich M A and Stepanov B I 1945 *Vibrations of Molecules. Volume 2* (Moscow)
[46] Ismailov N A 1976 *Electro Chemistry of Solutions* (Moscow: Chemistry)
[47] Gessner R V, Frederick C A, Quigley G J, Rich A and Wang A H 1989 *J.Biol.Chem.* **264** 7921-7935

Multiscale physics of ion-induced radiation damage

Eugene Surdutovich[1, 2] **and Andrey V. Solov'yov**[2,3]
[1] Department of Physics, Oakland University, Rochester, Michigan 48309, USA
[2] Frankfurt Institute for Advanced Studies, Ruth-Moufang-Str. 1, 60438 Frankfurt am Main, Germany
E-mail: surdutov@oakland.edu

Abstract. A multiscale approach to the physics of ion-beam cancer therapy, an approach suggested in order to understand the interplay of a large number of phenomena involved in radiation damage scenario occurring on a range of temporal, spatial, and energy scales, is being reviewed. The scenario is described along with a variety of effects that take place on different temporal, spatial, and energy scales and play major roles in the scenario of interaction of ions with tissue. The understanding of these effects leads to a quantitative assessment of relative biological effectiveness that relates the physical quantities, such as dose, to the biological values, such as the probability of cell survival.

1. Introduction

The development of ion-beam cancer therapy (IBCT) and other applications of ions boosted the scientific interest in obtaining of deeper understanding of radiation damage with ions. A number of important scientific questions related to the assessment of biological damage on the molecular level, have not yet been resolved. Therefore, this field has attracted much attention from the scientific community [1, 2, 3, 4, 5]. The COST [4] Action, "Nano-scale insights in ion beam cancer therapy (Nano-IBCT)" [6], approved in 2010, opened an opportunity for a wide exchange between scientists in different areas specifically devoted to the fundamental studies related to IBCT. Among these studies is the multiscale approach to the assessment of radiation damage induced by irradiation with ions. It is aimed at a phenomenon-based quantitative understanding of the scenario from the incidence of an energetic ion on tissue to the cell death. This method combines many spatial, temporal, and energy scales, and is therefore a truly multiscale approach.

The multiscale approach was formulated in Refs. [5, 7], where a scenario leading to DNA damage was suggested. Then, it was elaborated upon as different aspects of the scenario were added in a series of works [8, 9, 10, 11, 12, 13, 14, 15]. Its name emphasizes the fact that important interactions involved in the scenario happen on a variety of temporal, spatial, and energy scales. Right from the beginning, the approach was formulated as phenomenon-based and was aimed at elucidating the physical, chemical, or biological effects that are important or dominating on each scale in time, space, and energy.

The multiscale approach raised questions about the nature of the effects that define the biological outcome and predict the relative biological effectiveness (RBE) [1, 2] and explain

[3] On leave from A.F. Ioffe Physical Technical Institute, St. Petersburg, Russia.
[4] European framework for Cooperation in Science and Technology

the survival curves that relate the dose with the probability of cell survival. The main issues addressed by the multiscale approach are ion stopping in the medium, the production and transport of secondary electrons produced as a result of ionization and excitation of the medium, the interaction of secondary particles with biological molecules, the most important being DNA, the analysis of induced damage, and the evaluation of the probabilities of subsequent cell survival or death. These effects are happening on time scales ranging from 10^{-21} to 10^{-5} s, i.e., from nuclear to biochemical times. The aim of the physical part of the analysis is the calculation of the spatial distribution of primary DNA damage, including the degree of complexity of this damage.

The scenario of the radiation damage is described in Section 2, then the key effects on consequent time scales are reviewed in Sections 3-5.

2. Multiscale scenario of radiation damage

Radiation damage due to ionizing radiation is initiated by the projectiles incident on tissue. Initially they have energy ranging from a few to hundreds of MeV. In the process of propagation through tissue they transfer their energy in the processes of ionization, excitation, nuclear fragmentation, etc. Naturally, the radiation damage is associated with the transferred energy, and the dose (i.e., absorbed energy density) is a common indicator for the assessment of the damage [1, 2]. In the case of ion therapy, the profile of dose deposition along the ion's path is characterised with a plateau followed by a sharp Bragg peak. The position of this peak depends on the initial energy of the ion and marks the location of the maximum radiation damage. In the process of radiation therapy, a tumour is being "scanned" with the Bragg peak in order to deposit a large dose to the target and spare healthy tissues surrounding it.

However, the deposition of large doses in the vicinity of the Bragg peak does not explain how the radiation damage occurs, since projectiles themselves only interact with a few biomolecules on their way and this direct damage is only a small fraction of overall damage. It is commonly understood that the secondary electrons produced in the process of ionization of the medium with ions are largely responsible for the vast portion of biodamage. Let us discuss the scenario which explains it.

Secondary electrons are produced during a rather short time of $10^{-18} - 10^{-17}$ s following the ion's passage. These energy spectrum of these electrons has extensively been discussed in the literature [16, 17, 18, 8, 9] and the main result (relevant for this discussion) is that the most secondary electrons have energy below 50 eV and only a few (referred to as δ-electrons) have high (> 100 eV) energy. Moreover, this is true for a very large range of ion's energy. This has several important consequences. First, the ranges of propagation of these electrons in tissue are rather small, around 3 nm. Second, the angular distribution is largely uniform [19]; this allows one to consider their transport using random walk approach [5, 12, 20, 21, 22].

The next time scale $10^{-16} - 10^{-15}$ s corresponds to the propagation of secondary electrons in tissue. These electrons (which start with about 45-50 eV energy) are called ballistic. Their initial mean elastic and ionization free paths (in liquid water) are about 0.43 and 3.5 nm, respectively. This means that they ionize a molecule after about seven elastic collisions, while the probability of second ionization is small [8]. Thus the secondary electrons are losing their energy. This happens within 1-1.5 nm of the ion's path. After that they continue propagating elastically scattering with the molecules of the medium until they get bound or solvated electrons are formed. It is important to notice, that these low energy electrons remain important agents for biodamage since they can attach to biomolecules like DNA causing dissociation [23, 24]. The solvated electrons may play the main role in the damage scenario as well [25].

Yet, the energy lost by electrons during the previous stage is transferred to the medium, and as a result of this transfer, the medium becomes very hot [10]. This happens within about $1 - 1.5$-nm cylinder around the ion's path, which we refer to as the hot cylinder. The pressure

Journal of Physics: Conference Series **438** (2013) 012014 doi:10.1088/1742-6596/438/1/012014

inside this cylinder increases by a factor of about 10^3 compared to the atmospheric pressure in the medium outside the cylinder. This pressure builds up by about $10^{-14} - 10^{-13}$ s and it is a source of a cylindrical shock wave [11]. This shock wave propagates through in medium for about $10^{-13} - 10^{-11}$ s. Its relevance to the biodamage is as follows. If the shock wave is strong enough (the strength depends on the distance from the ion's path and the LET), it may inflict damage directly by breaking covalent bonds in a DNA molecule [14]. Besides, the radial collective motion that takes place during this time is instrumental in propagating the highly reactive species such as hydroxyl radicals, just formed solvated electrons, etc. to a larger distances (up to tens of nm) thus increasing the area of an ion's impact.

We assess the primary damage to DNA molecules and other parts of cells due to the above effects. This damage happens during the following 10^{-5} s and consists of various lesions on DNA and other biomolecules. Some of these lesions may be repaired by the living system, but some may not and the latter may lead to the cell death. The whole scenario is shown in Figure 1.

3. Ion stopping and the Bragg peak

The scenario starts with the ion's propagation in a medium. The Bragg peak in the stopping power of massive charged particles is obtained using a version of the Bethe - Bloch formula [26, 27, 28]. This formula provides the dependence of the stopping power on the energy of the ion and practically depends on the mean excitation energy. In the majority of works, liquid water was used as a tissue substitute because human cells on the average consist of 75% water. This energy for liquid water is chosen somewhere between 70 and 80 eV [29, 30]. Our approach to this problem was different; we used the singly-differentiated (with respect to the secondary electron energy) ionization cross sections of water molecules in the medium as a physical input [8, 9]. These cross sections were taken from experiments [31] with parameters tuned for liquid water [9]. The parameterizations were also modified to take into account relativistic effects in the beginning of the ion's path. Even a slight change in cross sections in the entrance channel (plateau region of LET curve) may significantly affect the position of the Bragg peak. Energy loss due to excitation of water molecules has also been included in the calculations [8]. The effect of charge transfer strongly affects the height of the Bragg peak, since it is proportional to the square of the effective charge of the projectile and the latter decreases as the projectile slows down because of the pick off electrons. Barkas's parameterization [32] was used for the calculation of the effective charge. The next effect, included in Ref. [8], was the scattering of projectiles that naturally occurs as they propagate in the medium. This causes the spread in longitudinal velocities or the so-called energy straggling. This widens and substantially diminishes the Bragg peak for the beam of projectiles. Another effect on this scale, not yet included in the multiscale approach, is the nuclear fragmentation that happens quite often when projectiles collide with the nuclei of the medium. This effect is deemed to be significant, especially for the calculations of the tail in the stopping power curve beyond the Bragg peak [30]. As a consequence, we were able to describe the Bragg peak in good agreement with simulations and experiments [8, 9, 33].

An important difference of our approach, relying on the singly-differentiated ionization cross section, allowed us to not only describe the Bragg peak, but also to use these cross sections to calculate the energy spectrum of secondary electrons, which turned out to be quite useful on the following scale of the electron transport and interactions of electrons with the medium.

A new development has been reported in Ref. [15], where a semiempirical model for calculating the electron emission from any organic compound after ion impact was employed in order to extend the calculations to a more realistic medium than liquid water. With the only input of density and composition of the target a quite accurate evaluation of ionization cross sections for such media became possible. Results for protons impacting in the most representative biological targets, such as water or DNA components, showed a good comparison with experimental data.

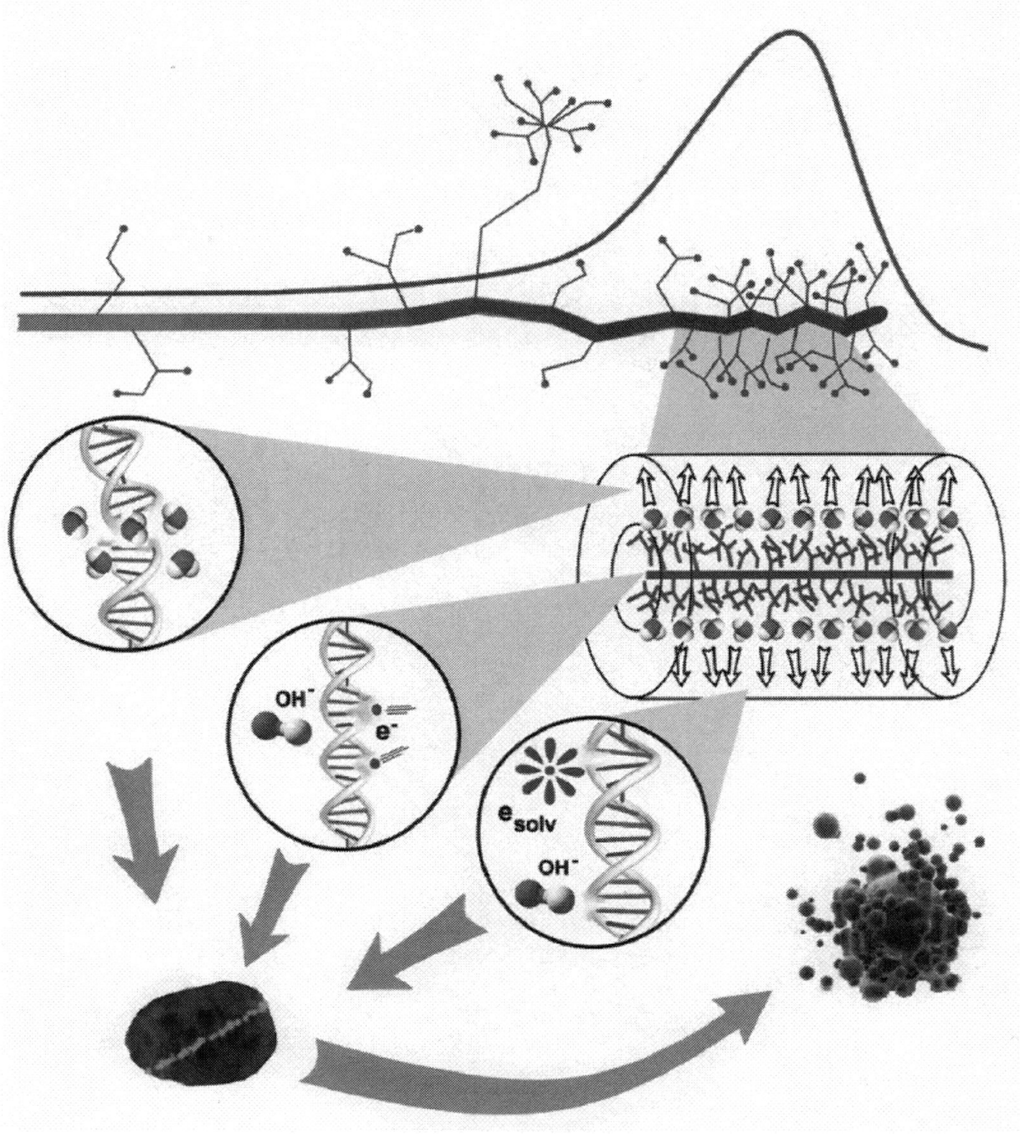

Figure 1. The scenario of biological damage with ions. Ion propagation is shown on top along with production of secondary electrons and the Bragg peak. Then the Bragg peak area is zoomed up to show propagation of secondary electrons within the hot cylinder and then a shock wave in liquid water expanding to the size of large cylinder. Three mechanisms of DNA damage are shown: with energetic water molecules of the shock wave, with ballistic electrons and radicals inside the hot cylinder, and with radicals and solvated electrons at the surface of the large cylinder. The next stage of biological repair is shown as repair protein foci on the background of the cell nucleus. Finally, the last stage is an apoptotic cell, representing a probable cell death.

Due to its simplicity and great predictive effectiveness, the method can be immediately extended to any combination of biological target and charged particle of interest in ion beam cancer therapy. As a result, the Bragg peak as well as the energy spectra of secondary electrons can be calculated for more complicated media than liquid water.

4. Transport of secondary electrons and DNA damage

The next scale in energy and space is related to the transport of the secondary particles, which has been considered in Refs. [5, 34, 12, 22]. These papers were built on the inference from the analysis of the energy distribution of secondary electrons that the average energy of these electrons is about 45-50 eV. At these and lower energies, the ionization cross sections are nearly isotropic, which allows one to employ a random walk approximation, i.e., assume that most of secondary electrons (excluding the more energetic δ-electrons) diffuse out from the ion's path. This diffusion was explored in a series of problems. First, we considered a random walk of electrons from the ion's path to a twist of DNA located at some distance from the path [5]. A single twist of a DNA molecule was chosen because it is an important unit for investigations of double strand breaks (DSBs), significant for the assessment of cell survival. In that study, the number of DSBs per μm of ion's path due to secondary electron action was estimated and compared with biophysical experiments [35].

A later development of that work in Ref. [12] explored a random walk approximation in order to assess the complex DNA damage. Complex DNA damage, investigated by biologists in Refs. [36, 37, 38, 39], is defined as a multiple number of primary DNA lesions happening on a length of two DNA twists. Such a correlation of lesions is called a cluster with a size equal to the number of singly-damaged sites. Different types of lesions may be qualified as effective, e.g., single strand breaks (SSBs), DSBs, base-damages, abasic sites, etc. The size of a cluster is related to the probability of its enzymatic repair. It is deemed that clusters of a sufficiently large size (larger than three or four lesions) are lethal for the cell. Thus, the problem of assessment of complex damage can, on the one hand, be tackled in the same fashion as the problem of DSB, described above; on the other hand, its solution is directly related to the assessment of cell survival as a result of irradiation with ions. The target for a complex damage site is a two-twist segment of a DNA molecule. The original idea for the calculation of complex damage was formulated in Ref. [40]. There, it was suggested that each lesion that counts in a cluster is due to an action of a single agent such as a secondary electron, a radical, etc. Then, if the average number of lesions per target N_i is calculated, the probability of occurrences of clusters of different size, $P_\nu(N_i)$, can be calculated using a Poisson distribution,

$$P_\nu = \exp\left(-N_i\right) \frac{N_i^\nu}{\nu!} \,, \tag{1}$$

where ν is the degree of complexity [40, 12]. The calculation of N_i, were solved using the random walk approach [21]. In that recent work, the random walk calculations of the fluence were successfully compared with the Monte Carlo simulation of transport of secondary electrons.

The radial distribution of clustered damage sites can be obtained using Eq. (1) and the radial dependence of N_i. Given that damage clusters with degree of complexity exceeding a certain number are lethal (this information comes from the biologists [39]), we can obtain the radial distribution of cell survival. Independently, the radial dose distribution around the ion's path can also be calculated using the random walk approach [20]. Then, the survival curve can be obtained as a parametric plot of cell survival on the radial dose [12].

This is an example of how the multiscale approach can be used for assessment of radiation damage. The involved quantities, the cell survival and the radial dose, are calculated independently based on either physical or biological effects. Their dependence is thus obtained phenomenologically rather than empirically.

5. The shock waves and their effects

Thermomechanical effects stem from the highly inhomogeneous dose deposition by ions. The ions lose most of their energy in the processes of ionization and excitation of the molecules of the medium. As a result a large number of electrons and holes are produced in the vicinity of the ion's path. These processes happen within 10^{-16} s. The electrons lose their energies in cascades of collisions and thermalize by about $10^{-15} - 10^{-14}$ s. At this time they transfer their energy to the molecules. The energy in the latter is distributed between vibrational and translational degrees of freedom and by $10^{-14} - 10^{-13}$ s the medium in the vicinity of the ion's trajectory becomes very hot. The inelastic thermal spike model predicts the temperature increases to be above 1000 K [10] for carbon ions in liquid water. Consequently, the pressure above 10 GPa is expected within a cylinder of 1 nm around the ion's path. This pressure initiates a cylindrical shock wave that starts at about 10^{-13} s after the ion's passage [11].

The pressure in the front of the shock wave gradually decreases with time. A simplified hydrodynamical treatment [11] predicted the propagation of these waves until 1 ns after the ion's passage. During this time, the wave front propagates for over 90 nm. In reality, the shock waves lose their strength sooner [13], but still their effects are expected at distances over 10 nm even for carbon ions. In relation to the radiation damage, we are interested in two effects. First is the rapid increase of pressure in the wave front with a consequent sharp decrease in the wake of the wave [11]. The second effect is the mass flow transferred by the shock wave.

The pressure on the wavefront is given by [11],

$$P(t) = \frac{2}{\gamma+1}\rho u_f^2 = \frac{\beta^2}{2(\gamma+1)}\frac{\sqrt{\epsilon\rho}}{t}\,, \tag{2}$$

where t is the time after the ion's traverse, $\gamma = C_P/C_V$ for the medium, u_f is the speed of the wave front, $\beta = 0.863$ is a dimensionless constant related to the geometry of the problem (see Ref. [11] for more detail), $\rho = 1$ g cm^{-3} (for liquid water) is the density of the medium, and ϵ is the fraction of the linear energy transfer (LET) that is carried by the shock wave. The radius, R, of the wave front is given by [11]

$$R(t) = \beta\sqrt{t}\left[\frac{\epsilon}{\rho}\right]^{1/4}\,. \tag{3}$$

From Eqs. (2) and (3) we can obtain the dependence of pressure on the radius of the wave front:

$$P = \frac{\beta^4}{2(\gamma+1)}\frac{\epsilon}{R^2}\,. \tag{4}$$

If a biomolecule, such as DNA, appears to be in the vicinity of the ion's path, forces proportional to the above pressure will act on its parts. These forces can be strong enough to break covalent bonds in the backbone of the DNA and thus contribute to the radiation damage. The question of whether the short-time action of such forces is sufficient to cause DNA strand breaks was analyzed in Refs. [13, 14] using a Molecular Dynamics (MD) simulation of a shock wave propagating through a nucleosome. At small values of LET, the shock waves are not sufficiently strong and the radiation damage is due to chemical effects such as interaction of DNA molecules with free or solvated electrons, free radicals, holes, etc. In Ref. [14], it is shown that for the bonds located about 2 nm from the ion's path, the critical value of LET at which the direct bond breaking by the shock wave starts to dominate over chemical effects is 5 keV/nm. Equation (4) tells us that the same (pressure) forces are achieved for the constant ratio of ϵ/R^2. This allows us to calculate the radius of dominance of the direct damage by shock waves [14]. A LET of about 5 keV/nm is achieved for Ar ions in the vicinity of the Bragg peak.

The value of LET is proportional to the square of the effective charge of the ion. This means that when the radiation damage is assessed for ions heavier than Ar, the direct effect has to be taken into account.

In order to study the transport effects of the shock wave on radiation damage, one should calculate the production of the reactive species such as free radicals and solvated electrons inside the hot cylinder and then calculate their transport by the wave in liquid water. The major anticipated effect will be the "inflation" of the 2-3-nm radius domain, dominated by ballistic electrons, to a cylinder of a radius of 20-30 nm. Damage mechanism predicted in Ref. [25] may be dominant in this region.

6. Cooonclusions

The main difference of the multiscale approach to the radiation damage from other methods of assessment of radiation damage is the in-depth focusing on physical effects. The main advantages of the multiscale approach follow from its architecture and are its fundamentality and versatility. The approach evaluates the relative contributions and significance of a variety of phenomena; it elucidates a complex multiscale scenario in sufficient detail and has a solid predictive power. It is structurally simple and inclusive, and allows for modifications and extensions by including new effects on different scales and improvements on the way.

The above effects lead to the calculation of primary biological damage as a result of irradiation of tissue with ions. At this time, there are many unknown parameters that enter these calculations, but the survival curves predicting the probability of cell survival can still be obtained [12]. There is a hope that the comparison with a variety of experiments will allow to build a reliable model for the predictive calculation of the RBE. Such a calculation will be based on the analysis of key effects that are parts of the scenario of radiation damage rather than on empiric coefficients taken from completely different scenario of damage with photons.

A new understanding of the scenario related to the shock wave effects seriously questions the validity of empirical approaches of assessment of biological effects. The notion of thermomechanical effects represents a paradigm shift in the understanding of radiation damage due to ions. These effects should be also considered for high-density ion beams, irradiation with intensive laser fields, and other conditions prone to causing high gradients of temperature and pressure on a nanometer scale. These predictions call for more studies, both theoretical and experimental. Further developments of radiation therapy including the use of sensitisers require similar understanding of a multiscale scenario.

Acknowledgments

We are grateful to the support of COST Action MP1002 "Nano-scale insights in ion beam cancer therapy."

References

[1] Surdutovich E and Solov'yov A 2012 *J. Phys.: Conf. Ser.* **373** 012001
[2] Schardt D, Elsässer T and Schulz-Ertner D 2010 *Rev. Mod. Phys.* **82** 383–425
[3] Kumar A and Sevilla M 2010 *Chem. Rev.* **110** 7002–7023
[4] Baccarelli I, Gianturco F, Scifoni E, Solov'yov A and Surdutovich E 2010 *Eur. Phys. J. D* **60** 1
[5] Solov'yov A, Surdutovich E, Scifoni E, Mishustin I and Greiner W 2009 *Phys. Rev.* **E79** 011909
[6] Nano-ibct: Nanoscale insights into ion beam cancer therapy (nano-ibct) accessed on 01/2013 URL `http://fias.uni-frankfurt.de/nano-ibct/`
[7] Surdutovich E and Solov'yov A 2009 *Europhys. News* **40/2** 21
[8] Surdutovich E, Obolensky O, Scifoni E, Pshenichnov I, Mishustin I, Solov'yov A and Greiner W 2009 *Eur. Phys. J. D* **51** 63–71
[9] Scifoni E, Surdutovich E and Solovyov A 2010 *Phys Rev. E* **81** 021903
[10] Toulemonde M, Surdutovich E and Solov'yov A 2009 *Phys. Rev. E* **80** 031913
[11] Surdutovich E and Solov'yov A 2010 *Phys. Rev. E* **82** 051915

[12] Surdutovich E, Gallagher D C and Solov'yov A V 2011 *Phys. Rev. E* **84** 051918
[13] Yakubovich A V, Surdutovich E and Solov'yov A V 2012 *Nucl. Instr. Meth. B* **279** 135–139
[14] Surdutovich E, Yakubovich A V and Solov'yov A V 2013 *Sci. Rep.* **3** 1289
[15] de Vera P, Garcia-Molina R, Abril I and Solov'yov A V 2013 *Phys. Rev. Lett.* **110** in press
[16] Pimblott S and LaVerne J 2007 *Rad. Phys. Chem.* **76** 1244–1247
[17] Pimblott S and Siebbeles L 2002 *Nucl. Inst. Meth. B* **194** 237–250
[18] Pimblott S, LaVerne J and Mozumder A 1996 *J. Phys. Chem* **100** 8595–8606
[19] Nikjoo H, Uehara S, Emfietzoglou D and Cucinotta F A 2006 *Radiat. Meas.* **41** 1052
[20] Surdutovich E and Solov'yov A V 2012 *Eur. Phys. J. D* **66** 245
[21] Bug M, Surdutovich E, Rabus H, Rosenfeld A B and Solov'yov A V 2012 *Eur. Phys. J. D* **66** 291
[22] Surdutovich E and Solov'yov A V 2012 *Eur. Phys. J. D* **66** 206
[23] Park Y, Li Z, Cloutier P, Sanche L and Wagner J 2011 *Radiat. Res.* **175** 240–246
[24] Sanche L 2010 Low-energy electron interaction with dna: Bond dissociation and formation of transient anions, radicals, and radical anions *Radical and radical ion reactivity in nucleic acid chemistry* ed Greenberg M (New York: J. Wiley & Sons, Inc.) p 239
[25] Smyth M and Kohanoff J 2012 *J. Am. Chem. Soc.* **134** 9122–9125
[26] Bethe H 1930 *Ann. Phys.* **397** 325–400
[27] Bloch F 1933 *Z. Phys. A: Hadrons Nucl.* **81** 363–376
[28] Bloch F 1933 *Ann. Phys.* **408** 285–320
[29] Abril I, Garcia-Molina R, Denton C, Kyriakou I and Emfietzoglou D 2011 *Radiat. Res.* **175** 247–255
[30] Pshenichnov I, Mishustin I and Greiner W 2008 *Nucl. Inst. Meth. B* **266** 1094–1098
[31] Rudd M E, Kim Y K, Madison D H and Gay T 1992 *Rev. Mod. Phys.* **64** 441
[32] Barkas W H 1963 *Nuclear Research Emulsions I. Techniques and Theory* vol 1 (New York, London: Academic Press)
[33] Surdutovich E, Scifoni E, and Solov'yov A 2010 *Mutat. Res.* **704** 206–212
[34] Scifoni E, Surdutovich E and Solov'yov A 2010 *Eur. Phys. J. D* **60** 115
[35] Tobias F, Durante M, Taucher-Scholz G and Jakob B 2010 *Mutat. Res.* **704** 54–60
[36] Ward J 1988 *Prog. Nucleic Acid. Res. Mol. biol.* **35** 95–125
[37] Ward J 1995 *Radiat. Res.* **142** 362–368
[38] Malyarchuk S, Castore R and Harrison L 2009 *DNA Repair* **8** 1343–1354
[39] Sage E and Harrison L 2011 *Mutat. Res.* **711** 123–133
[40] Surdutovich E, Yakubovich A and Solov'yov A 2010 *Eur. Phys. J. D* **60** 101

Journal of Physics: Conference Series **438** (2013) 012015 doi:10.1088/1742-6596/438/1/012015

Ionization of biomolecular targets by ion impact: input data for radiobiological applications

Pablo de Vera[1,2], Isabel Abril[1], Rafael Garcia-Molina[3] and Andrey V. Solov'yov[2,4]

[1] Departament de Física Aplicada, Universitat d'Alacant, E-03080 Alacant, Spain
[2] Frankfurt Institute for Advanced Studies, Ruth-Moufang-Str. 1, 60438 Frankfurt am Main, Germany
[3] Departamento de Física – Centro de Investigación en Óptica y Nanofísica, Universidad de Murcia, E-30100 Murcia, Spain
[4] On leave from: A.F. Ioffe Physical-Technical Institute, Politechnicheskaya 26, 194021 St. Petersburg, Russia

E-mail: pablo.vera@ua.es

Abstract.

In this work we review and further develop a semiempirical model recently proposed for the ion impact ionization of complex biological media. The model is based on the dielectric formalism, and makes use of a semiempirical parametrization of the optical energy-loss function of bioorganic compounds, allowing the calculation of single and total ionization cross sections and related quantities for condensed biological targets, such as liquid water, DNA and its components, proteins, lipids, carbohydrates or cell constituents. The model shows a very good agreement with experimental data for water, adenine and uracil, and allows the comparison of the ionization efficiency of different biological targets, and also the average kinetic energy of the ejected secondary electrons.

1. Introduction

Ion beam cancer therapy is a powerful emerging tool for treating cancer, specially deep-seated tumours. It makes use of the characteristic energy deposition pattern of ion beams, mainly governed by the electronic stopping force, which is maximum at energies around 100 keV/u, i.e., at the end of the ion's trajectory, giving place to the Bragg peak, a sharp and narrow maximum of the dose near the ion's range. This particular depth-dose profile maximizes the damage produced in tumoral regions, while sparing the energy deposition in surrounding healthy tissues [1].

Although the basic working principles of ion beam cancer therapy are well known, many questions still remain regarding the detailed understanding of the biological outcomes that the physical and chemical stages of the ion propagation in biological media produce. Hadron therapy is, in fact, a complex multiscale problem [2], which involves many different processes in several energy, space and time scales, ranging from nuclear fragmentation reactions until the physical and chemical damage and repair mechanisms of macromolecules such as DNA or proteins, being an important stage the direct electronic excitations and ionizations of these biological molecules by ion impact.

Nowadays, the energy deposition by swift ions in materials of interest in radiotherapy is well known. Particularly, calculations performed within the dielectric formalism [3] have shown to be able of providing accurate values of the electronic stopping force (including many-body and phase effects) in both inorganic (liquid water, metals, semiconductors) [4, 5, 6, 7] and organic (DNA, polymers) materials [8, 9]. These electronic stopping data are needed for detailed simulations of the ion beam propagation through these targets [7, 10]. Nonetheless, one limiting factor of the dielectric formalism is that it includes both electronic excitations and ionizations together, being difficult to extract detailed information on the secondary electron emission. In fact, this information is very relevant, since the major part of the biological damage at a microscopic level is produced by secondary electrons, even the very low energetic ones [11, 12]. Although the dielectric approach has been adapted for describing the electron emission from liquid water induced by ion beams [13, 14], its extension to other relevant biological targets only was achieved very recently [15]. Until then, only methods applicable to atomic targets or small molecules, or limited to some specific targets, projectiles, or energy ranges were available.

In this work we review the main features of the semiempirical method proposed to apply the dielectric formalism to calculate the ionization of complex biological media [15] (section 2), and then we apply it to obtain proton impact ionization data, such as single differential and total cross sections and average kinetic energies of ejected electrons (section 3) of a selection of relevant biological targets. The final conclusions and remarks are given in section 4.

2. Theoretical approach

Several theoretical and semiempirical methods are available to study the electron ejection from atoms and molecules generated by ion impact [16]. Among them, the Binary Encounter Apprximation (BEA) [17] or the Rudd formula [18] are widely used. Nonetheless, these models are restricted to some particular targets or energy ranges: for example, the BEA is quite accurate for atoms or small molecules, whereas its application is difficult for large macromolecules. Moreover, it is not valid for very low ion energies (where two-center effects in the collision are not taken into account) and for high energies (where dipole interactions play a major role and are not considered). On the other hand, the Rudd formula is valid for a wide energy range, but its application is limited to some selected targets for which its parameters have been fitted to experimental data [16]. Regarding other more sophisticated theoretical models, such as quantum models like the Continuum-Distorted-Wave–Eikonal-Initial-State (CDW-EIS) method [19, 20], or the Classical Trajectory Monte Carlo (CTMC) method [21], their application to large biological molecules is computationally quite complex and expensive.

The dielectric formalism (based on the Plane Wave Born Approximation, PWBA) [3] provides a first perturbative theoretical framework in which the electronic magnitudes can be analytically treated for a wide energy range, for various charged projectiles such as protons, heavier ions or electrons, and where many-body and condensed phase effects are taken into account.

2.1. Ionization of the outer shell electrons: dielectric formalism

According to the dielectric formalism, the macroscopic single differential cross section (SDCS), or inverse mean free path, for the ejection of an electron with kinetic energy W from the electronic i-shell of the target, by a projectile of kinetic energy T, mass M_1 and charge Z_1 is [22]

$$\left.\frac{\mathrm{d}\Lambda(T,W)}{\mathrm{d}W}\right|_i = \frac{e^2}{\pi\hbar^2}\frac{M_1 Z_1^2}{T}\int_{k_-}^{k_+}\frac{\mathrm{d}k}{k}\mathrm{Im}\left[\frac{-1}{\epsilon(k,B_i+W)}\right]_i , \tag{1}$$

where E and $\hbar k$ are, respectively, the energy and momentum transfers. We have used $E = B_i + W$, being B_i the binding energy of the i-shell. The excitation spectrum of the target enters in this equation through the quantity known as the energy-loss function (ELF),

Table 1. Physical properties of the biological targets studied in this work. Atomic compositions have been converted from mass percentage compositions from Ref.[27] (except otherwise stated), Z_2 and A_2 are the sum of the atomic and mass numbers of all the atoms in each formula, while $\mathcal{Z}_2$ is Z_2 divided by the number of atoms in the formula. Densities and $\overline{B}$ values were set equal to 1.00 g/cm^3 and 20 eV, respectively, when they were unknown.

Material (formula)	Z_2	A_2	$\mathcal{Z}_2$	ρ (g/cm^3)	$\overline{B}$ (eV)
Liquid water (H_2O)	10	18.02	3.33	1.00	18.13 [13]
Adenine ($C_5H_5N_5$)	70	135.13	4.67	1.35 [24]	20.44 [37]
Guanine ($C_5H_5O_1N_5$)	78	151.13	4.87	1.58 [24]	21.09 [37]
Thymine ($C_5H_6O_2N_2$)	66	126.11	4.40	1.36 [31]	20.29 [37]
Cytosine ($C_4H_5O_1N_3$)	58	111.10	4.46	1.30 [24]	20.47 [37]
Uracil ($C_4H_4O_2N_2$)	58	112.09	4.83	1.40 [31]	20.73 [37]
DNA backbone ($C_5H_{10}O_5P_1$)	95	181.10	4.52	1.00	19.14 [37]
DNA ($C_{20}H_{27}O_{13}N_7P_2$) [28]	330	635.41	4.78	1.35 [28]	20.00
Protein ($C_{142.56}H_{209.97}O_{44.09}N_{38.92}S_1$)	1706.49	3206.50	3.91	1.35	20.00
Lipid ($C_{9.44}H_{17.18}O_1$)	81.87	146.78	2.96	0.92	20.00
Carbohydrate ($C_{1.20}H_{2.00}O_1$)	17.21	32.45	4.10	1.56	20.00
Cell nucleus ($C_{60.07}H_{843.06}O_{371.77}N_{18.31}P_{6.73}S_1$)	4422.75	8016.25	3.40	1.00	20.00
Cytoplasm ($C_{1926.40}H_{8105.31}O_{2724.83}N_{138.76}P_1S_{5.80}Na_{3.71}Cl_{3.49}K_{4.16}$) [29]	42720.60	77434.50	3.31	1.00	20.00

$\mathrm{Im}\left[-1/\epsilon(k, B_i + W)\right]$, where $\epsilon(k, B_i + W)$ is the complex dielectric function of the target. The sub-index i in eq.(1) refers to the contribution of the i electronic shell to the ELF. The integration limits, imposed by conservation laws, are $k_\pm = \sqrt{2M_1}(\sqrt{T} \pm \sqrt{T-E})$. The macroscopic cross section, Λ, is related to the microscopic one, σ, through $\Lambda = \mathcal{N}\sigma$, being $\mathcal{N}$ the molecular density of the target.

Once the SDCS is known, other key radiobiological quantities can be calculated from it, such as the total ionization cross section (TICS)

$$\Lambda(T) = \int_0^\infty \frac{\mathrm{d}\Lambda(T,W)}{\mathrm{d}W}\mathrm{d}W \; , \tag{2}$$

and the average energy of the emitted electrons

$$W_{\mathrm{average}}(T) = \frac{1}{\Lambda(T)}\int_0^\infty W\frac{\mathrm{d}\Lambda(W)}{\mathrm{d}W}\mathrm{d}W \; . \tag{3}$$

The advantage of the dielectric formalism is that, provided that the ELF is experimentally known, many-body interactions and target physical-state effects are naturally included in the subsequent calculations. Therefore, the key magnitude to compute the ionization cross sections is the target ELF over the whole energy and momentum transfer plane (the so-called Bethe surface). In the following section we discuss how to obtain the ELF of biological materials in a simple and efficient manner. Once the ELF is known, we need a criterion to split it in excitations and ionizations, which will be explained in section 2.3.

2.2. The energy-loss function of bioorganic targets

The ELF for all the electronic excitations and ionizations of a target can be experimentally determined from optical data (zero momentum transfer, $k = 0$) or X-ray and electron energy-loss experiments ($k \neq 0$). Optical data are available for a large number of materials, among which only a few are organic or biological compounds.

Regarding the latter ones, Tan and co-workers [23, 24] noticed that the optical-ELF of 13 bioorganic condensed compounds and liquid water at low energies (corresponding to the excitation of the outer shell electrons) are rather similar, with an intense peak around 20–25 eV, and sometimes with other small peaks at lower energies. This fact led them to the idea that the experimental optical ELF of the outermost electrons of organic compounds can be parametrized with a single-Drude function

$$\mathrm{Im}\left[\frac{-1}{\epsilon(k=0,E)}\right] = \frac{a(Z_2)E}{[E^2 - E_\mathrm{p}^2(\mathcal{Z}_2)]^2 + \gamma^2(\mathcal{Z}_2)E^2}\,, \tag{4}$$

where $a(Z_2)$ (in eV^3), $E_\mathrm{p}(\mathcal{Z}_2)$ (in eV) and $\gamma(\mathcal{Z}_2)$ (in eV) represent the intensity, position and width of the single-Drude ELF. Using the experimental optical ELF of the 14 bioorganic compounds, $E_\mathrm{p}(\mathcal{Z}_2)$ and $\gamma(\mathcal{Z}_2)$ were parametrized as a function of the mean atomic number of the target, $\mathcal{Z}_2$ (i.e., the number of electrons per formula divided by the number of atoms) [23],

$$E_\mathrm{p} = 19.927 + 0.9807 \cdot \mathcal{Z}_2\,, \tag{5}$$
$$\gamma = 13.741 + 0.3215 \cdot \mathcal{Z}_2\,. \tag{6}$$

The remaining parameter, $a(Z_2)$, can be simply obtained by imposing the accomplishment of the f-sum rule [25], linked to the number of electrons in the target, Z_2 (i.e., the sum of the atomic numbers of all the atoms in the formula); of course, the inner shells also contribute to the number of electrons. In this parametric approach, eq.(4) is used for transferred energies below 40 eV, and above 50 eV ELF is obtained from X-ray atomic scattering factors [26], ensuring that the inner-shell electrons are considered for the calculation of the parameter $a(Z_2)$. Between 40 and 50 eV, a parabola is used to link Eq.(4) with the X-ray data (see Ref.[23] for a detailed explanation).

This method allows the prediction of the optical ELF for an arbitrary bioorganic compound in the optical limit ($k = 0$), provided that the atomic composition (for calculating $\mathcal{Z}_2$) and density (which affects the f-sum rule) are known. This information is easily accessible for any material. Specifically, the average composition and density of a large number of biological targets is compiled in the ICRU Report 46 [27]. We summarize in Table 1 the physical properties of the representative biological materials studied in this work, which have been obtained from Ref. [27], except for DNA [28] and the cytoplasm [29]. The average mass composition of each target has been converted to number of atoms per formula, assigning 1 atom to the less abundant component, in order to calculate the values of Z_2, A_2 and $\mathcal{Z}_2$. In Fig. 1 we show the predicted optical ELFs for liquid water and solids dry DNA, adenine, guanine, thymine and uracil, together with the available experimental data [28, 30, 31, 32, 33]. As it can be seen, the agreement is very good in practically all cases, except for liquid water, where the asymmetry of the experimental ELF can not be reproduced with a single-Drude function.

Once the ELF is known in the optical limit, we need to extend it to $k \neq 0$ in order to obtain the whole Bethe surface. There are several methods available to obtain the ELF at finite values of the momentum transfer [34]. Among them, we use here a simple dispersion relation introduced by Ritchie and Howie, derived from the similarity between the Drude function (eq.(4)) at $k = 0$ and $\gamma = 0$ with the Lindhard function for the free electron gas [35], which simply consists in replacing E_p in eq.(4) by

$$E_\mathrm{p}(k) = E_\mathrm{p}(k=0) + \alpha\frac{\hbar^2k^2}{2m_\mathrm{e}} \approx E_\mathrm{p}(k=0) + \frac{\hbar^2k^2}{2m_\mathrm{e}}\,, \tag{7}$$

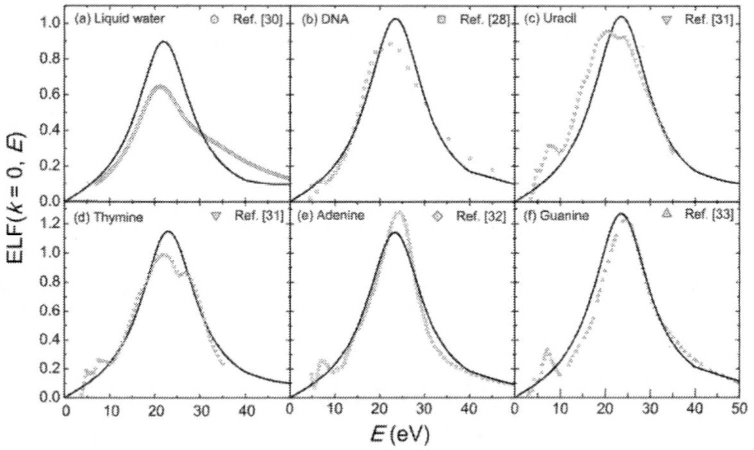

Figure 1. Comparison of the optical-ELF obtained through the parametric model of Eq.(4) (solid lines) with available experimental data (symbols) for: (a) liquid water [30], (b) solid DNA [28], (c) solid uracil [31], (d) solid thymine [31], (e) solid adenine [32], and (f) solid guanine [33].

where m_1 is the electron mass and $\alpha = 6E_{\rm Fermi}/5E_{\rm plasmon}$, being $E_{\rm Fermi}$ and $E_{\rm plasmon}$ the nominal Fermi and plasmon energies of the free electron gas [36]. $\alpha \approx 1$ for liquid water [34], and we have kept this value for the rest of biomaterials for the sake of simplicity. No dispersion relation was assumed for the damping coefficient γ. Although this extension of the ELF to $k \neq 0$ is only an approximation, its accuracy is in the order of other approximations made in this work.

2.3. Ionization of the outer-shell and inner-shell electrons

The SDCS given by eq.(1) is not only valid for the ionization of the i-shell, but it can also account for excitations. It can describe the ionization of an electronic shell i provided that we know the contribution of the ionization of this shell to the total ELF. Since the ELF was obtained in the previous section from an empirical parametrization to experimental ELFs, which include all the possible excitations and ionizations of the target, we need some criterion to distinguish the ionizations from the excitations and evaluate the electron production cross sections.

One case already studied is liquid water: its ELF has been parametrized in contributions of excitations and ionizations of different shells in the works by Dingfelder [13] and Emfietzoglou [14]. Basically, these authors used the experimental excitation spectrum, represented by the imaginary part of the dielectric function, and fitted its different peaks by normal and derivative Drude functions in order to identify and reproduce each resonance, making use of information on the excitation and ionization energies of each shell.

This procedure is satisfactory for water, which only has four outer molecular orbitals. Nonetheless, the generalization of such a parametrization for large and complex macromolecules is not a trivial issue. For these reason, we introduce here a further approximation. If one looks at the existing parametrizations performed for liquid water [13, 14], it can be seen that the excitations are practically confined at low transferred energies, while at high energies only ionizations remain. Therefore, we can estimate a mean binding energy, $\overline{B}$, from the ionization thresholds of all the outer electronic shells, and assume that ionizations will only occur at energies above this threshold (the electron being ejected with an energy $W = E - \overline{B}$), and excitations only below it. This mean binding energy is 18.13 eV for liquid water [13]; in fact, a few electronvolts above this energy excitations practically vanish, while only ionizations remain. Moreover, the

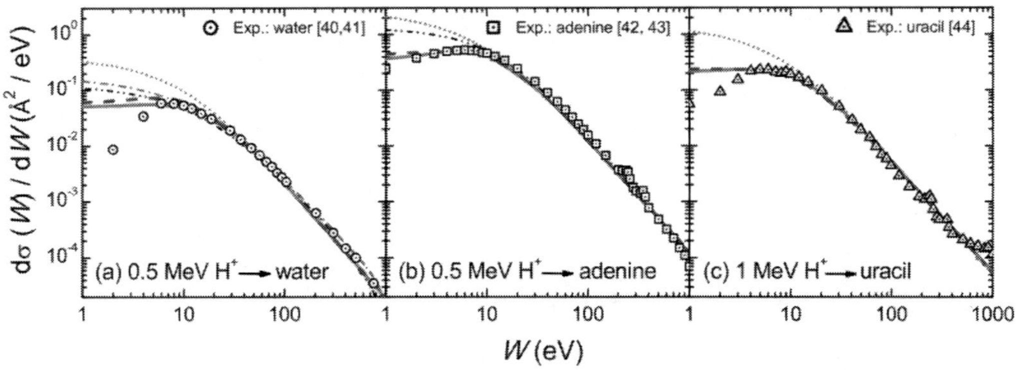

Figure 2. SDCS for ionization of (a) water, (b) adenine and (c) uracil by protons of a given energy. Symbols represent experimental data [40, 41, 42, 43, 44], while solid (dashed) lines are our calculations within the dielectric formalism using the experimental (parametrized) optical ELF of the target, as described in the text. Dotted lines are calculations performed within the BEA model, while the dash-dotted line is the result of the Rudd formula for liquid water. The dash-dot-dotted lines are other calculations for the water molecule (Classical Trajectory Monte Carlo method [21]) and adenine molecule (quantum mechanical calculations [20]).

use of such a mean binding energy is justified, since the difference between the actual binding energies of each shell and the mean value will be of the order of a few electronvolts, which is negligible for the further transport of the secondary electrons. Finally, it will be seen in section 3 that this approximation yields good results in comparison with the available experimental data.

Information on the ionization energies of the different shells of biological molecules can be easily obtained from the literature, from databases or from quantum chemistry calculations. In particular, there is information available for liquid water [13], DNA components [37], amino acids [38] and other organic molecules [39]. From these data we have calculated the mean binding energies of the targets studied in this work, which are collected in Table 1. Since the values of $\overline{B}$ are always close to 20 eV, we have set this value for the complex targets for which we do not have direct information, such as the DNA molecule, proteins or the cell nucleus.

Although the major contribution to the ionization cross section comes from the outer shells, the inner shells can also have noticeable effects in the calculation of some quantities, such as the average energy of ejected electrons, eq.(3). So far we have dealt only with outer shell electrons. For the inner shells, which can be regarded as isolated cores, where chemical and phase effects are less important, simpler theoretical models such as the BEA can be used instead. For describing the SDCS of the inner shells, in this work we have used a simple BEA equation [16], where the binding and mean kinetic energies of each shell, B_i and U_i respectively, have been obtained from a database by the National Institute of Standards and Technology (NIST) [39].

3. Results and discussion

We have calculated ionization cross sections for proton impact in several representative biological targets using the models described above, i.e., the dielectric formalism with the ELF parametrization of bioorganic compounds and the mean ionization energy for the outer shell electrons, and the BEA for the inner shell electrons. The compositions, densities and mean binding energies of the targets are summarized in Table 1.

Figure 2 shows the SDCS calculated with the proposed model for protons of a given energy in liquid water, solid adenine and solid uracil, compared with available experimental data for

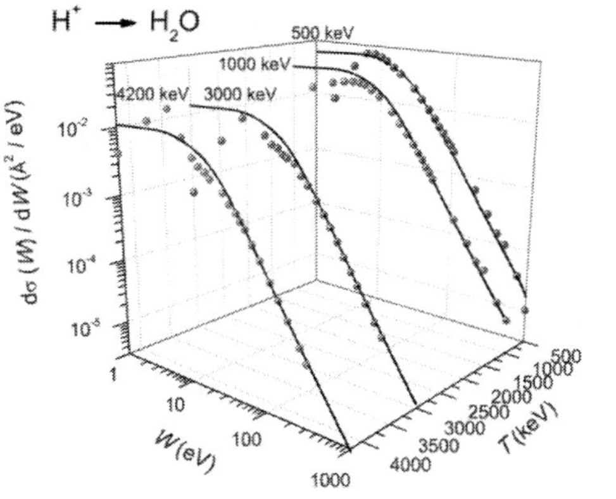

Figure 3. Single differential cross sections (SDCS) for ionization of liquid water by protons of several energies obtained with the present model. Symbols represent experimental data for water vapour [40, 41].

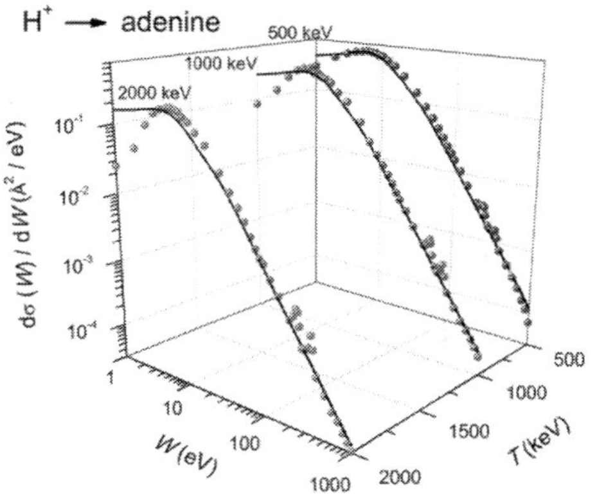

Figure 4. Single differential cross sections (SDCS) for ionization of solid adenine by protons of several energies obtained with the present model. Symbols represent experimental data for adenine vapour [42, 43].

the corresponding gas phases [40, 41, 42, 43, 44]. Although the calculations are performed for condensed phases, we do not know experimental data for condensed targets with which we can compare. Nonetheless, the agreement with the gas data is excellent at almost all ejection energies, except for some differences at very low electron energies. These differences could be due to phase effects or to the extreme difficulty of experimentally determining these energies.

In Fig. 2 we plot by solid lines the calculations performed using the experimental optical ELF of these materials, while dashed lines depict the calculations done using the predicted ELF by the parametric model of Ref. [23]. The comparison of both calculations allows us to estimate the error coming from the ELF prediction. As it can be seen, the differences are very small for adenine and uracil. Larger differences are observed in the case of liquid water at very low energies, where the parametric model is not able of reproducing the asymmetry of the experimental ELF. In this case, the error in the maximum of the SDCS is around 30 %, which is not so big taking into account the predictive power of the method. In any case, it is preferable to use the experimental ELF when it is available, like in the cases shown in Fig. 1, being the method in this case more reliable. We show in Figs. 3 and 4 our calculated SDCS for proton beams at several energies in liquid water and solid adenine, respectively, using their experimental ELF, finding an overall agreement with experiments.

The comparison of our calculations with other models also gives support to our methodology. In Fig. 2 we also show by dotted lines the SDCS calculated for the gas phases within the BEA model, while the dash-dotted line of Fig. 2(a) is the SDCS obtained with the Rudd formula for water vapour and the dash-dot-dotted lines of Fig. 2(a) and (b) are *ab initio* calculations, within the Classical Trajectory Monte Carlo method for water [21], and quantum mechanical calculations for adenine [20]. The agreement between all the models and experiments is rather good at high energies, but at low energies, say below 10–20 eV, our calculations seem to agree better with the experimental data. The causes of these discrepancies (especially with the *ab initio* calculations) deserve more attention, but a possible explanation of our better agreement is that we are considering a more realistic target excitation spectrum through its ELF.

In Fig. 5 we show our calculated TICS (eq. (2)) for all the materials of Table 1: Fig. 5(a) shows the macroscopic cross sections (i.e., inverse mean free paths) for liquid water, DNA, protein, lipid, carbohydrate, cytoplasm and the cell nucleus, while Fig. 5(b) shows the microscopic cross sections (i.e., TICS per molecule) of the DNA/RNA molecular components adenine, thymine, cytosine, guanine, uracil

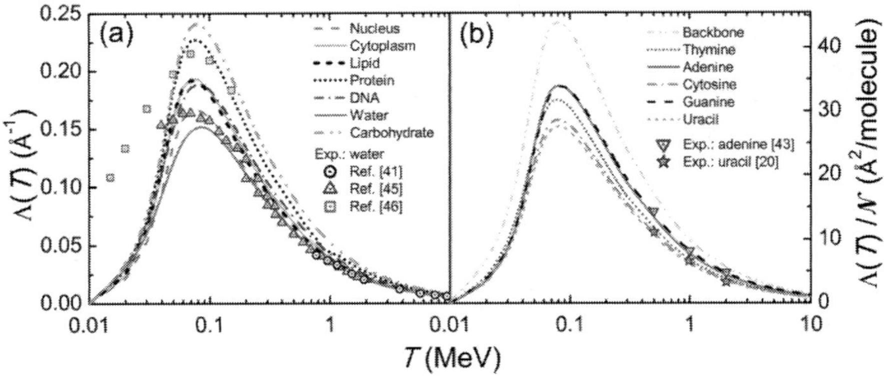

Figure 5. (a) Macroscopic TICS (inverse mean free path), and (b) microscopic TICS per molecule for protons in several biological materials in the condensed phase, compared with experimental data in water vapour [41, 45, 46], adenine vapour [43] and uracil vapour [20].

and sugar-phosphate backbone. Also shown in these figures are experimental data for the water, adenine and uracil molecules in the gas phase [20, 41, 43, 45, 46]. As it can be seen, our calculations nicely agree with the experiments, except for some differences around 100 keV in the case of water measurements from Refs. [45] and [46]; the slight disagreement with the measurements by Rudd *et al.* [45] could be due to phase effects, and also to the fact that the dielectric formalism (which is based on the first Born approximation) starts to lose validity at these low energies. In any case, the dispersion between the measurements by Rudd *et al.* [45] and by Bolorizadeh *et al.* [46] is remarkable.

The information depicted in Fig. 5 is quite interesting, since it shows the different sensitivities of each target to ion impact ionization. Among the macroscopic targets shown in Fig. 5(a), it seems that carbohydrates and protein has the largest TICS, probably influenced by their bigger densities (see Table 1). Regarding the cell nucleus, it has a TICS which is practically identical to that of liquid water in almost all the energy range, but with slight differences in the maximum. Nonetheless, these differences are too small (within the uncertainties of the ELF prediction for liquid water) that it is not possible to conclude that the cell nucleus is more sensible than liquid water itself. Nonetheless, the cytoplasm, lipid and DNA show TICS larger than liquid water and the cell nucleus, a fact that could indicate a larger ionization probability of these targets. It is remarkable the possible effect of the larger density of the DNA, but here lipids and the cytoplasm are considered to have densities similar to that of liquid water. Among the microscopic TICS shown in Fig. 5(b), now the density is not a factor (TICS per molecule are shown), and we still observe differences in their cross sections, being remarkable that the DNA sugar-phosphate backbone shows the largest ionization probability. Perhaps, in this case, the dominant factor is the number of electrons per molecule.

Other interesting quantity which can be calculated for several biological media is the average energy of secondary electrons. According to eq.(3), the high energy transfers will be more relevant in this calculations, so ionization of the inner shells (introduced here through the BEA) will play a noticeable role. In Fig. 6 we show our results for seven representative biological targets: liquid water, dry DNA, protein, lipid, cell nucleus, cytoplasm, and carbohydrate. We also include other calculation for liquid water from Ref. [47], which agrees quite well with our water curve. Our results indicate that the differences in W_{average} between these targets is not very significant, spanning from the maximum of liquid water and the minimum of proteins, which are around 55 and 45 eV, respectively, at 1 MeV proton impact.

4. Summary and conclusions

In this work we review, explain in more detail and expand our semiempirical method proposed to evaluate the ion impact ionization cross sections for condensed complex biological targets based on the dielectric

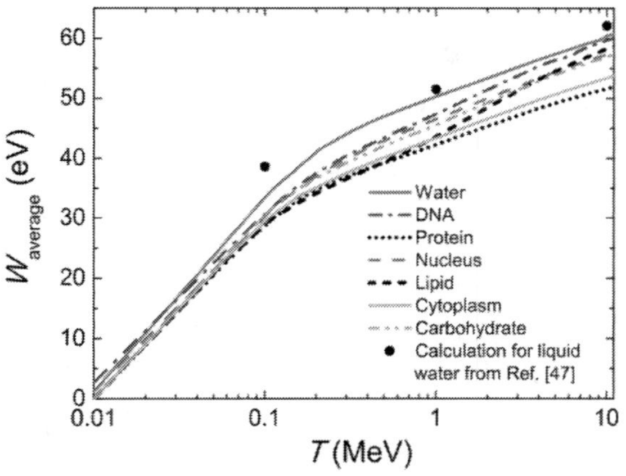

Figure 6. Average kinetic energy of the secondary electrons ejected from different biological materials by proton impact, calculated with our semiempirical model for the outer shell electrons, and with the BEA for the inner shells.

formalism and a semiempirical parametrization of the experimental optical ELF of organic compounds [15]. Calculations are extended to a broad selection of representative biological targets (liquid water, DNA, proteins, lipids, carbohydrates, the cell nucleus and cytoplasm, and the DNA/RNA molecular components adenine, guanine, cytosine, thymine, uracil and sugar-phosphate backbone), including the contribution from the inner shells to properly perform the calculation of the average energy of secondary electrons, besides the single differential and total cross sections. Our calculations show an excellent agreement with experimental data in the gas phase for water, adenine and uracil, and they also agree with other semiempirical, theoretical and *ab initio* calculations at high ejection energies. At low energies, our model seems to agree better than the others with the experimental data. We evaluate ionization cross sections and average energies of secondary electrons for all the cited targets, showing some differences between their ionization efficiency, but a quite similar behaviour of the average energy of secondary electrons. This method will be extended in future works to account for heavier ions in different charge states, and also for electron impact.

Acknowledgments

The authors want to thank fruitful discussions and helpful data provided by Drs. E. Surdutovich, E. Scifoni, D. Emfietzoglou and C. Champion. We recognize the financial support from the Spanish Ministerio de Economía y Competitividad and the European Regional Development Fund within the Project FIS2010–17225 and from the Conselleria d'Educació, Cultura i Esport de la Generalitat Valenciana for granting PdV within the VALi+d program. Support from the COST Action MP1002 NanoIBCT is greatfully acknowledged.

[1] Kraft G 2000 *Progr. Part. Nucl. Phys.* **45** S473.
[2] Solov'yov A V, Surdutovich E, Scifoni E, Mishustin I and Greiner W 2009 *Phys. Rev. E* **79** 011909.
[3] Lindhard J 1954 *K. Dan. Vidensk. Selsk. Mat. Fys. Medd.* **28** (8).
[4] Heredia-Avalos S, Garcia-Molina R, Fernández-Varea J M and Abril I 2005 *Phys. Rev. A* **72** 052902-1.
[5] Heredia-Avalos S, Moreno-Marín J C, Abril I and Garcia-Molina R 2005 *Nucl. Instr. Meth. B* **230** 118.
[6] Garcia-Molina R, Abril I, Denton C D, Heredia-Avalos S, Kyriakou I and Emfietzoglou D 2009 *Nucl. Instr. Meth. B* **267** 2647.
[7] Garcia-Molina R, Abril I, Heredia-Avalos S, Kyriakou I and Emfietzoglou D 2011 *Phys. Med. Biol.* **56** 6475.
[8] Abril I, Garcia-Molina R, Denton C D, Kyriakou I and Emfietzoglou D 2011 *Rad. Res.* **255** 247.
[9] de Vera P, Abril I and Garcia-Molina R 2011 *J. Appl. Phys.* **109** 094901.

[10] de Vera P, Abril I and Garcia-Molina R 2013 *Appl. Rad. Isot.*, in press. DOI: 10.1016/j.apradiso.2013.01.023
[11] Boudaïffa B, Cloutier P, Hunting D, Huels M A and Sanche L 2000 *Science* **287** 1658.
[12] Pan X, Cloutier P, Hunting D and Sanche L 2003 *Phys. Rev. Lett.* **90** 208102.
[13] Dingfelder M, Hantke D, Inokuti M and Paretzke H G 1998 *Rad. Phys. Chem.* **53** 1.
[14] Emfietzoglou D 2003 *Rad. Phys. Chem.* **66** 373.
[15] de Vera P, Garcia-Molina R, Abril I and Solov'yov A V 2013 *Phys. Rev. Lett.*, in press (2013).
[16] International Commission on Radiation Units and Measurements 1996 *Secondary Electron Spectra from Charged Particle Interactions* (ICRU Report 55) (Bethesda, MD: ICRU).
[17] Vriens L 1967 *Proc. Phys. Soc. London* **90** 935.
[18] Rudd M E, Kim Y-K, Madison D H and Gay T J 1992 *Rev. Mod. Phys.* **64** 441.
[19] Crothers D S F and McCann J F 1983 *J. Phys. B: At. Mol. Phys.* **16** 3229.
[20] Galassi M E, Champion C, Weck P F, Rivarola R D, Fojón O and Hanssen J 2012 *Phys. Med. Biol.* **57** 2081.
[21] Errea L F, Illescas C, Méndez L and Rabadán I 2013 *Phys. Rev. A* **87** 032709.
[22] Emfietzoglou D 2012 Inelastic scattering of charged particles in condensed media: a dielectric theory perspective, Section III in *Interaction of Radiation with Matter*, edited by H. Nikjoo, S. Uehara, and D. Emfietzoglou (Boca Raton: CRC Press).
[23] Tan Z, Xia Y, Zhao M, Liu X, Li F, Huang B and Ji Y 2004 *Nucl. Instr. Meth. B* **222** 27.
[24] Tan Z, Xia Y, Zhao M and Liu X 2006 *Nucl. Instr. Meth. B* **248** 1.
[25] Altarelli M and Smith D Y 1974 *Phys. Rev. B* **9** 1290.
[26] Henke B L, Gullikson E M and Davis J C 1993 *At. Data Nucl. Data Tables* **54** 181. The ASCII files for the scattering factors of the different elements can be downloaded from `http://henke.lbl.gov/optical_constants/asf.html`
[27] International Commission on Radiation Units and Measurements 1992 *Photon, Electron, Proton and Neutron Interaction Data for Body Tissues* (ICRU Report 46) (Bethesda, MD: ICRU).
[28] Inagaki T, Hamm R N, Arakawa E T and Painter L R 1974 *J. Chem. Phys.* **61** 4246.
[29] Byrne H L, McNamara A L, Domanova W, Guatelli S and Kuncic Z 2013 *Phys. Med. Biol.* **58** 1251.
[30] Hayashi H, Watanabe N, Udagawa Y and Kao C C 2000 *Proc. Nat. Acad. Sci. USA* **97** 6264.
[31] Isaacson M 1972 *J. Chem. Phys.* **56** 1803.
[32] Emerson L C, Williams M W, Tang I'lan, Hamm R N and Arakawa E T 1975 *Radiat. Res.* **63** 235.
[33] Arakawa E T, Emerson L C, Juan S I, Ashley J C and Williams M W 1986 *Photochem. Photobiol.* **44** 349.
[34] Garcia-Molina R, Abril I, Kyriakou I and Emfietzoglou D 2012 Energy Loss of Swift Protons in Liquid Water: Role of Optical Data Input and Extension Algorithms, Chapter 15 in *Radiation Damage in Biomolecular Systems*, edited by G. García Gómez-Tejedor and M. C. Fuss (Dordrecht: Springer).
[35] Ritchie R H and Howie A 1977 *Philos. Mag.* **36** 463.
[36] Ashcroft N W and Mermin N D 1976 *Solid State Physics* (Fort Worth: Saunders College Publishing).
[37] Bernhardt Ph and Paretzke H G 2003 *Int. J. Mass Spectr.* **223-224**.
[38] Peudon A, Edel S and Terrisol M 2006 *Radiat. Prot. Dos.* **122** 128.
[39] Kim Y-K, Irikura K K, Rudd M E, Ali M A, Stone P M, Chang J, Coursey J S, Dragoset R A, Kishore A R, Olsen K J, Sansonetti A M, Wiersma G G, Zucker D S and Zucker M A 2004 *Electron-Impact Ionization Cross Section for Ionization and Excitation Database* (version 3.0) (Gaithersburg, MD: National Institute of Standards and Technology). Available online: `http://www.nist.gov/pml/data/ionization/index.cfm`
[40] Toburen L H and Wilson W E 1977 *J. Chem. Phys.* **66** 5202.
[41] Wilson W E, Miller J H, Toburen L H and Manson S T 1984 *J. Chem. Phys.* **80** 5631.
[42] Iriki Y, Kikuchi Y, Imai M and Itoh A 2011 *Phys. Rev. A* **84** 032704.
[43] Iriki Y, Kikuchi Y, Imai M and Itoh A 2011 *Phys. Rev. A* **84** 052719.
[44] Iriki Y, Kikuchi Y, Doi K, Imai M and Itoh A 2012 *J. Phys.: Conf. Series* **388** 102041.
[45] Rudd M E, Goffe T V, DuBois R D and Toburen L H 1985 *Phys. Rev. A* **31** 492.
[46] Bolorizadeh M A and Rudd M E 1986 *Phys. Rev. A* **33** 888.
[47] Pimblott S M, LaVerne J A 2007 *Radiat. Phys. Chem.* **76** 1244.

International Conference on Dynamics of Systems on the Nanoscale (DySoN 2012) IOP Publishing
Journal of Physics: Conference Series **438** (2013) 012016 doi:10.1088/1742-6596/438/1/012016

Radiation damage of biomolecules (RADAM) database development: current status

S Denifl[1], G Garcia[2,3], BA Huber[4], BP Marinković[5], N Mason[6], J Postler[1], H Rabus[7], G Rixon[6], AV Solov'yov[8,*], E Suraud[9], AV Yakubovich[8,*]

[1]Institut für Ionenphysik und Angewandte Physik, Technikerstr. 25 / 3, A-6020 Innsbruck

[2]Instituto de Física Fundamental, Consejo Superior de Investigaciones Científicas,(CSIC) Serrano 113-bis, 28006 Madrid, Spain

[3]Centre for Medical Radiation Physics, University of Wollongong, NSW 2522, Australia

[4]Centre de Recherche sur les Ions, la Matière et la Photonique(CIMAP), Bd Henri Becquerel, 14070 Caen Cedex 05, France

[5]Institute of Physics, University of Belgrade, Pregrevica 118, 11080 Belgrade and Advanced School for Electrical Engineering and Informatics, 11030 Belgrade, Serbia

[6]Open University, Faculty of Science, Walton Hall, Milton Keynes, MK7 6AA, UK

[7]Physikalisch-Technische Bundesanstalt, Bundesallee 100, 38116 Braunschweig, Germany

[8]Frankfurt Institute for Advanced Studies, Ruth-Moufang Str. 1, 60438, Frankfurt am Main, Germany

[9]Laboratoire de Physique théorique, IRSAMC, CNRS, Université Paul Sabatier, 118 Routre de Narbonne, 31062 Toulouse cedex, France

[*]On leave from A.F. Ioffe Physical-Technical Institute, Polytechnicheskaya 26, 194021, St. Petersburg, Russia

E-mail: nano-ibct@fias.uni-frankfurt.de

Abstract. Ion beam therapy offers the possibility of excellent dose localization for treatment of malignant tumours, minimizing radiation damage in normal tissue, while maximizing cell killing within the tumour. However, as the underlying dependent physical, chemical and biological processes are too complex to treat them on a purely analytical level, most of our current and future understanding will rely on computer simulations, based on mathematical equations, algorithms and last, but not least, on the available atomic and molecular data. The viability of the simulated output and the success of any computer simulation will be determined by these data, which are treated as the input variables in each computer simulation performed. The radiation research community lacks a complete database for the cross sections of all the different processes involved in ion beam induced damage: ionization and excitation cross sections for ions with liquid water and biological molecules, all the possible electron - medium interactions, dielectric response data, electron attachment to biomolecules etc. In this paper we discuss current progress in the creation of such a database, outline the roadmap of the project and review plans for the exploitation of such a database in future simulations.

1. Introduction

The damaging effect of ionizing radiation has been known for many years. It being commonly accepted that high-energy tracks formed by α, β, γ radiation and atomic ions ionize cell components along the track, thereby leading to various dissociation channels and to the formation of damaging radicals. This knowledge has triggered intensive research on radiation protection and the development of intensive biomedical uses of different radiation, generally called radiotherapy, used especially for tumour-based diseases. In this field, clinical trials and technological advances often develop more rapidly than our understanding of the underlying physico-chemical processes. Thus curative protocols are established on an empirical basis based on experience and expertise of the personnel involved, rather than on pure scientific knowledge.

Ion Beam Cancer Therapy (IBCT, or hadron therapy) represents a new and effective method for providing high-dose delivery directly into tumors, thereby maximizing killing of the cancer cells while simultaneously minimizing the radiation damage to surrounding healthy tissue.However, the full potential of Ion Beam Cancer Therapy can only be realized by a better understanding of the physical, chemical and biological mechanisms, on a range of time and space scales that lead to cell death under ion irradiation. The COST Action MP 1002 NANO-IBCT aims to combine, using a multiscale approach, the unique experimental and theoretical expertise available within Europe to acquire a greater insight into radiation damage induced by ion impact at the nanoscopic and molecular level. For more information, visit **http://fias.uni-frankfurt.de/nano-ibct.**

2. Current Status

Theoretical investigations of the effect of ionizing radiation are usually conducted using computer simulations, since the underlying molecular processes are too complex to treat them purely analytically. Several computer codes for radiation transport (e.g. Geant4: http://geant4.cern.ch/, FLUKA: http://www.fluka.org/, MCNP: http://mcnp-green.lanl.gov/) and track structure analysis (KURBUC, PARTRAC, LEPTS, OREC, SHERBROOKE) have been developed for the simulation of interactions of ionising radiation with various media. MBN Explorer [1] (http://www.mbnexplorer.com/) is a multi-purpose computer code, which is designed to study molecular systems of various degrees of complexity. A broad variety of interatomic potentials implemented in the MBN-Explorer allows a simulation of the structure and dynamics of different molecular systems, such as atomic clusters, fullerenes, nanotubes, proteins, DNA, composite systems, nanofractals, etc. A recent project exploits MBN-Explorer to focus on the thermo-mechanical pathways of DNA damage as a consequence of irradiation in ion beam cancer therapy [2,3].

However, such simulations are strongly dependent upon the quality of the input data. At a recent workshop organized by the COST Action Nano-IBCT in Vienna February 24-26, 2012 the need for a new database for the data on the processes relevant for the interaction of ionizing radiation with organic molecules became obvious. It was decided that this database entitled RADAM should exploit the experience of the VAMDC project [http://www.vamdc.eu/], which is devoted to the creation of an interoperable e-Infrastructure for the exchange of atomic and molecular data. The RADAM database could be extended with the data from simulated outputs (using a database network model), to compare the capabilities of different computer codes.

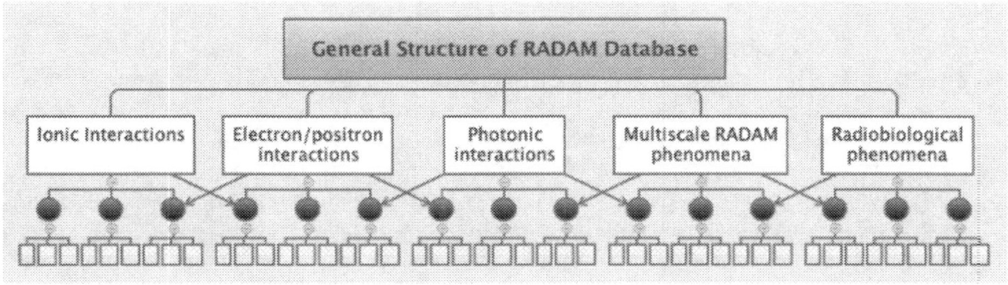

Figure 1. The principal structure of the RADAM database as it was elaborated at the Nano-IBCT workshop in Vienna, February 2012 and Nano-IBCT database workshop in Innsbruck, February 21-22 2013. Blue dots show different RADAM database nodes. Small squares at the bottom represent sets of data stored at each node.

The general structure of the RADAM Database is to consist of five principal areas: Ionic Interactions, Electron/Positron Interactions, Photonic Interactions, Multiscale RADAM phenomena and Radiobiological phenomena, see figure 1.

Each of the aforementioned areas will be represented by at least one database node, where the corresponding information will be stored. Data stored at all the nodes should be accessed through a single user portal (RADAM DB portal), which will provide the user interface for creating database queries using a unified query language for all the databases. This query language is currently the VAMDC SQL subset 2 (VSS2), but might be extended in the future as necessities arise.

The RADAM DB nodes will be also connected to the VAMDC portal in order to provide the possibility for users to search for data compatibility with other VAMDC databases.

Currently the following nodes are being created for the RADAM Database:

2.1. Cross sections for dissociative electron attachment

The Nano-Bio-Physics group at the University of Innsbruck has established a database node containing cross sections for dissociative electron attachment (DEA) processes in the gas phase. The node is called "Innsbruck DEA Database" (IDEADB) and is fully compatible with VAMDC standards version 11.12 (the current, production release) and 12.07 (the forthcoming release). Additionally, a simple web interface was developed in order to allow easy inspection and comparison of data (http://ideadb.uibk.ac.at)

IDEADB currently contains gas phase measurements of the electron energy dependence of cross sections for DEA processes to various molecules, some of which are biomolecules. For each process, the originating species and resulting anion fragments are described in as much detail as possible (e.g. InChI for molecules where possible) and the electron energy resolution used for the measurement is given along with comments taken from the original publication. Since all data is taken from publications and DOIs are given for each process, the data can be cited easily and the additional information given by a full publication can be accessed. For further investigations of data, each cross section can be exported to an ASCII file.

The system is based on the Python implementation of the Node Software developed by the VAMDC project. It uses nginx as a reverse proxy and gunicorn as an application server to power the Django framework. Since the amount of data is rather small, the VAMDC performance goals can be met easily. The system is hosted on dedicated hardware at the University of Innsbruck and therefore well connected to scientific networks.

2.2. Electron and positron interactions with biomolecules

2.2.1 Low energy (0-10 keV) electron and positron interaction

Low energy (0-10 keV) electron and positron interaction cross sections and energy loss in biomolecular systems are being collated in a database prepared by G Garcia and co-workers at CSIC Madrid. Cross section data for electrons and positrons derived from the IAM-SCAR [4,5] calculation procedure are stored together with a critical compilation of available theoretical and experimental data. This node is connected with the Centre for Antimatter-Matter Studies (CAMS, Australia) and the Laboratory for Atomic Collisional Processes of Institute of Physics University of Belgrade (see subsection 2.2.2).

2.2.2 Electron/positron interactions with biomolecules

Electron/positron interactions with biomolecules node is created in Belgrade at the Laboratory for Atomic Collisional Processes of Institute of Physics University of Belgrade, IPB ,(mail.ipb.ac.rs/~centar3/acp.html). Primarily data measured at the own laboratory will be stored in the node. Those include absolute electron differential (in angle and energy) cross sections for elastic scattering and excitation of biomolecules and molecules of biological interest, but will also include targets such as metal atoms. Secondary data sets will be assembled by collating calculated data for the same targetsmeasured at IPB. All data are in compliance with VAMDC standards and the xsams scheme for the collisional processes. Data sets are presented as 3D surfaces (cross sections versus electron impact energy and scattering angle). Data error analysis is included as absolute error bars for each data point.

2.3. Ion interactions with biomolecules and clusters

The CIMAP group in Caen intends to create a node in which data concerning the interaction of singly and multiply charged ions with molecules and clusters of biological interest will be stored. Initially cross section data for the elastic and inelastic (ionization and charge transfer) scattering of ions from biomolecular systems are being collated. These data will be presented as a function of the scattering angle and the collision energy in tabular and in graphical form. Targets will include small biomolecules such as halogenated nucleobases, nucleosides and nucleotides, amino acids and (poly) peptides as well as clusters of these biomolecules. Where possible, experimental data will be compared with theoretical results, for example as presented for uracil in [6]. Furthermore, the RADAM database will include ion-induced relative fragmentation yields as a function of the charge state and kinetic energy of the projectile (see for example [7]). In a second step, the database will be extended to include information on secondary particle production (e.g. electrons and radicals) induced by ion impact.

2.4. Interaction of ionizing particles with all constituents of DNA

In an ongoing research program at PTB, cross section data for the interaction of ionizing particles with all constituents of the DNA or of model molecules for these building blocks (such as tetrahydrofuran for deoxyribose or pyrimidine for the nucleic bases cytosine and thymine) are determined. For electrons with kinetic energies between 20 eV and 1 keV, total scattering cross sections as well as single differential elastic and double differential inelastic cross sections are determined, where the experimental apparatus covers a range of scattering angles between 15° and 135° [8]. The experimental data are extrapolated to forward and backward scattering angles via phenomenological models that are supported by theoretical calculations based on, for example, the modified independent-atom model [9]. Total elastic and single differential inelastic electron cross sections are calculated by numerical integration, and the latter are interpolated as a function of energy using the binary-encounter Bethe model [10].

A second experiment that determines the absolute cross sections for the fragmentation of biomolecules after impact ionization has recently started operation and is currently used for measurements of

electron impact on the DNA molecular building blocks tetrahydrofuran and pyrimidine. The sample preparation system is based on a specially designed supersonic jet that allows also non volatile substances like the nucleic bases to be studied.
A third experiment for measuring double differential ionisation cross sections of DNA constituents for impact of protons and light ions with energies typical for the Bragg-peak region close to the end of the ion track is under construction. This experiment will also allow the determination of cross sections for charge-transfer between these projectiles and target molecules. The data will be evaluated in terms of analytical models for the energy dependence of the cross sections such as the HKS model [11].
As an input to the data base, the following data will be provided: the experimental values including an uncertainty estimate; a report on the evaluation for consistency; and description of the parameters of the evaluated models with an uncertainty estimate for this best-fit interpolation that takes correlations into account.

2.5. Multiscale RADAM phenomena
The Multiscale RADAM phenomena node has been created in Frankfurt in the Meso-Bio-Nano Science group of Prof. Dr. Andrey Solov'yov (fias.uni-frankfurt.de/mbn). The node will store experimental and theoretical data on RADAM related phenomena such as stopping power, diffusion coefficients, Bragg peak characteristics, thermo-mechanical damage, beams defocusing/spreading, complex dielectric functions, molecular fragmentation cross sections, chemical reaction rates. See subsection 4 in the appendix for details. The initial version of Multiscale RADAM phenomena node was created in collaboration with the group of Prof. Bratislav Marinković at Institute of Physics, Belgrade. Currently the node stores experimental data on the linear energy transfer for carbon ions with different initial energies in water. The corresponding links to the publications, where the data was reported, are provided for each query.

3. Roadmap of the RADAM DB project.
The roadmap of the RADAM DB project is presented in figure 2. In order to create the database it is necessary to define its principal logic structure. Working group leaders responsible for each database component have proposed structures for each of the topical areas and these are shown in the appendix.

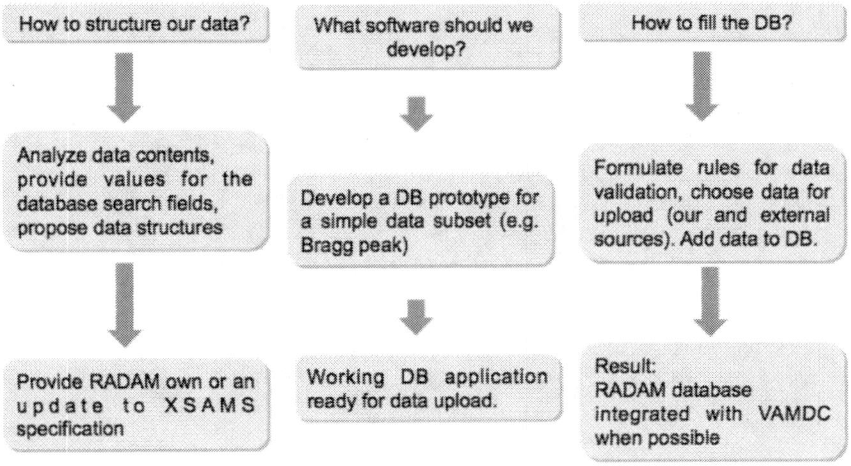

Figure 2. The roadmap of RADAM DB project. Creation of the database implies elaboration of the following questions: structuring, validation of the data and creation of the software, physical location of the nodes, their maintenance.

At the current stage of the development we will include only data, which has published in peer-reviewed journals. However, in certain cases unpublished but thoroughly checked data can be also of high interest to the community. We plan to develop protocols for data verification at later stages of the project in collaboration with the SUP@VAMDC project (**www.sup-vamdc.vamdc.org**).

4. Conclusions and future plans

The RADAM database being created within the framework of the COST Action Nano-IBCT will have a multi-node expandable infrastructure. Each node will be devoted to a certain RADAM data domain. Several nodes devoted to similar data domains can be maintained by different groups.

A dedicated RADAM data portal for constructing queries to the nodes will be created. The RADAM portal will support all types of data stored at the nodes. The nodes will be also accessed by the queries from the VAMDC portal. In case of data compatibility with VAMDC XSAMS standards the reply to the VAMDC queries will be created. Further modifications of VAMDC SQL subset will include extensions towards broader coverage of RADAM data by VAMDC queries.

Any group willing to provide data for RADAM DB or setting up a personal node is cordially invited to participate in the project. RADAM DB developers within the COST Action will provide the joining groups with necessary software for the node installation as well as help and support on configuration and maintenance of the node. The RADAM DB developers can be accessed by e-mail to nano-ibct@fias.uni-frankfurt.de or by direct contact with the working group leaders.

5. Acknowledgements

Parts of this work were supported by the Fonds zur Förderung der wissenschaftlichen Forschung (FWF), I978 and 22665, Wien.

AY, AS and BM acknowledge the support from bilateral project between Germany (DAAD ID 54394840) and Serbia (#680-00-00095/2012-09/9; OI 171020). BM acknowledges the support from VAMDC.

The authors also acknowledge the support and inspiration from COST Action 1002 Nano-IBCT under whose auspices the RADAM DB has been developed.

References

[1] Solov'yov IA, Yakubovich AV, Nikolaev PV, Volkovets I, Solov'yov AV 2012 *J. Comp. Chem.* **33** 2412
[2] Yakubovich AV, Surdutovich E, Solov'yov AV 2011, *AIP Conf. Proc.* **1344** 230
[3] Yakubovich AV, Surdutovich E, Solov'yov AV 2012 *Nucl. Instr. Meth. B* **279** 135
[4] Blanco F and García G 2003 *Phys. Rev. A* **67** 022701
[5] Blanco F and Garcia G 2004 *Phys. Lett. A* **330** 230
[6] Champion C et al. 2012 *JPCS* **373** 012004
[7] Maclot S. et al. 2011 *Chem. Phys. Chem,* **11**, 930
[8] Baek WY, Bug M, Rabus H, Gargioni E. and Grosswendt B 2012 *Phys. Rev. A* **86** 032702,
[9] Hayashi S and Kuchitsu K, 1976 *J. Phys. Soc. Japan* **41** 1724
[10] Kim YK.and Rudd ME 1994 *Phys. Rev. A* **50** 3954
[11] Bernal MA and Liendo JA *Nucl. Instr. Meth. Phys. Res. B* 2006 **251** 171

Appendix. RADAM DB structure and topical areas

1. *Ionic interactions*

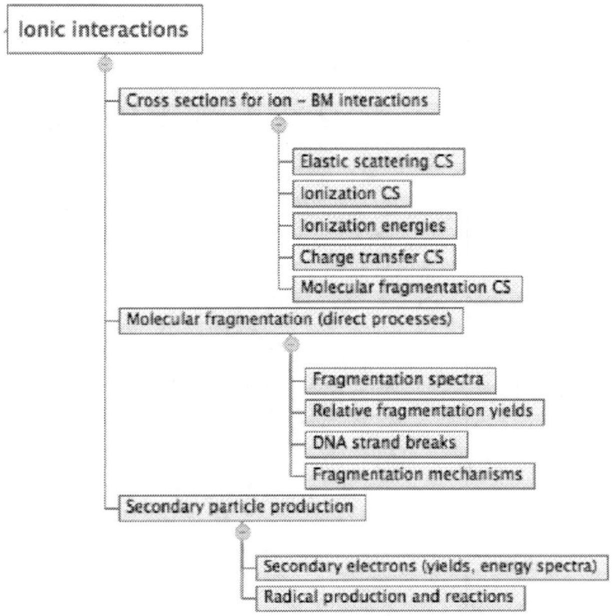

2. *Electron/positron interactions*

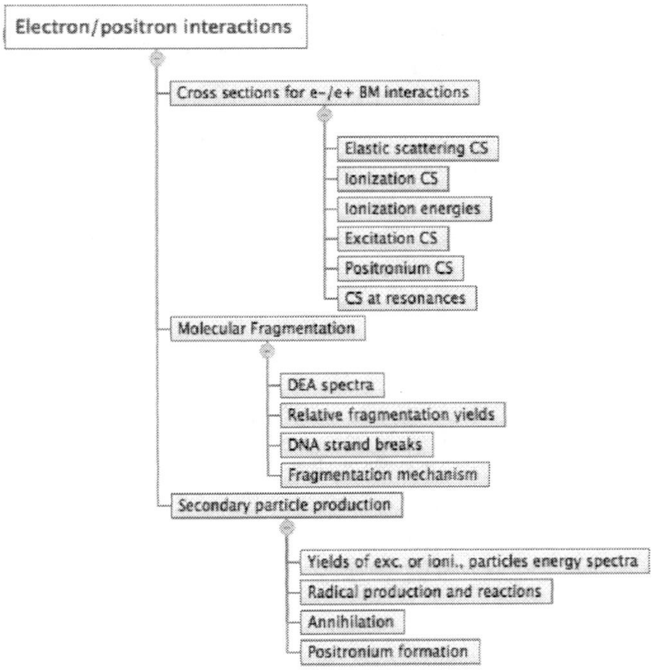

3. *Photonic interactions*

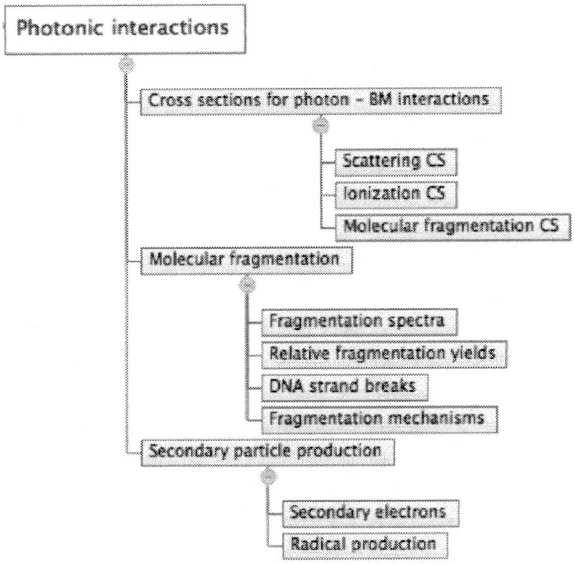

4. *Multiscale RADAM phenomena*

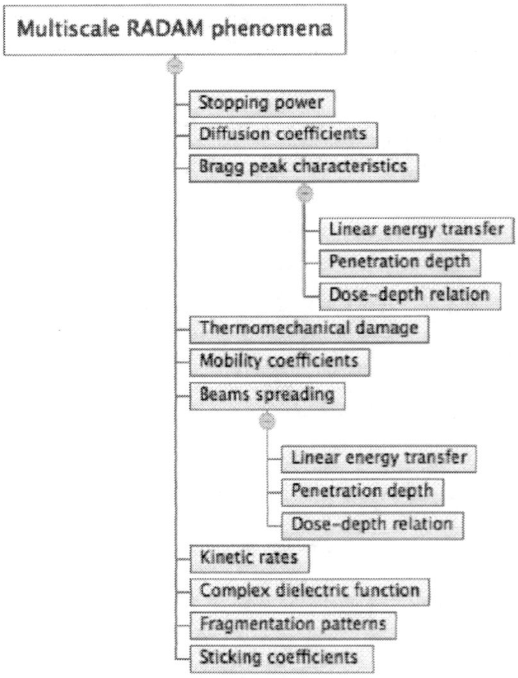

5. *Radiobiological phenomena*

Channeling and Radiation of Electrons in Silicon Single Crystals and $Si_{1-x}Ge_x$ Crystalline Undulators

H Backe[1], D Krambrich[1], W Lauth[1],
K K Andersen[2], J Lundsgaard Hansen[2] and and Ulrik I Uggerhøj[2]
[1] Johannes Gutenberg-University Mainz, Institute for Nuclear Physics, D-55099 Mainz, Germany
[2] University of Aarhus, Department of Physics and Astronomy, DK-8000 Aarhus C, Denmark
E-mail: backe@kph.uni-mainz.de

Abstract. The phenomenon of channeling and the basic features of channeling radiation emission are introduced in a pedestrian way. Both, radiation spectra as well as dechanneling length measurements at electron beam energies between 195 and 855 MeV feature quantum state phenomena for the (110) planar potential of the silicon single crystals. Radiation from a crystalline undulator, produced at the Aarhus University (UAAR), has been investigated at the Mainz Microtron electron accelerator facility MAMI. The 4-period epitaxially grown strained layer $Si_{1-x}Ge_x$ undulator had a period length $\lambda_u = 9.9$ μm. At a beam energy of 375 MeV a broad excess yield around the theoretically expected photon energy of 0.132 MeV has been observed. Model calculations on the basis of synchrotron-like radiation emission suggest that evidence for a weak undulator effect has been observed.

1. Introduction

In order to understand the phenomenon of planar channeling, it suffices to envisage a single crystal as composed of planes (sheets) in the average potential of wich the transverse motion of a particle can be bound. This way the particle can be steered over long longitudinal distances. The situation is quite similar as for people walking between grape-vines, see Fig. 1.[1] In this article we focus on channeling of electrons. Electrons have, in contrast to positrons, a much greater chance to interact with positively charged atomic nuclei comprising a plane. Consequently they have a much shorter dechanneling length. The latter characterizes the mean distance a particle is bound in the potential well. Imagine in Fig. 1 a violent interaction of a person with a grape-vine tree. It probably will finish the walk.

In general the particle oscillates in the potential well as shown in Fig. 2, right panel, and thereby so called channeling radiation is emitted. Although the potential depths are in the laboratory frame for positrons or electrons only in the order of some tens of eV, the oscillation may result in photons with energies in the range between keV to multi MeV if the relativistic particle energies are in the range between a few MeV to some GeV, respectively. How this feature comes about is explained in section 2.1 of this paper.

[1] For the Romans Mogontiacum, first mentioned by Tacitus (about 58-120 AD), now Mainz, was a military base to protect the empire against barbarians living in the north of the river Rhine. Vine was an important beverage for the troops. To avoid the cumbersome transport of vine in amphoraes over the Alps, Romans started in the third century to cultivate vine-growing also in the region around Mainz.

Figure 1. Left: A colorful way to demonstrate channeling. Picture captured by one of us (H.B.) at the vineyard Baiken at Rauenthal, Rheingau, Germany. Right: Channeling of a positively charged particle. Shown are two strings of atoms composing crystal planes.

If the crystal planes are periodically bent at the direction the particle moves, undulator-like radiation in the hundreds of keV up to the MeV region may be emitted in addition to channeling radiation. The possibility to produce very intensive γ-ray sources in the MeV range at channeling of electrons or positrons has been discussed theoretically since the mid 70th of the last century [1, 2, 3]. Nowadays this field of research has been matured and detailed theoretical predictions on the feasibility of a γ-ray laser on the basis of micro undulators have been made, for an overview see Ref. [4], and the recent text book of A. V. Korol, A. V. Solov'yov, and W. Greiner [5], and references cited therein. However, the experimental demonstration of such devices hampers essentially from two obstacles. On the one hand, high quality positron beams in the GeV range are not easily available, in contrast to electron beams, and, on the other hand, crystal-undulator production and characterization of their features are still in a rather rudimentary stage. Because of this situation, experiments have been commenced at the Mainz Microtron MAMI, where a high quality electron beam with energies of up to 855 MeV is available. The aim is to characterize epitaxially produced $Si_{1-x}Ge_x$ superlattices [6, 7] [2] designed as crystal undulators. Periodically undulating (110) planes can be obtained by a variation of the concentration x, for details see e.g. Ref. [9, 11].

Preliminary experimental results with a 4-period undulator with period length of 9.9 μm, grown at the Aarhus University, have been described in recent papers [10, 11]. Clear enhancements of intensities were observed in the energy region where the undulator radiation peak is expected. However, only weak evidence for a small peak was observed at the beam energy of 270 MeV [11]. Simulation calculations of the undulator radiation spectra have been performed on the basis of synchrotron-like radiation emission which suggest a very small coherence effect. To figure out the reason for the obviously very small dechanneling length, we focus in this contribution on the analysis of another measurement with the same undulator at a somewhat higher beam energy of 375 MeV. Experiment and model calculations are described in sections 3 and 4. In addition, we reanalyzed dechanneling length measurements already published in Ref. [12] under refined model assumptions. The results are presented in subsection 2.3.

For readers only interested to understand principles, a pedestrian introduction to the phenomenon of channeling radiation emission is presented in subsections 2.1 and 2.2.

[2] For earlier references on the structure, properties and applications of $Si_{1-x}Ge_x$ strained layers and superlattices see the review article Ref. [8].

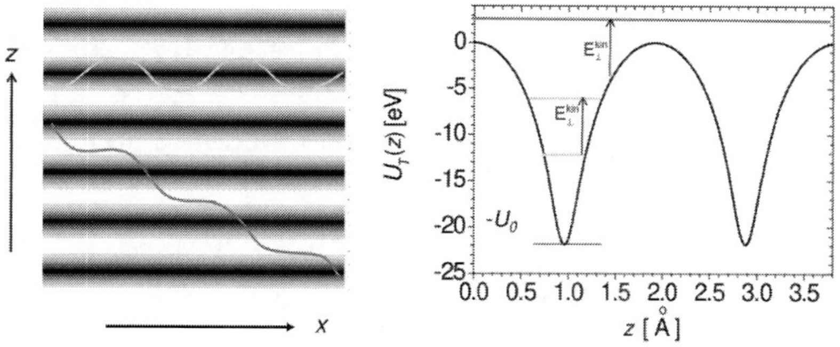

Figure 2. Left: Density plot of the mean potential of planes in a single crystal. White corresponds to a low and black to high potential for a positive probe charge. The transverse motion of an electron in a bound state is called planar channeling, and in an unbound state of low continuum energy planar quasi-channeling. Right: Cut through the potential along the z axis. The energy is given by the potential at the point of impact to which the transverse kinetic energy $E_\perp^{kin} \approx pv \cdot \psi_\perp^2/2$ must be added, with $\psi_\perp$ the tilt angle of the impinging electron with respect to the (110) planes and pv the momentum of the particle. For further details see also Ref. [22].

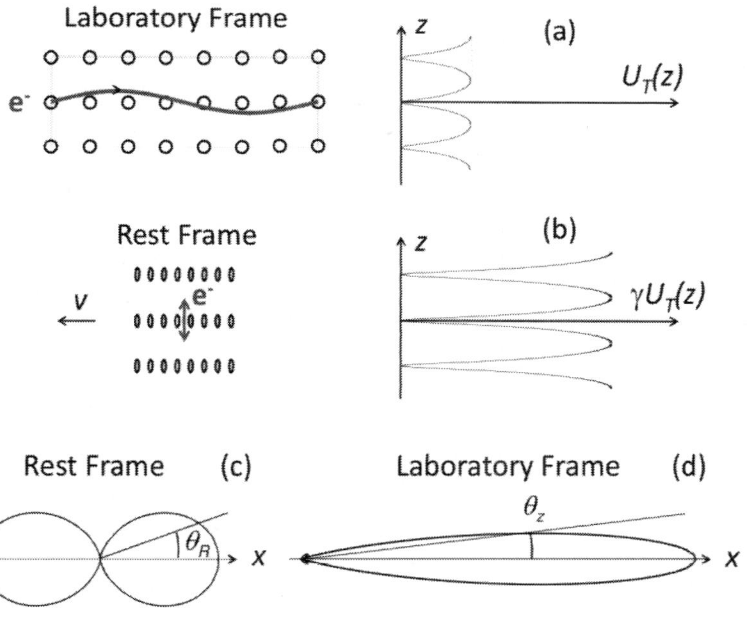

Figure 3. Explaining the basic features of channeling radiation. Shown are the potential for an electron at (110) planar channeling in a silicon single crystal in the laboratory frame (a), and in the rest frame of the electron (b), angular distribution of the emitted radiation in the dipole approximation in the rest frame (c), and in the laboratory frame (d). In this example the Lorentz factor is $\gamma = 3$.

2. Channeling and channeling radiation

The phenomena of channeling and channeling radiation emission have been studied by a number of authors, both theoretically as well as experimentally. In the following only the most important facts are compiled for the purpose that an interested non-specialized reader may be able to qualitatively understand the subject of this article. Comprehensive overview articles have been written to which advanced readers in this field of research are referred to [13, 14, 15, 16, 17, 18, 19, 20, 21].

2.1. Basic features of channeling radiation emission

Let us first understand how the high photon energies observed in the laboratory system come about. As shown in Fig. 3 (a) and (b) the electron sees in its rest frame a Lorentz contracted lattice. This fact results in an increase of the potential depth by the Lorentz factor $\gamma = (1-\beta^2)^{-1/2}$, with $\beta = v/c$ the reduced velocity of the electron in the laboratory system, and, accordingly, also of the emitted photon energy. The emission characteristics for a dipole transition $dn_R/d\Omega_R \propto \cos^2\theta_R$ in the rest frame of the electron is depicted in Fig. 3 (c). Note that the angle θ_R is the angle between the x axis and the emission direction, and not between the dipole moment and the emission direction. The angular distribution in the laboratory frame, given by the equation [23]

$$\frac{dn}{d\Omega} = \frac{dn_R}{d\Omega_R}\frac{d\Omega_R}{d\Omega} = \frac{dn_R}{d\Omega_R}\frac{1}{\gamma^2(1-\beta\cdot\cos\theta)^2}, \tag{1}$$

is strongly forward directed, see Fig. 3 (d).

So far channeling has been treated as a pure classical phenomenon. However, in reality quantum states exist in the potential well and a transition may result in emission of monochromatic radiation. Deexcitation photons emitted in the rest frame of the electron are boosted in the laboratory system to an energy [14]

$$\hbar\omega = \frac{\hbar\omega_R}{\gamma(1-\beta\cdot\cos\theta)} \cong \frac{2\gamma^2}{1+(\theta\gamma)^2}\hbar\omega_0. \tag{2}$$

At the right side of this equation $\hbar\omega_R = \gamma\hbar\omega_0$ has been substituted with the transition energy in the potential well in the laboratory frame $\hbar\omega_0$, see Fig. 3 (a).

2.2. Quantum states and channeling radiation

To calculate the energy levels in the potential well, we follow Ref. [19, Eqn. (2.189)]. The potential has been approximated by $U_T(z) = U_0/\cosh^2(z/b)$ with $U_0 = 22.9$ eV, $b = 0.303$ Å for which the one dimensional Schrödinger equation can be solved analytically. The solution for the eigenstates reads

$$\varepsilon_n = -\frac{(\hbar c)^2}{2\gamma m_e c^2 b^2}\left(-\frac{1}{2}+\sqrt{\frac{1}{4}+\frac{2\gamma m_e c^2 U_0 b^2}{(\hbar c)^2}}-n\right)^2, \quad 0 \le n \le -\frac{1}{2}+\sqrt{\frac{1}{4}+\frac{2\gamma m_e c^2 U_0 b^2}{(\hbar c)^2}} \tag{3}$$

with the quantum number n being 0 or a positive integer. As an example, the experimental photon spectra in Fig. 4 are the result of (110) planar channeling of electrons in a silicon single crystal. Shown are also corresponding calculated level schemes. The photon energy for a transition $\varepsilon_i - \varepsilon_f$ between two levels in the potential well is given according to Eqn. (2) by $\hbar\omega \simeq 2\gamma^2(\varepsilon_i - \varepsilon_f)$ at $\theta = 0$. It is interesting to notice that $\varepsilon_i - \varepsilon_f$ scales for low principal quantum numbers about proportional to $\gamma^{-1/2}$ similar as for an harmonic oscillator, see Ref. [14, p. 458 f]. This behavior originates from the fact that the potential at the minimum can well be approximated by a harmonic oscillator potential.

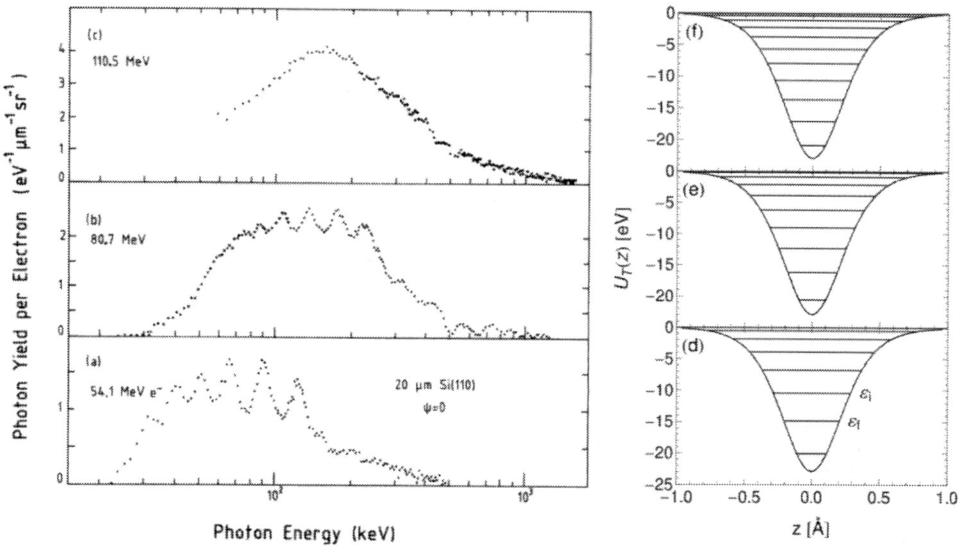

Figure 4. Radiation spectra for 54.1 (a), 80.7 (b), and 110.5 MeV (c) electrons incident along the (110) planar direction of a 20 μm silicon single crystal. Spectra at random orientation of the crystals were subtracted. Figure taken from Ref. [26]. Panels (d), (e), and (f) show corresponding level schemes. The potential is an approximation which allows to solve the one-dimensional Schrödinger equation analytically [19, Eqn. (2.189)].

The line structure disappears at the electron energy of 110.5 MeV for which the level density is so high that the lines overlap. The line width originates mainly from incoherent scattering of the electrons on lattice atoms resulting in the creation or absorption of phonons (also frequently called in literature thermal scattering) [3], as well as from multiple scattering. The latter is effective only for the dimension parallel to a crystallographic plane, since perpendicular to a plane the electron is bound and a scattering process results in a transition between levels in the potential well or into the continuum. Coherence lengths calculations were presented in a number of papers for various transitions and crystals, however, an disentanglement between incoherent scattering, leading to a Lorentzian line shape, and multiple scattering with a Gaussian line shape was first performed in Ref. [27] for diamond single crystals. As for diamond, the coherence length is expected also for silicon to be in the order of a μm decreasing slowly as function of increasing beam energy, see Ref. [27, TABLE I]. Indeed, at a beam energy of 110.5 MeV an assumed coherence length $l_{coh} \simeq 0.5$ μm results for $\varepsilon_1 - \varepsilon_0 = 3.8$ eV in a relative line widths $\Gamma/\hbar\omega = \hbar c/[l_{coh}(\varepsilon_i - \varepsilon_f)] = 10$ % (FWHM). At the corresponding photon energy at observation in forward direction $\hbar\omega = 355$ keV the width amounts to $\Gamma = 37$ keV which can hardly be resolved at a typical energy separation of the photon lines of also 10 %. At still higher electron energies only smooth structures are observed for the same reasoning. However, it is an open question up to which beam energy the lowest bound states cannot be treated classically, meaning that for the dechanneling length additional considerations are required. In trying to answer this question, dechanneling length measurements performed for the relevant beam energies between 195 and 855 MeV at the Mainz Microtron MAMI [12] have been reanalyzed.

[3] We mention that the line broadening effect due to incoherent scattering has some similarity with the well known collisional line broadening in atomic physics.

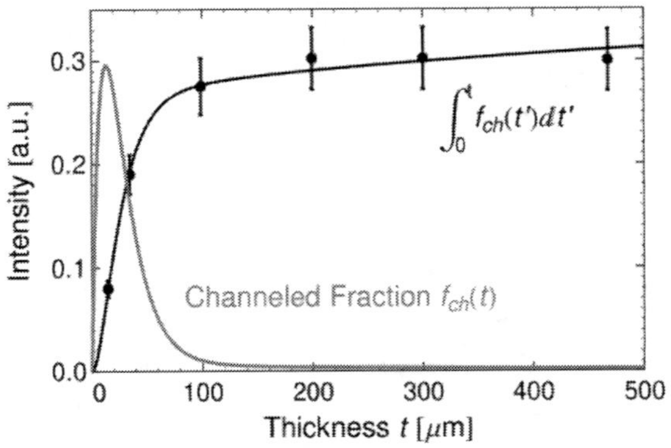

Figure 5. Dechanneling length measurement at a beam energy of 855 MeV. Error bars indicate the experimental signal as function of the crystal thickness. The signal was generated by electrons which have lost about 50 % of their primary energy. For experimental details see Ref. [32]. The red curve assigned with "Channeled Fraction f_{ch}" is the occupation probability according to Eqn. (4) with $a = 0.01$ and $t_0 = 6$ μm. The black curve assigned with $\int_0^t f_{ch}(t')dt'$ is the integrated channeled fraction fitted to the data.

2.3. Dechanneling length in the quantum state picture

One of the very important prerequisites for experimental studies of undulator-like radiation in strained-layer $Si_{1-x}Ge_x$ superlattices is the knowledge of the dechanneling length of electrons for the (110) crystallographic plane. Unfortunately, even for undistorted plane crystals only little is known experimentally. Because of this fact, we have performed dechanneling length measurements at various beam energies for plane crystals. The results were published in Ref. [32]. In Fig. 5 a typical measurement is depicted here once more.

The most striking feature is the approximative signal saturation as function of the crystal thickness t. Such a behavior is not at all expected in the classical picture, based on the Fokker-Planck equation, for the fraction of electrons residing in the potential pocket. A strong rechanneling probability causes a much steeper increase of the total probability as function of increasing crystal thickness, see Ref. [25, Fig. 16]. Therefore, it might be tempting to speculate that the signal originates from the lowest quantum state in the potential pocket. With the probability density maximum at atomic planes a large probability for collisions with small impact parameters may result which probably are required for the large energy loss of the electron of about 50 % of their primary energy. However, there is indication for a small rechanneling probability at larger crystal thicknesses which was neglected in our previous analysis. In the latter it was assumed that the probability density $f_{ch}(t)$ is proportional to $\exp(-t/L_d)$ meaning that the occupation probability has its maximum already at $t = 0$ from where it decays exponentially with the characteristic dechanneling length L_d. This assumption is questionable in the quantum picture since the lowest quantum state might also be occupied by radiative transitions.

This discussion led us for the probability density $f_{ch}(t)$ to a differential equation of the type

$$\frac{df_{ch}(t)}{dt} + \frac{f_{ch}(t)}{L_d} - r_0 \cdot r(t)(1 - f_{ch}(t)) = a \exp\left[-\frac{t}{t_0}\right] \quad (4)$$

$$g_{ch}(t) = C \int_0^t f_{ch}(t')dt' \quad (5)$$

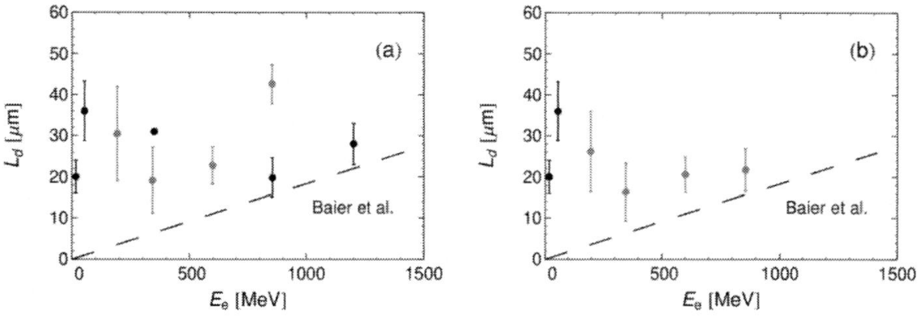

Figure 6. Dechanneling lengths for (110)-planar channeling of electrons at silicon single crystals as function of the electron energy. The experimental values assigned in black are for 17 MeV and 54 MeV the 1/e occupation lengths derived from the 1→ 0 radiative transition [28]. The point at 350 MeV is taken from Ref. [29] for which no error bar has been quoted. The value at 855 MeV is a reanalysis of the measurement published in Ref. [25, Fig. 15] with allowance for rechanneling, and the value at 1200 MeV is taken from Ref. [30]. The reanalysis of measurements at MAMI [32] are shown with red error bars for two models: Model (I) assumes prompt feeding taking into account rechanneling (a), model (II) assumes delayed feeding (b). The dashed line assigned with "Baier et al." is a calculation according to Ref. [31, Eqn. (10.1)] with $U_0 = 22$ eV.

which must be solved to obtain the fit function Eqn. (5). The latter is the integral over the crystal thickness t of the solution of Eqn. (4). The solution of the truncated differential equation $df_{ch}/dt + f_{ch}/L_d = 0$ is simply the already mentioned exponential function. The third term on the left hand side of the differential equation (4) takes into account rechanneling. It has been assumed that it is proportional to the occupation probability for an electron in the potential pocket obtained from the solution of the Fokker-Planck equation for 855 MeV electrons [25] for which the function $r(t) = \sqrt{t_s/(t+t_s)}$ with $t_s = 2.68$ μm is a good approximation. The function $\exp[-t/t_0]$ on the right hand side takes into account a possible delayed feeding of the quantum state.

For the best fit procedure two model assumptions have been tested. For model (I) the homogeneous differential equation (4) with $a = 0$ has been solved. The dechanneling length L_d, the rechanneling parameter r_0 and the common overall factor C at the integral for $g_{ch}(t)$ were taken as free fit parameters assuming for the initial occupation $f_{ch}(t \to 0) = 0.5$. This procedure is equivalent to a prompt feeding of the bound level. The fit results are shown in Fig. 6 (a). In model (II) a delayed feeding of the lowest bound state has been assumed by solving the inhomogeneous differential equation with $a \neq 0$ and choosing $f_{ch}(t \to 0) = 0$ as initial occupation, selecting a proper t_0 with the boundary condition of a constant product $a \cdot t_0 = 0.06$, and treating for the best fit procedure again the three above mentioned parameters as free. The results are shown in Fig. 6 (b).

2.4. Discussion of the dechanneling length measurements

From the classical model of Baier et al. [31, Eqn. (10.1)] a linear increase of the dechanneling length as function of the electron energy is expected. However, the experimental data suggest a significant larger dechanneling length for our measurement at 195 MeV in accord with the data points at 17 and 54 MeV taken from Ref. [28]. We interpret this finding is indication for a quantum state effect at dechanneling also for the beam energy of 195 MeV. Comparing the results for the measurements at 195, 345 and 600 MeV depicted in Fig. 6 (a) and (b) we see that the two model assumptions result in more or less the same dechanneling lengths. This is not surprising since the parameter $t_0 = 1$ μm had to be chosen rather small (with one exception).

A special case are the two measurements at 855 MeV for which model (I) gives quite different results. A significant better fit to the upper red data point was obtained with model (II) by choosing $t_0 = 6$ μm, this example is shown in Fig. 5, resulting in the shown decrease of the dechanneling length in Fig. 6 (b). Indeed, with this choice of the delayed feeding parameter t_0 it does not contradict the reanalyzed measurement from Ref. [25, Fig. 15] with allowance for rechanneling, see black error bar in Fig. 6 (a). However, the analysis procedures are inconsistent since for this black data point we found no indication for a delayed feeding.

In summary of this part, the nearly saturated signals as function of the crystal thickness, see for an example Fig. 5, as well as the reanalysis for the dechanneling lengths both provide evidence for quantum state effects. It manifests itself for the measurement at 195 MeV in a significant larger dechanneling length in comparison with the classical model of Baier et al. [31, Eqn. (10.1)]. For the measurement at 855 MeV we obtained inconsistent results. Obviously, the experimental data basis is not good enough for a definite assignment of the dechanneling length. Much more data points should be taken, in particular at crystal thicknesses less than about 20 μm including measurements with very thin crystals in the order of only a few μm.

Let us finish this subsection with a qualitative discussion how a longer quantum state dechanneling length emerges in comparison with the classical picture. As already mentioned, at the crystal entrance face well separated low lying quantum states may be populated either directly or delayed. In both cases the excitation of an electron from the ground state level requires a single scattering of the electron with a minimum scattering angle $\Delta\psi_\perp^2 \approx 2\Delta E_\perp^{kin}/(pv)$, with $\Delta E_\perp^{kin} \simeq 60$ eV$/\sqrt{\gamma}$ the transverse energy transfer for the transition to the first excited state. For this approximation $\gamma \gtrsim 200$ must be fulfilled. At an electron energy of 855 MeV an angle $\sqrt{\Delta\psi_\perp^2} = 59$ μrad results. To reach the region with high level density, this angle might even be larger by a factor of about 3-4. Such single scattering events might have a low probability even if taken into account that the angular integration extends to infinity and the probability density of the electron at the crystal plane is high. As a result, a longer "quantum state" dechanneling length might be the consequence, even at a beam energy as high as 855 MeV, in contrast to a weakly bound classical state. For the low beam energies of 17 and 54 MeV "quantum state" effects are well known. The term "occupation length" used in Ref. [28] is actually a synonym for our dechanneling length. [4]

3. Experiments with the crystalline undulator at 375 MeV beam energy

In the following we report on an experiment performed with a 4-period $Si_{1-x}Ge_x$ superlattice undulator crystal with period length of 9.9 μm produced at the Aarhus University (UAAR). The high quality electron beam with an energy of 375 MeV of the Mainz Microtron MAMI has been employed. We remind the reader on the experiments already performed with this undulator at beam energies of 270 and 855 MeV [10, 11]. A Fourier analysis of the amplitude distribution for the (110) plane of the undulator yields 4.79 Å, see Ref. [11, Fig. 1], from which an undulator parameter $K = 0.223$ and an expected peak energy of the undulator radiation of 0.132 MeV can be deduced.

The experimental setup is shown in Fig. 7. The undulator crystal was mounted on goniometers with which rotations around three axes could be accomplished. Details on the goniometers, the procedure to align the crystal, etc., are described in Ref. [25]. Photon spectra are shown in Fig. 8. The calculated bremsstrahlung contributions, corresponding to the crystal tuned to a random off-channeling position, are shown in panel (a) as light blue and red lines. These contributions have been subtracted in panel (b). Further details of the analyzing procedure

[4] That an increased "quantum state" dechanneling length might still be observable at beam energies as large as a few hundred MeV was brought to our attention at a meeting at FIAS [33].

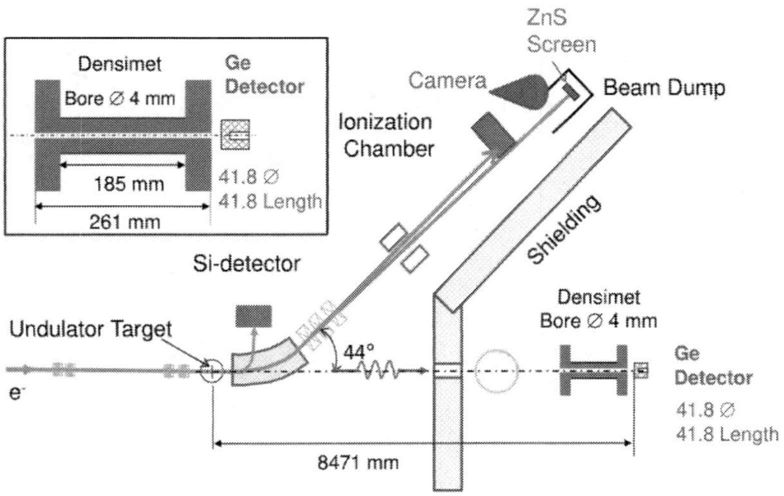

Figure 7. Experimental setup at MAMI. Downstream the undulator target the beam is deflected horizontally by a 44°-bending magnet and thereafter vertically by a second 7.2°-bending magnet. The ionization chamber and the Si detector are monitor counters for the detection of channeling. Photon spectra are recorded with a Ge(i) detector, see inset (bore: 7.8 mm Ø and 30 mm length). The photon beam from the target is collimated with Densimet 176 (density of 17.6 g/cm^3, 92.5% tungsten, 5% nickel, 2.48% iron). The diameters are 52 mm for the inner part and 96 mm for the outer parts. Photons proceed 8.109 m in vacuum and 0.362 m in air just in front of the Ge(i) detector.

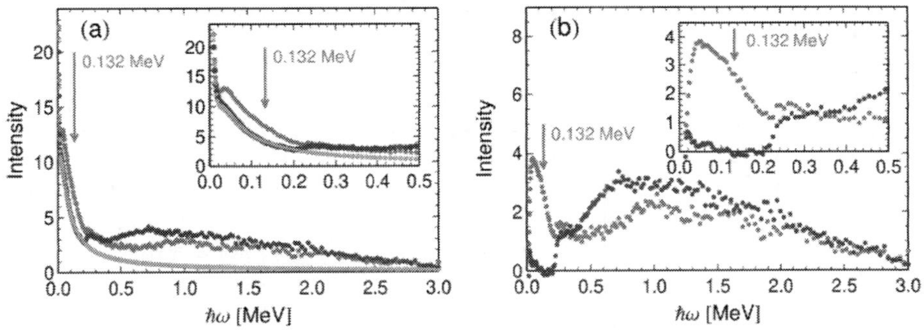

Figure 8. Panel(a): Deconvoluted photon number spectra at (110)-planar channeling of 375 MeV electrons for the flat (blue) and the undulator crystal (red) with effective thicknesses of 49.5 μm and 39.6 μm, respectively. Data were taken with the Ge detector, see Fig. 7. - Panel (b): Same as panel (a) with the bremsstrahlung background subtracted.

are described in Ref. [11].

In the next section we will refer to a model with the aim to understand the origin of the low energy excess yield of the undulator crystal at least qualitatively.

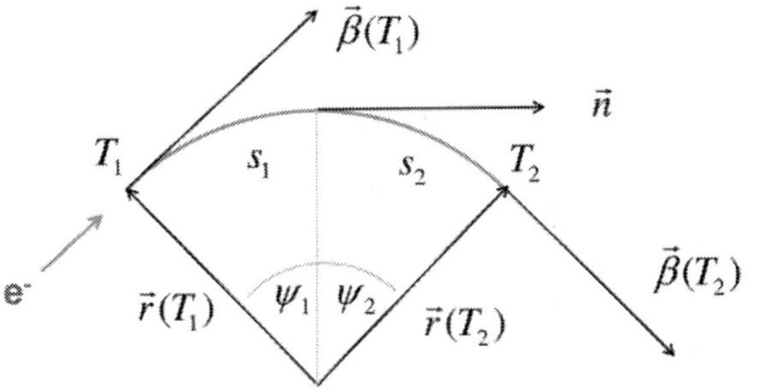

Figure 9. Synchrotron radiation emission from a single finite arc element of the undulator. At times T_1 and T_2 the electron has reduced velocity vectors $\vec{\beta}(T_1)$ and $\vec{\beta}(T_2)$, respectively. For the corresponding radius vectors holds $|\vec{r}(T_1)| = |\vec{r}(T_2)| = \rho$. The vector $\vec{n}$ denotes the observation direction in the plane of the circular arc with bending radius ρ the electron moves.

4. Model calculations

The most striking feature of the spectra shown in Fig. 8 is the enhancement of the radiation intensity for the undulator crystal at low photon energy around 0.1 MeV. We will show that this excess yield is most probably related to synchrotron-like radiation emission.

4.1. Synchrotron-like radiation emission

As has been described in detail in Ref. [11], the undulator consists in a very good approximation of totally eight circular arcs with a bending radius of 6.60 mm. Therefore, we consider synchrotron-like radiation emission as the basic process. On a first glance this picture seems to fail since the critical energy for synchrotron radiation emission $\hbar\omega_c \simeq (3/2)\gamma^3\hbar c/\rho = 17.7$ keV turns out to be much too low. However, the prerequisite for Schwinger's approach with which this formula has been derived, see e.g. Ref. [34, ch. 14.6], is not at all valid for our undulator. We find an arc angle of $\psi_1 = \psi_2 = 0.375$ mrad, see Fig. 9, which is significantly smaller in comparison with $1/\gamma = 1.36$ mrad. This immediately tells that Schwinger's approach with integration from $-\infty$ to $+\infty$ must fail. Because of this fact the emitted time impulse is much shorter and the spectral distribution much broader than calculated with Schwinger's formalism.

The radiation amplitudes from the eight arcs of the undulator must be superimposed coherently taking into account de- and rechanneling. While in a bent crystal strong dechanneling is expected, enhanced by centrifugal forces, at the same time also a significant rechanneling may occur since after half an undulator period all electrons will find a proper direction for rechanneling. This situation is quite dissimilar in comparison with a plane crystal for which the fraction of electrons which scatter back into the nominal direction continuously decreases as function of the traversed crystal depth. It has been assumed in our model that rechanneled electrons again emit a coherent wave train but, of course, without the contribution from the first pole. The corresponding intensity of the second wave train has been added incoherently to the intensity of the first one. This procedure continues until the electron has reached the last pole. Consequently, the irradiated photon number intensity is given by the first term of the formula

$$\frac{d^2N}{d\Omega d\hbar\omega} = \frac{d^2N_S}{d\Omega d\hbar\omega}\sum_{n=1}^{2n_u} p_n|S_n(\hbar\omega)|^2 + \frac{d^2N_W}{d\Omega d\hbar\omega}\sum_{n=1}^{2n_u}(1-p_n) \tag{6}$$

with $d^2N_S/d\Omega d\hbar\omega$ the radiation contribution from a single pole and $S_n(\hbar\omega)$ the structure factor, both described in Ref. [11], n_u the number of undulator periods, and p_n the channeling probability at pole n. The remaining fraction $(1 - p_n)$ of electrons which does not channel are assumed to produce an incoherent radiation contributions $d^2N_W/d\Omega d\hbar\omega$ at each pole of the undulator. It has been assumed that the electrons emit radiation either by performing a sort of "orbiting" along a crystal channel in a rather tiny fraction of the arc, or in a volume reflection process. In the latter, see, e.g. Ref. [35] and references cited therein, electrons are deflected by the crystal potential without being channelled. We have not estimated whether the corresponding radiation component is of significance for our experiment, or not. In case it would be, one might assume that the orbit in the turning point can be approximated also by a circular arc and, as a consequence, synchrotron-like radiation emission from volume reflection can approximately also be subsumed into the second term of Eqn. (6).

4.2. Description of the fitting procedure

The shape of the radiation spectra depends on several parameters which must be adapted in a fitting procedure. Most sensitively enters the dechanneling length L_d via the structure factor $S_n(\hbar\omega)$ which controls the shape of the undulator peak. The channeling probabilities p_n are described by the recursion relation

$$p_{n+1} = p_n \cdot (1 - p_{de}) + [1 - p_n \cdot (1 - p_{de})] \cdot p_{re}. \tag{7}$$

The p_n depend on the three parameters p_1, p_{de}, and p_{re} which are the initial channeling probability at the first pole, the dechanneling, and rechanneling probabilities, respectively. The rechanneling probability has been assumed to be independent on the pole number what is a rough approximation. The dechanneling probability is assumed to be connected with the dechanneling length by the relation $p_{de} = 1 - \exp[-(\lambda_U/2)/L_d]$. The synchrotron-like radiation intensity for a single undulator pole $d^2N_S/d\Omega d\hbar\omega$ has been calculated for a full arc, and no additional parameter enters. The single incoherent de- or reflection intensity $d^2N_W/d\Omega d\hbar\omega$ depends on an additional parameter ψ_W, assuming $\psi_1 = \psi_2$, see Fig. 9. This parameter controls the intensity and the slope of the spectral component extending to higher photon energies.

The analytical model has totally fife parameters which have been varied by a trial and error method until an acceptable fit of the photon spectra below about 200 keV was achieved. In this region the channeling radiation component for the undulator crystal is assumed to be small, as it is for the plane crystal. Real sensitive are only four parameters, L_d, p_{re}, ψ_W, and the plasma frequency $\hbar\omega_p$. For silicon at random orientation the latter is $\hbar\omega_{p,mean} = 31.06$ eV. However, at channeling it should be somewhat larger because of an enhanced overlap of the electron with an atomic plane where the charge density is larger than in the mean. Indeed, the fit improved significantly by choosing a value 1.2 $\hbar\omega_{p,mean} = 37.27$ eV. It was found that L_d, p_{re} are strongly correlated to each other.

Care has been taken to reproduce the total spectrum which includes the bremsstrahlung contribution. The latter has been calculated for our experimental geometry with well known formulas [36, Eqn. (10.2)] taking into account also the density effect [37, ch. 3] and transition radiation emission. The latter should be absent at on-axis observation as in our experiment. However, scattering of the electron in the target results in a contribution from the rear surface. These contributions are of importance in the low energy part of the spectrum, below about 30 keV. The calculated undulator contribution was added to the calculated bremsstrahlung spectrum and the resulting shape of the sum spectrum was compared with the measured one.

5. Results and discussion

The result of the best fit procedure described in the previous section is depicted in Fig. 10. We find a reasonable good overall agreement with the parameters quoted in the figure caption. The arc length, responsible for emission of the incoherent part, is $L_d' \approx (\psi_W/y'_{max})\lambda_U/2 = 0.99$ μm

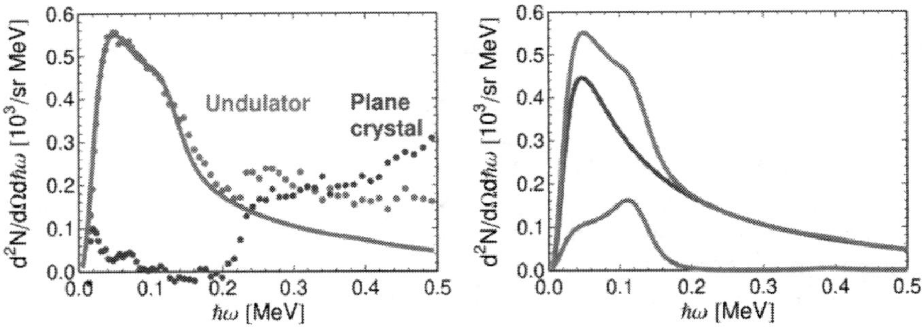

Figure 10. Left panel: Result of the best fit. The data points are taken from Fig. 8 right panel (inset) which were scaled by a factor of 0.145. The full red line represents the calculated undulator radiation line shape for which the following set of parameters were used: $L_d = 4.0\ \mu$m ($p_{de} \approx 0.71$), $\psi_W = 0.075$ mrad, $p_1 = 0.02$, and $p_{re} = 0.01$. The channeling probability drops from 0.02 at the first pole to about 0.014 at the last pole of the undulator. Right panel: Decomposition of the calculated spectrum. The blue spectrum represents the incoherent synchrotron-like radiation contribution, second term of Eqn. (6), the violett the coherent contribution, first term of Eqn. (6). The red curve is the sum of both.

which is rather small. It might be interpreted as dechanneling length of weakly bound electrons in the crystal potential. Whatever the explanation might be, also radiation in the volume reflection process may not be excluded, these rather sharp pulses in the time domain describe already the experimentally observed gross structure quite well, as can be concluded from Fig. 10, right panel (blue curve). It should be mentioned that the deconvolution procedure introduces some systematic errors which could well be in the order of 20 %.

The hump at the expected peak energy of 0.132 MeV can only be reproduced by the coherent term in Eqn. (6) originating from a small fraction of only about 2 % of electrons with a dechanneling length of about 4.0 μm, see Fig. 10, right panel (violet curve). In view of what has been discussed in section 2.3 it might be tempting to speculate that this fraction are electrons which are deeply bound in the crystal potential, i.e. occupying a quantum state.

Let us discuss why the coherence length turns out to be only about 4.0 μm while a larger value of about 15 μm is expected for a plane crystal, see Fig. 6 (b), although with a rather large error bar. We estimated that centrifugal forces diminish the potential barrier from 22.6 eV to 17.6 eV. It is hardly imaginable that by this lowering by only 22 % the dechanneling length of deeply bound electrons will be reduced to such a small value. Another explanation may be found in a break down of the strain by formation of misfit dislocations in the crystal. Finally, if looking to the $Si_{1-x}Ge_x$ crystal structure microscopically, see Fig. 11, still another explanation may be conceivable. In such a view an isolated germanium atom is surrounded by many silicon atoms resulting in a locally perturbed lattice structure. It is quite possible that a plane in which the electron channels looks microscopically bumpy, with the consequence of a locally fluctuating bending radii. In addition, also the transverse potential may be lowered, similar as for thermal vibrations at increasing temperature. Both effects are expected to reduce the dechanneling length.

It should be mentioned that the following alternative explanation for the coherent fraction must be considered as well. During subsequent emission of the short pulses the phase information might not completely be lost resulting in a partial coherence for the second term in Eqn. (6). This possible explanation could not be ruled out during the preparation of this paper, and presently we cannot claim that the observed interference structure originates from quantum

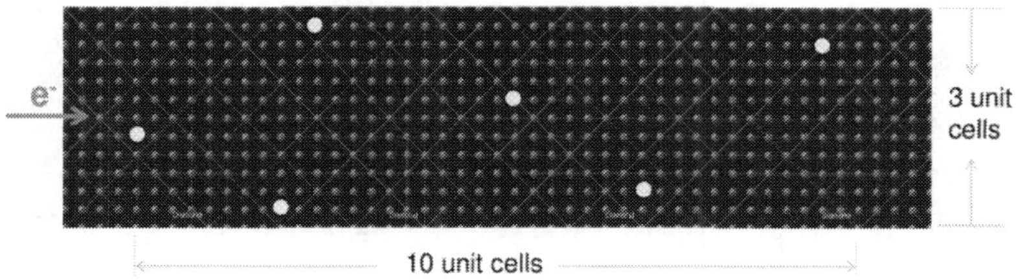

Figure 11. Ge Doping of a silicon single crystal. Shown are 10 × 3 × 1 unit cells with 240 atoms. At 2.5 % doping there are 6 Ge atoms present, i.e. in the mean 0.5 Ge atoms in every of the 12 rows drawn.

state channeling.

In summary, based on a rather simple model with mean trajectories evidence for a weak undulator effect may be concluded also for the experiment at 375 MeV electron beam energy. This finding is in accord with an experiment at 270 MeV [11]. However, there are a number of open questions which might be elucidated by Monte Carlo simulation methods. Various studies are presently underway aiming in a better theoretical understanding of radiation spectra emitted from a crystalline undulator under realistic conditions [38, 39, 40, 42, 43, 41]. In future such simulations may also be of some help for the design of a strong γ-ray source or even a laser on the basis of positrons. At present the reliability of those simulation codes have been or will be tested with the results of our channeling experiments with electrons at MAMI. This fact underlines the importance of the experiments with electrons as described in this paper.

6. Conclusions

Already published dechanneling length measurements [12], employing as signal electrons which have lost about 50 % energy of their primary energy, have been reanalyzed under more realistic model assumptions. The results suggest that low lying quantum states in the channeling potential pocket play an important role. In addition, an experiment with an epitaxially grown strained-layer $Si_{1-x}Ge_x$, 4-period, $\lambda_U = 9.9\ \mu$m undulator has been performed at an electron beam energy of 375 MeV. As for previous experiments at 270 MeV and 855 MeV [11], also in this experiment a clear enhancement of radiation in the expected low energy domain has been observed. Analysis with a simple analytical model suggests that the gross structure can be explained by synchrotron-like radiation emission from small arc elements of the undulator. In order to explain the hump at the expected peak of 0.132 MeV energy, a coherent contribution must be added the origin of which is presently not fully understood. A fraction of about 2 % of the electrons which channel with a dechanneling length of 4.0 μm would explain the experimental observation. Whatever the final explanation will be, the coherent radiation component is rather small. Why this is the case is presently also an open question. Crystal defects may not be excluded. To shed light on this puzzling situation, additional experiment will be analyzed which have been performed at beam energies of 195 and 855 MeV. In addition, experiments have been performed with an undulator for which the Ge content has been reduced by about a factor of two.

Acknowledgments

We gratefully acknowledge fruitful discussions with A.V. Korol, A.V. Solov'yov, and A. Kostyuk, as well as with A. Mazzolari, D. De Salvador, and W. Wagner on experimental details.

This work was supported by the Deutsche Forschungsgemeinschaft DFG under contract BA 1336/2-1.

In part, this work was supported also by the European Commission IRSES CUTE 269131 project (Crystalline Undulator: Theory and Experiment).

References

[1] Kumakhov M A 1976 *Phys. Lett.* **A 57** 17
[2] V.V. Kaplin, S.V. Plotnikov and S.A. Vorob'ev 1980 *Zh. Tekh. Fiz.* **50** 1079-81 [English translation: *Sov. Phys. - Tech. Phys.* **25** 650-51]
[3] Baryshevsky V G, Dubovskaya I Ya and Grubich A O 1980 *Phys. Lett.* **A 77** 61
[4] Korol A V, Solov'yov A V and Greiner W 2004 *Int. J. Mod. Phys.* E **13** 867
[5] Korol A V, Solov'yov A V and Greiner W 2013 *Channeling and Radiation in Periodically Bent Crystals* (Springer Series on Atomic, Optical and Plasma Physics 69)
[6] H. Ikezi, Y. R. Lin-Liu and T. Ohkawa 1984 *Phys. Rev.* **B 30** 1567
[7] Mikkelsen U and Uggerhøj E 2002 *Nucl. Instr. Meth. in Phys. Res.* **A 483** 455
[8] S. C. Jaint and W Hayes 1991 *Semicond. Sci. Technol.* **6** 547-76
[9] Krause W, Korol A V, Solov'yov A V and Greiner W 2002 *Nucl. Instr. Meth. in Phys. Res.* **A 483** 455
[10] Backe H, Krambrich D, Lauth W, Lundsgaard Hansen J and Uggerhøj Ulrik I 2011 *IL NUOVO CIMENTO* **34** C 157-65
[11] Backe H, Krambrich D, Lauth W, Lundsgaard Hansen J and Uggerhøj Ulrik I 2013 Radiation Emission at Channeling of Electrons in a Strained Layer $Si_{1-x}Ge_x$ Undulator Crystal in *Proceedings of the 5th International Conference on Charged and Neutral Particles Channeling Phenomena - Channeling 2012, Alghero, Italy, Sept. 23-28, 2012, Nucl. Instr. and Meth.* **B** in print
[12] W. Lauth, H. Backe, P. Kunz and A. Rueda 2008 *Channeling Experiments with Electrons at the Mainz Microtron MAMI, in Charged and Neutral Particles Channeling Phenomena, Channeling 2008, Proceedings of the 51st Workshop of the INFN ELOISATRON Project, Erice, Italy, 25 October - 1 November 2008, ed. by Sultan B Dabagov & Luigi Palumbo, A. Zichichi* (New Jersey: World Scientific 2010) 335-42.
[13] Beloshitsky V V and Komarov F F 1982 *Phys. Rep.* **93** 117
[14] Andersen J U, Bunderup E and Pantell R H 1983 *Ann. Rev. Nucl. Sci.* **33** 453
[15] Kimball J C and Cue N 1985 *Phys. Rep.* **125** 69
[16] Beloshitsky V V, Komarov F F and Kumakhov M A 1986 *Phys. Rep.* **139** 293
[17] Sáens A W and Überall H 1985 *Topics in Current Physics, Coherent Radiation Sources* (Berlin: Springer)
[18] Carrigan R A and Ellison J A 1987 *Relativistic Channeling, NATO ASI series B: Physics; Vol 165, Proceedings of a NATO Advanced Research Workshop on Relativistic Channeling, Maratea, Italy, March 31 - April 4, 1986* (New York: Plenum Press)
[19] Kumakhov M A and Komarov F F 1989 *Radiation From Charged Particles in Solids* (New York: AIP American Institute of Physics)
[20] Kumakhov M A and Wedell R 1991 *Radiation of Relativistic Light Particles during Interaction with Single Crystals* (Heidelberg, Berlin, New York: Spektrum, Akademischer Verlag)
[21] Berman B L 1991 *Radiation Effects and Defects in Solids* **122-123** 277
[22] Backe H, Lauth W, Rueda A, El-Ghazaly M, Kunz P, Picard A, Scharafutdinov A, Sossalla A and Weber T 2005 *Proceedings of the Memorial Symposium for Gerhard Soff, Frankfurt am Main, Germany, April 25-26, 2005, Topics in Heavy Ion Physics, ed. W. Greiner and J. Reinhardt* (Budapest: EP Systema Bt.) 1-17.
[23] Backe H 2006 *Hyp. Int.* **171** 93-107
[24] Breese M B H 1997 *Nucl. Instr. Meth. in Phys. Res.* B **132** 540-47
[25] Backe H, Lauth W, Kunz P and A. Rueda 2008 *Nucl. Instr. Meth. in Phys. Res.* **B 266** 3835-51
[26] Gouanere M, Sillou D, Spighel M, Cue N, Gaillard M J, Kirsch R G, Poizat J-C, Remillieux J, Berman B L, Catillon P, Roussel L, Temmer M 1988 *Phys. Rev.* **B 38** 4352.
[27] Azadegan B, Wagner W, Pawelke J. 2006 *Phys. Rev.* **B 74** 045209
[28] Kephart J O, Pantell R H, Berman B L, Datz S, Park H and Klein R K 1989 *Phys. Rev.* **B 40** 4249
[29] Komaki K, Ootuka A, Fujimoto F, Horikawa N, Nakanishi T, Gao C Y, Iwata T, Fukui S, Mutou M and Okuno H 1984 *Nucl. Instr. Meth. in Phys. Res.*, **B 2** 71
[30] Adeishvili D I, Bochek G L, Vit'kov V I, Kovalenko G D and Shramenko B I 1985 *Rad. Eff. Lett.* **87** 135
[31] Baier V N, Katkov V M and Strakhovenko V M 1998 *Electromagnetic Processes at High Energies in Oriented Single Crystals* (Singapore, New Jersey, London, Hongkong: World Scientific)
[32] Backe H, Lauth W, Kunz P and Rueda A 2008 *Charged and Neutral Particles Channeling Phenomena, Channeling 2008, Proceedings of the 51st Workshop of the INFN ELOISATRON Project, Erice, Italy 15 October - 1 November 2008, ed. Sultan B Dabagov & Luigi Palumbo, A. Zichichi* (New Jersey: World Scientific) 335-42

[33] Solov'yov A V, Korol A V, Kostyuk A, Greiner W 2008 *"Electron based undulator"*, talk presented at the PECU meeting, January 18, 2008.
[34] Jackson J D 1999 *Classical Electrodynamics, Third Edition* (New York, etc.: John Wiley&Sons, Inc.)
[35] Guidi V. and Bandiera L 2012 *Phys. Rev.* **A 86** 042903
[36] Akhiezer A I, Shulga N F and Truten V I 1998 *Physics Review* Vol **19**
[37] Ter-Mikaelian M L 1972 *High-Energy Electromagnetic Processes in Condensed Media* (New York, London, Sydney, Toronto: Wiley Interscience)
[38] Kostyuk Andriy, Korol Andrei, Solovyov Andrey and Greiner Walter 2011 *J. Phys. B. At. Mo. Opt. Phys.* **44** 075208
[39] Kostyuk A, Korol A V, Solovyov A V and Greiner W 2012 arXiv: 1104.3890v2 [physics.acc-ph]
[40] Kostyuk A 2013 Recent Progress in the Theory of Crystalline Undulator in *Proceedings of the 5th International Conference on Charged and Neutral Particles Channeling Phenomena - Channeling 2012, Alghero, Italy, Sept. 23-28, 2012, Nucl. Instr. and Meth.* **B** in print
[41] Kostyuk A 2013 *Phys. Rev. Lett.* **110** 115503
[42] 2013 Sushko G B, Bezchastnov V G, Solov'yov I A, Korol A V, Greiner W and Solov'yov A V *Simulation of Ultra-Relativistic Electrons and Positrons Channeling in Crystals with MBN Explorer*, Frankfurt Institute of Advanced Studies, Preprint 2013, submitted to Journal of Computational Physics
[43] Baryshevsky V G and Tikhomirov V V 2013 *Crystal Undulators: from the Prediction to the Mature Simulations in Proceedings of the 5th International Conference on Charged and Neutral Particles Channeling Phenomena - Channeling 2012, Alghero, Italy, Sept. 23-28, 2012*, Nucl. Instr. and Meth. B, in print

Sub-GeV Electron and Positron Channeling in Straight, Bent and Periodically Bent Silicon Crystals

G B Sushko¶, A V Korol¶§, Walter Greiner¶, A V Solov'yov§ [1]

¶ Frankfurt Institute for Advanced Studies, Ruth-Moufang-Str. 1, 60438 Frankfurt am Main, Germany

§ St. Petersburg State Maritime University, Leninsky ave. 101, 198262 St. Petersburg, Russia

E-mail: korol@fias.uni-frankfurt.de

Abstract. Preliminary results of numerical simulations of electron and positron channeling and emission spectra are reported for straight, uniformly bent and periodically bent silicon crystal. The projectile trajectories are computed using the newly developed module [1] of the MBN Explorer package [2, 3]. The electron and positron channeling along Si(110) and Si(111) crystallographic planes are studied for the projectile energies 195–855 MeV.

1. Introduction

The basic effect of the channeling process in a crystal is in an anomalously large distance which particle can penetrate moving along a crystallographic plane (the planar channeling) or an axis (the axial channeling) and experiencing the collective action of the electrostatic field of the lattice ions [4]. The field is repulsive at small distances for positively charged particles and, therefore, they are steered into the inter-atomic region, while negatively charged projectiles move in the close vicinity of ion strings or planes.[2]

Channeling of charged particles is accompanied by the channeling radiation [6] which arises due to the transverse motion (the channeling oscillations) of the particle inside the channel under the action of the interplanar or axial field. The intensity of the radiation depends on the type of the projectile, on its energy, and on the type of crystal and crystallographic plane (axis). In a bent crystal, the synchrotron radiation appears as a result of circular motion of the channeling particle along the bent crystallographic planes. The condition of stable channeling in bent crystals [7, 8] inplies the bending radius R to be much smaller than the (typical) curvature radius of the channeling oscillations. Therefore, the synchrotron radiation modifies the soft-photon part of the emission spectrum. In a periodically bent crystal, additional mechanism of radiation appears to the undulating motion of channeling particles which follow the periodic bending of crystallographic planes. The concept of a crystalline undulator (CU) was formulated quite recently [9, 10]. By means of CU it is feasible to produce monochromatic undulator-like radiation in the hundreds of keV up to the MeV photon energy range. The intensity and

[1] On leave from A.F. Ioffe Physical-Technical Institute, St. Petersburg, Russia

[2] Channeling effect can be discussed not only for crystals but, in principle, for any structured material which provides "passages", moving along which a projectile has much lower value of the mean square of the multiple scattering angle than when moving along any random direction. The examples of such materials are nanotubes and fullerites, for which the channeling effects has been also investigated, see, e.g., Ref. [5].

characteristic frequencies of the radiation can be varied by changing the type of channeling particles, the beam energy, the crystal type and the parameters of periodic bending. (see recent review [11] for more details).

In recent years, several experiments were carried out to measure the channeling parameters and the characteristics of emitted radiation of sub-GeV light projectiles. These include the attempts made [12] or planned to be made [13] to detect the radiation from a positron-based CU. More recently, a series of the experiments with straight, bent and periodically bent crystals have been carried out with 195–855 MeV electron beams at the Mainz Microtron (Germany) facility [15, 16]. The CUs, used in the experiment, were manufactured in Aarhus University (Denmark) using the molecular beam epitaxy technology to produce strained-layer $Si_{1-x}Ge_x$ superlattices with varying germanium content as described in [17,18]. Another set of experiments with diamond CUs is scheduled for the year 2013 at the SLAC facility (USA) with 10...20 GeV electron beam [19].

Theoretical support of the ongoing and future experiments as well as accumulation of numerical data on channeling and radiative processes of ultra-relativistic projectiles in crystals of various content and structure must be based on an accurate procedure which allows one to simulate the trajectories corresponding to the channeling and non-channeling regimes. Recently, a universal code to simulate trajectories of various projectiles (positively and negatively charged, light and heavy) in an arbitrary scattering medium, either structured (straight, bent and periodically crystals, superlattices, nanotubes etc) or amorphous (solids, liquids) has been developed as a new module [1] of the MBN Explorer package [2, 3]. To simulate propagation of particles through media the algorithms used in modern molecular dynamics (MD) codes were applied. Verification of the code against available experimental data as well as against predictions of other theoretical models were carried out for 6.7 GeV and 855 MeV electrons and positrons in Si(110) as well as in amorphous Si [1]. In the cited paper, critical analysis was carried out of the underlying physical model and the algorithm implemented in the recent code for electron channeling described in Refs. [20]. It was shown, that the specific model for electron–atom scattering leads in a noticeable overestimation of the mean scattering angle. As a results, the data on the dechanneling lengths presented in [20] are incorrect.

In the present paper we present new results on electron and positron channeling and emission spectra in straight, uniformly bent and periodically bent silicon crystal. The electron and positron channeling along Si(110) and Si(111) crystallographic planes are studied for the projectile energies 195–855 MeV.

2. Channeling Radiation of Electrons in Si (110)

A relevant benchmark for our simulations are the channeling spectra of 195...855 MeV electrons in Si(110) that have been addressed in previous experimental [14–16, 21] studies. To this end, we have performed extensive calculations of the particles trajectories and the emitted radiation spectra formed in $L = 50$ μm crystalline silicon. Two examples of our calculations are presented in Fig. 1.

In both panels, the curves represent the spectral enhancement factor, i.e., the spectral distribution $dE/d(\hbar\omega)$ of the radiation emitted in the crystalline medium normalized by that in the amorphous silicon. The latter was calculated within the Bethe-Heitler approximation [22].

For 270 MeV incident energy (left panel) the presented spectrum, which corresponds to the aperture 0.24 mrad, was averaged over $N_0 = 3500$ simulated electron trajectories. At the crystal entrance, the initial velocity for each trajectory was parallel to the Si(110) plane and aligned along the $[10, -10, 1]$ crystallographic direction. The error bars reflect the statistical uncertainties which correspond to the probability 0.999.

For 855 MeV electrons (right panel) the spectra were computed for the aperture 0.21 mrad and for three different directions of the initial velocity as indicated. The numbers of simulated

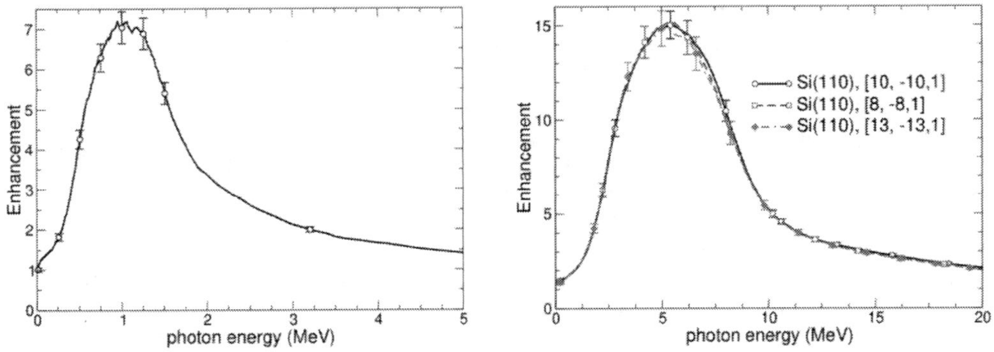

Figure 1. Enhancement factor for 270 MeV (left panel) and for 855 MeV (right panel) electron channeling along Si(110) plane in $L = 50$ μm thick straight crystal. The 270 MeV data refer to the maximum emission angle $\theta_{\max} = 0.24$ mrad, the 855 MeV data – to $\theta_{\max} = 0.21$ mrad. Three curves on the right panel correspond to three different directions of the incident beam, as indicated.

trajectories are: $N_0 = 3120$ for the $[10, -10, 1]$ direction, $N_0 = 1919$ for $[8, -8, 1]$, and $N_0 = 1679$ for $[13, -13, 1]$. The goal of these calculations was to analyze the sensitivity of the channeling radiation to the initial (random) direction along the (110) plane. We can state that all spectra are indistinguishable within statistical errors.

The theoretical results for the enhancement factors were compared to the experimental data [21]. The analysis of the theory-versus-experiment will be published elsewhere. Here we just note that perfect correspondence, both in the peak position and height, was found for 270 MeV incident electron beam. For 855 MeV electrons, theoretical curves reproduce the peak position but underestimate the intensity of the channeling radiation by approximately 25-30 per cent. The analysis of the source of this discrepancy is currently under way.

3. 855 MeV Electrons and Positrons in Straight and Bent Si (111)

The structure of a silicon single crystal implies that a Si(111) planar channel contains two (111) planes separated by the distance $d_{\rm n} = 0.784$ Å . Total width of the channel is $d = 3.136$ Å.

The presence of two planes of crystal atoms in a single channel leads to specific features of the channeling oscillations for both negatively and positively charged projectiles. These features are absent in the case of channeling along (100) or/and (110) planes.

To discuss qualitatively the channeling oscillations of an electron and a positron we refer to Figure 2, which presents the Si(111) interplanar potential U calculated in the continuous approximation [4] with the use of the Molière atomic potential. Let us stress, that the results of numerical simulations presented further in this section, were not based on the continuous approximation. The latter is used for illustrative considerations only.

In the case of electron channeling (left panel in Figure 2) the interplanar potential has two wells symmetrically separated with respect to the midplane, where the potential has a local maximum U_0. At the boarders of the channel, i.e. at the distances $\pm d/2$, the potential has maxima $U_{\max}$ which exceed U_0. As a result, if the transverse energy $\varepsilon_{\perp}$ of an electron satisfies the condition $\varepsilon_{\perp} < U_0$ then the channeling oscillations are restricted to one of the wells. In the case $\varepsilon_{\perp} > U_0$ the particle oscillates with larger amplitude within $[-d/2, d/2]$. The electron interplanar potential is strongly anharmonic, therefore, the period of oscillations depends on the amplitude. On average, the large-amplitude oscillations are slower than the small-amplitude ones in the vicinity of the local minima.

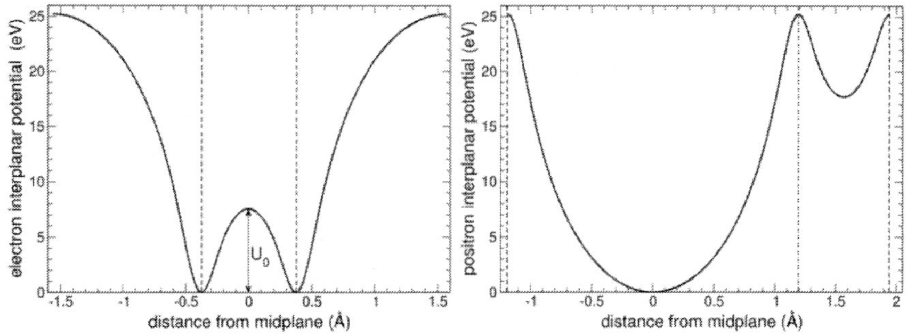

Figure 2. Electron (left panel) and positron (right panel) *continuous* Molière interplanar potential for Si(111). Vertical dashed lines mark the positions of the atomic planes. The curves correspond to the crystal temperature 300 K.

The positron interplanar potential is presented on the right panel of Fig. 2. In this case, the potential also has two wells although strongly asymmetric. Both of the wells can be approximated by parabolic dependencies. The frequency of channeling oscillations in the narrow (and shallow) well is approximately 2 times larger than that in the wide well.

Two types of channeling oscillations for electrons and positrons manifest themselves in the emission spectra calculated for small emission angle and discussed below in the paper.

The motion of 855 MeV electrons and positrons collimated at the entrance along Si(111) planes is illustrated in Fig. 3 by sets of randomly chosen simulated trajectories. The data refer to the straight crystal of the length $L = 100$ μm.

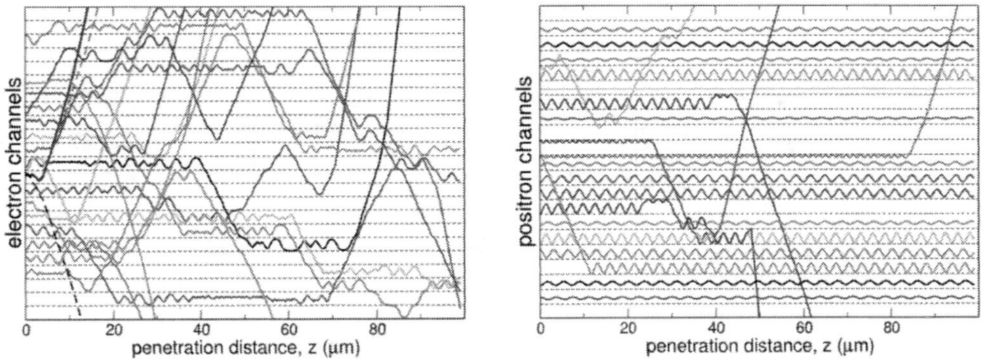

Figure 3. Channeling of 855 MeV electrons (left panel) and positrons (right panel) in a 100 μm thick silicon crystal. The plots show randomly chosen trajectories of the particles initially collimated along Si(111) crystallographic planes. Horizontal dashed lines indicate the (111) planar channels separated by the distance $d = 3.136$ Å.

For positrons, noticeable are nearly harmonic oscillations. Two types of oscillations, occurring in the wide part of the channel and in the narrow as well, are clearly seen in the presented trajectories. Another feature of positron channeling through a $L = 100$ μm thick crystal is a comparatively small number of the dechanneling events. This is also not surprising if one compares the crystal size with the dechanneling length $L_d \approx 700$ μm for a 855 MeV positron in Si(111). The latter value can be obtained using Eq. (1.50) from [8] with the correction for a light projectile introduced in [23]. Therefore, it is not surprising that most of the incident particles traverse the crystal in the channeling mode.

Much less regular are the channeling oscillations of electrons, see the left panel of Fig. 3. The electron trajectories exhibit a broader variety of features: channeling motion, over-barrier motion, rechanneling process, rare events of hard collisions etc. First, let us note that the dechanneling length of a 855 MeV electron in Si(111), estimated with the help of Eq. (10.1) from [24], is $L_{\rm d} \approx 23$ μm. Therefore, it is not surprising that the events of channeling through the whole crystal are quite rare. On the other hand, the events of rechanneling, i.e., capture to the channeling mode of an over-barrier particle, are quite common for electrons. Even the multiple rechanneling events are not rare. This phenomenon has been already noted in the simulations of the electron channeling [1,20] with a qualitative explanation provided [20] of the difference in the rechanneling rate for positively and negatively charged projectiles. Also it is worth noting a visible anharmonicity in the channeling oscillations of electrons which is a direct consequence of a strong deviation of the the electron interplanar potential from a harmonic shape. As a result, the period of the oscillations varies with the amplitude. Similar to the positron channeling, two types of oscillations, related to the two wells structure of the interplanar potential (see Fig. 2 left), are clearly seen in the presented trajectories.

The simulations of the trajectories were performed for straight and bent Si(111) channels. In the former case, two crystal lengths along the incident beam direction were considered: $L_1 = 50$ μm and $L_2 = 100$ μm. For the bent crystal (the uniform bending with the bending radius $R = 33$ mm) the simulations were carried out for $L_1 = 50$ μm in accordance with the experimental conditions at the Mainz Microtron facility [21].

The simulated trajectories were used to estimate the dechanneling length (in the case of the electron channeling) and to calculate spectral distribution of the emitted radiation.

To quantify the electron dechanneling process we calculated two penetration lengths introduced in Ref. [1]. The first one, notated below as $L_{\rm p1}$ was found as a mean value of the primary channeling segments, which started at the entrance and lasted till the dechanneling point somewhere inside the crystal. Generally speaking, this quantity is dependent on the angular distribution of the particles at the entrance. The $L_{\rm p1}$ values quoted below were obtained for a zero-emittance beam collimated initially along the (111) planar direction. The second penetration depth, $L_{\rm p2}$, is defined as a mean value of all channeling segments, including those which appear due to the rechanneling. In the rechanneling process an electron is captured into the channeling mode having, statistically, an arbitrary value of the incident angle ψ not greater than Lindhard's critical angle. Therefore, $L_{\rm p2}$ mimics the penetration depth of the beam with a non-zero emittance $\approx \psi_{\rm L}$.

In addition to $L_{\rm p1}$ and $L_{\rm p2}$ we calculated the total channeling length $L_{\rm ch}$, defined as an average length of all channeling segments per trajectory.

The results for $L_{\rm p1}$, $L_{\rm p2}$ and $L_{\rm ch}$, together with the calculated values of the channel acceptance $\mathcal{A} = N_{\rm acc}/N_0$ (where N_0 and $N_{\rm acc}$ are numbers of the incident and the accepted particles, respectively), are summarized in Table 1.

For a straight crystal, it is instructive to compare the obtained values $L_{p1} = 18.70 \pm 0.69$ and $L_{p2} = 15.92 \pm 0.40$ μm with the dechanneling lenghts for the initial beam, $L_{\rm d0} = 13.57 \pm 0.12$ μm, and for the rechanneled particles, $L_{\rm d} = 13.69 \pm 0.07$ μm obtained in Ref. [20]. The calculations performed in the cite paper were based on the peculiar model of the elastic scattering of an ultra-relativistic projectile from the crystal constituents. The model substitutes the atom with its "snapshot" image: the atomic electrons are treated as point-like charges placed at fixed positions around the nucleus. The interaction of an ultra-relativistic projectile (e.g., an electron) with each atomic constituent is treated in terms of the classical Rutherford scattering. In Ref. [1] it was demonstrated, that such a "snapshot" model noticeably overestimates the mean scattering angle in the process of elastic scattering in a single electron-atom collision. The mean square

Table 1. The penetration lengths L_{p1}, L_{p2} and $L_{\rm chan}$ for $\varepsilon = 855$ MeV electrons in straight and bent Si (111) crystals (the bending radius is 33 mm). The crystal lengths are $L_1 = 50$ μm and $L_2 = 100$ μm. Also indicated are: the number of simulated trajectories N_0, the number of accepted particles $N_{\rm acc}$, and the acceptance $\mathcal{A} = N_{\rm acc}/N_0$.

Crystal	N_0	N_a	$\mathcal{A}$	L_{p1} (μm)	L_{p2} (μm)	$L_{\rm chan}$ (μm)
straight L_1	3467	2560	0.74	18.37 ± 0.82	15.48 ± 0.52	24.72 ± 0.83
straight L_2	2160	1632	0.76	19.22 ± 1.24	16.38 ± 0.61	38.71 ± 1.76
straight L_1 & L_2	5627	4192	0.75	18.70 ± 0.69	15.92 ± 0.40	—
bent L_1	3603	2471	0.69	16.62 ± 0.77	15.96 ± 0.67	19.15 ± 1.11

angle for a single scattering is a very important quantity in the multiple-scattering region, where there is a large succession of small-angle deflections symmetrically distributed about the incident direction. It was noted in Ref. [1] that the "snapshot" approximation underestimates the dechanneling length of 855 MeV electrons in straight Si (110) by approximately 30 per cent. Similar to this, the Si(111) data from [20], undervalues the dechanneling length presented in Table 1: $L_{\rm d0}$ is less than L_{p1} by 37 ± 5 % whereas $L_{\rm d}$ is 17 ± 3% smaller than L_{p2}.

Let us note that the obtained length $L_{p1} = 18.70 \pm 0.69$ μm is in agreemeent with the value 18.8 μm evaluated recently in Ref. [25] from the solution of the Fokker-Plank equation.

The simulated trajectories were used to compute spectral distribution of the emitted radiation following the formalism and algorithm described in detail in Ref. [1]. The results are presented in figures 5 – 6. The calculated spectral intensities are normalized to the Bethe-Heitler values (see, for example, Ref. [22]) and, thus, are plotted as the enhancement factors over the bremsstrahlung spectrum in amorphous silicon. Statistical uncertainties due to the finite number ($\approx 3000 \dots 4000$) of the analyzed trajectories are indicated by the error bars. The calculations were performed for two detector apertures: $\theta_{\max} = 0.21$ and 2 mrad. The first value, which is close to the aperture used in the experiments with the 855 MeV electron beam [14, 15, 21], is much smaller than the natural emission angle $\gamma^{-1} \approx 0.6$ mrad. Therefore, the corresponding spectra refer to nearly forward emission. On the contrary, the second angle greatly exceeds γ^{-1}, so that the cone $\theta_{\max}$ collects nearly all emitted radiation.[3]

Figures 4 and 5 present the enhancement of radiation in straight silicon crystals.

First, we note that for both electrons and positrons the intensity of radiation in the oriented crystal greatly exceeds (by more than an order of magnitude) the bremsstrahlung background. The enhancement comes from the particles moving along quasi-periodic channeling trajectories, which bear close resemblance with the undulating motion. As a result, constructive interference of the waves emitted from different but similar parts of the trajectory are added coherently. For each value of the emission angle θ the coherence is most pronounced for the radiation into harmonics, which frequencies can be estimated as (see, e.g., [24]):

$$\omega_n = \frac{2\gamma^2\,\Omega_{\rm ch}}{1 + \gamma^2\theta^2 + K_{\rm ch}^2/2}\, n, \quad n = 1, 2, 3, \dots, \qquad (1)$$

where $\Omega_{\rm ch}$ is the frequency of channeling oscillations and $K_{\rm ch}^2 = 2\gamma^2 \left\langle v_\perp^2 \right\rangle / c^2$ is the mean square of the undulator parameter related to them. Within the framework of continuous potential

[3] The intensity of channeling radiation of 855 MeV electrons in Si(111) crystals of various thickness was calculated in [25] for large apertures. In the cited paper the channeling process was modeled on the basis of the Fokker-Plank equation.

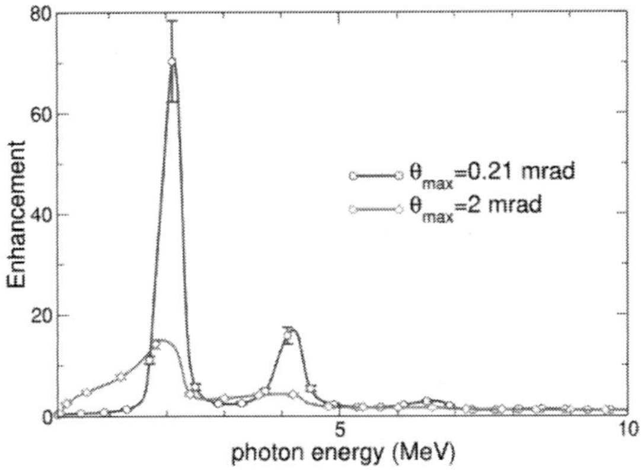

Figure 4. Enhancement factor for 855 MeV *positrons* channeled in $L = 50$ μm crystalline target along Si (111) planes calculated for two values of the maximum emission angle as indicated.

approximation, these quantities are dependent on the magnitude of the transverse energy which, in turn, determines the amplitude of oscillations.

Different character of channeling by positrons and electrons results in differences in the spectra of the channeling radiation.

The nearly perfect sine-like channeling trajectories of positrons lead to the emission spectrum close to that of the undulator radiation with $K^2 \ll 1$. Two peaks in the positron spectrum, see Fig. 4, is due to two types of channeling oscillations mentioned above. The peak at ≈ 2 MeV is due to the emission in the fundamental harmonic ($n = 1$) from the trajectories corresponding to the channeling motion in the wide well of the positron Si(111) channel (see Fig. 2 right). It is more pronounced for the smaller aperture, since in this case a strong inequality $(\gamma\theta)^2 \ll 1$, valid for all angles $\theta \le \theta_{\max}$, ensures the independence of ω_1 on the emission angle. The second, less accented peak, corresponds to the emission in the first harmonic due to the channeling motion in the narrow part of the channel. In this case, the amplitudes are smaller (this result in the decrease of the intensity) but the channeling frequencies are higher leading to the higher value of ω_1. For the larger aperture, a big part of the energy is radiated into the cone $\gamma^{-1} < \theta < \theta_{\max}$. For these relatively large emission angles the first harmonic energy decreases with θ. As a result, the peaks become broader and less intensive.

Due to strong anharmonicity of the electron channeling oscillations, the peaks in the spectrum of channeling radiation are less pronounced even for the small aperture, see Fig. 5 left. For the large aperture $\theta_{\max} = 2$ mrad the second peak is completely smeared out. The right panel in the figure illustrates the differences in the emission spectra (for the small aperture) for electron channeling in Si (111) and Si (110) channel (the calculations of the latter were performed in [1]). The Si(110) channel can be modeled as a single-well interplanar potential which leads to a single peak in the emission spectrum.

Enhancement of radiation emitted by 855 MeV positrons and electrons in uniformly bent Si (111) channels are presented in Figs. 6 and 7. The data correspond to the crystal length $L = 50$ μm and the bending radius $R = 33$ cm. In a bent crystal, the channeling condition implies the centrifugal force $F_{\rm cf} \approx \varepsilon/R$ to be smaller than the maximum interplanar force $U'_{\max}$ [7]. In our case, the value of the bending parameter $C = F_{\rm cf}/U'_{\max}$ [11] is much less than one, $C \approx 0.045$, so that the channeling condition is well-fulfilled. This is reflected, in particular, by small deviation of the values of the penetration lengths $L_{\rm p1}$, $L_{\rm p2}$ and $L_{\rm ch}$ calculated for an electron channeled in the bent Si(111) from those for the straight channel, see Table 1.

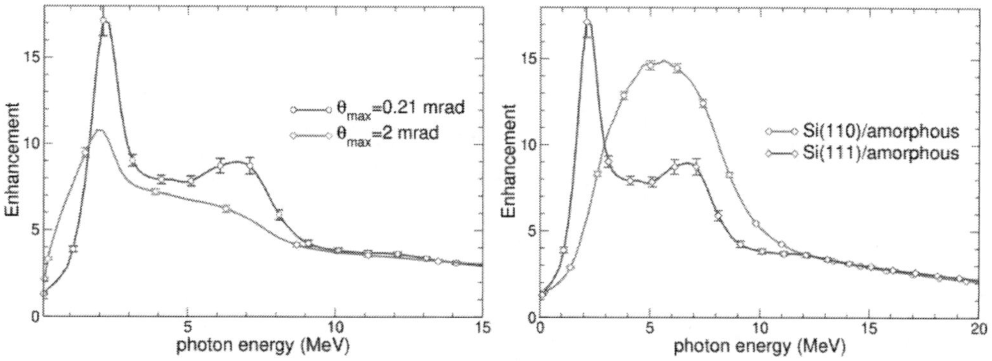

Figure 5. *Left panel.* Enhancement factors, calculated for two indicated apertures $\theta_{\max}$, for 855 MeV *electron* channeling in $L = 50$ μm straight Si crystal along (111) planes
Right panel. Enhancement factor for 855 MeV *electrons* channeled in $L = 50$ μm straight silicon crystal along (110) and (111) planes. The data refer to $\theta_{\max} = 0.21$ mrad.

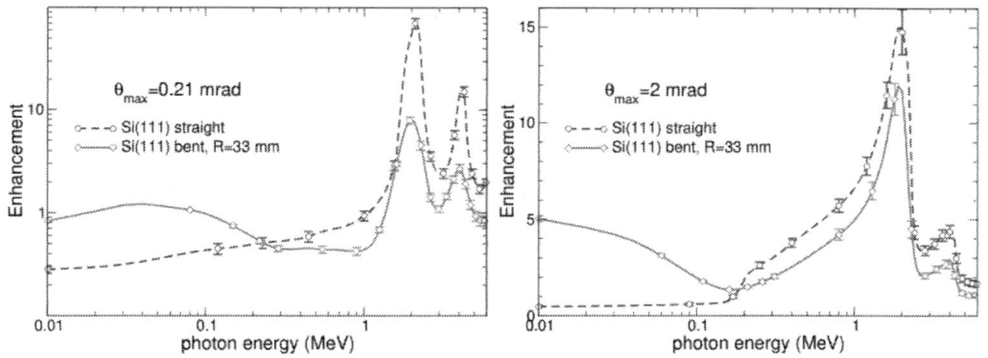

Figure 6. Enhancement factor calculated for two apertures $\theta_{\max}$ for 855 MeV *positrons* channeled along straight (dashed line) and bent (solid line) Si (111) plane. The bending radius $R = 33$ mm, the crystal length $L = 50$ μm. Note the double log scale in the left graph.

Let us note two features of the emission spectra formed in bent silicon crystal. First, the bending gives rise to the synchrotron radiation, since the channeled particle experiences the circular motion in addition to the channeling oscillations. This leads to the increase of the intensity in the photon energy range $\lesssim 10^2$ keV. For these energies the excess of the radiation yield from the bent channel over that from the straight channel is clearly seen in the figures (compare the solid and the dashed curves). Second, it is seen that the decrease in the intensity of the channeling radiation in the bent channel is more pronounced for the small aperture, $\theta_{\max} = 0.21$ mrad than for the large one equal to 2 mrad. This feature can be understood by comparing $\theta_{\max}$ to the bending angle $\Theta = L/R \approx 1.5$mrad $\gg \gamma^{-1}$. Since the emission angle is defined with respect to the direction of the incident beam, the emission within the cone $\theta_{\max} = 0.21$ mrad $\ll \Theta$ occurs, effectively, only from the initial parts of the channeling trajectories. In the opposite limit, $\theta_{\max} > \Theta$, the radiation from nearly all the trajectory is detected.

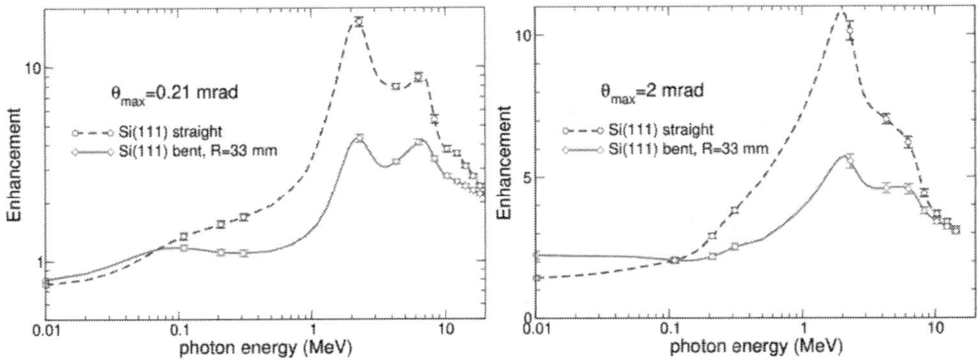

Figure 7. Same as in Fig. 6 but for 855 MeV *electrons.*

4. Channeling and Radiation in Crystalline Undulators

In this section we present some of the results obtained recently with the help of the newly developed code [1] for the channeling phenomenon and radiation emitted in a crystalline undulator (CU). In CU, in addition to the channeling radiation, the undulator-type radiation appears due to the undulating motion of channeling particles which follow the periodic bending of crystallographic planes [9–11]. Two types of harmonic periodic bending of the channel centerline, which correspond to the sine and to the cosine profiles, can be considered:

$$y(z) = a \sin\left(2\pi z/\lambda_{\rm u}\right), \qquad y(z) = a \cos\left(2\pi z/\lambda_{\rm u}\right) . \qquad (2)$$

Here, the coordinate z is measured along the straight channel, the y axis is perpendicular to the straight plane. The quantities a and $\lambda_{\rm u}$ are the bending amplitude and period and they satisfy the relation $d < a \ll \lambda_{\rm u}$ where d is the interplanar distance (see the cited papers for more details on the description of the CU concept).

We have performed simulation of the trajectories, the quantitative analysis of the channeling motion and computation of the spectral intensities of the radiation formed by ultra-relativistic *electrons and positrons* within the energy range 195 . . . 855 MeV in the CU with the parameters used in the experiments at at the Mainz Microtron (Germany) facility [15, 16]. The 4-periods CUs were manufactured in Aarhus University (Denmark) using the molecular beam epitaxy technology to produce strained-layer $Si_{1-x}Ge_x$ superlattices with varying germanium content as described in [17, 18].

The following values of the CU parameters were used in the calculations:

- Channeling plane: Si(110) (the interplanar distance $d = 1.92$ Å)
- Crystal length: $L = 39.6$ μm
- Bending period: $\lambda_{\rm u} = 9.9$ μm
- Bending amplitude: $a = 3 \ldots 5$ Å

In full the results of our calculations will be published elsewhere. Here we announce only few results for 355 and 855 MeV electrons which are presented in Fig. 8. The left panel compares the spectral enhancement of radiation emitted by 355 MeV electrons channeled in straight Si (110) and in the sine-shaped CU with two bending amplitudes as indicated. All three dependencies exhibit powerful maximum at about 1.7 MeV which corresponds to the channeling radiation. In the case of the undulating crystals the maxima are lower is due to the decrease in the allowed amplitude of the channeling oscillations in periodically bent channel in comparison with the

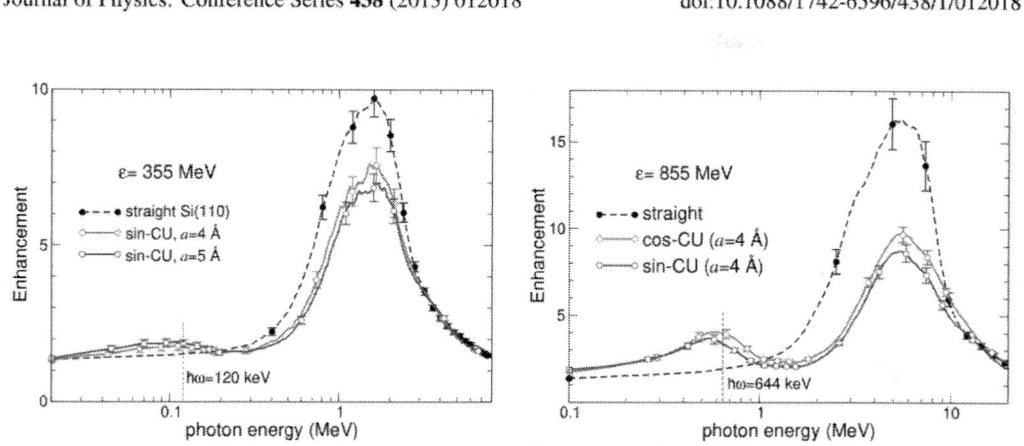

Figure 8. Enhancement of radiation emitted in straight Si (110) (dashed lines) and in CU (solid lines) over the Bethe-Heitler spectrum. Left panel: the data refer to $\varepsilon = 355$ MeV electrons. Two solid lines present the dependencies for the sine-like CUs of the indicated amplitudes. Righ panel: the data refer to $\varepsilon = 855$ MeV electrons. Two solid lines stand for the spectra emitted from the sine- and cosine-like CUs of the same amplitude. The maximum emission angle is $\theta_{\max} = 0.21$ mrad for both panels.

straight one [23]. However, the CU undulator radiation manifests itself as a hump in the photon energy range 40...100 keV (the vertical dashed line marks the first harmonic of the radiation in the forward direction). Hence, our simulations indicate that it is possible to observe the CU radiation even for comparatively low energies of the electron beam.

With the increase of the electron beam energy the CU radiation peak becomes more accented, as it is illustrated by the right panel. Here, the maximum at about 600 keV is seen for both sine- and cosine-shaped CUs.

5. Conclusion

Using the newly developed code [1], which was implemented as a module in the MBN Explorer package [2], we have performed the Monte Carlo simulations of trajectories of ultra-relativistic electrons and positrons in oriented straight, bent and periodically bent single Si crystals.

The simulated trajectories were used as the input data for numerical analysis of the intensity of the emitted radiation. In the case of straight crystals the reduces to the channeling radiation emitted atop the incoherent bremsstrahlung background. In a bent channel the spectrum is enriched by the synchrotron radiation due to the circular motion of the projectile along the bent centerline. In a periodically bent crystal, in addition to the channeling radiation, the undulator-type radiation appears due to the periodicity of the bending. In this paper, for the first time the characteristics of the crystalline undulator radiation were computed on the basis of the Monte Carlo simulation of electron and positron trajectories.

The calculation of the spectra as well as the numerical analysis of channeling conditions and properties (acceptance, dechanneling length) have been carried out in a broad range of the beam energies, 195 ... 855 MeV. The obtained and presented results are of interest in connection with the ongoing experiments with electron beams at Mainz Microtron [15] and with possible experiments with the positron beam [13].

Acknowledgments

We are grateful to Hartmut Backe and Werner Lauth for fruitful and stimulating discussions. The work was supported by the European Commission CUTE (the CUTE-IRSES project, grant

GA-2010-269131). The possibility to perform complex computer simulations at the Frankfurt Center for Scientific Computing is gratefully acknowledged.

References

[1] Sushko G B, Bezchastnov V G, Solov'yov I A, Korol A V, Greiner W and Solov'yov A V 2013 *J. Comp. Phys.* (in print)
[2] Solov'yov I A, Yakubovich A V, Nikolaev P V, Volkovets I and Solov'yov A V 2012 *J. Comp. Chem.* **33** 2412–2439
[3] http://www.mbnexplorer.com/
[4] Lindhard J 1965 *K. Dan. Vidensk. Selsk. Mat. Fys. Medd.* **34** 1-64
[5] Artru X, Fomin S P, Shulga N F, Ispirian K A, Zhevago N K 2005 *Phys. Rep.* **412** 89-189
[6] Kumakhov M A 1976 *Phys. Lett.* **57A** 17–18
[7] Tsyganov E N 1976 Fermilab Preprint TM-682, TM-884 (Fermilab, Batavia)
[8] Biryukov V M, Chesnokov Yu A and Kotov V I 1996 *Crystal Channeling and its Application at High-Energy Accelerators* (Springer-Verlag, Berlin, Heidelberg)
[9] Korol A V, Solov'yov A V and Greiner W 1998 *J. Phys. G: Nucl. Part. Phys.* **24**, L45–L53
[10] Korol A V, Solov'yov A V and Greiner W 1999 *Int. J. Mod. Phys.* E **8** 49-100
[11] Korol A V, Solov'yov A V and Greiner W 2013 *Channeling and Radiation in Periodically Bent Crystals*, (Springer-Verlag Berlin Heidelberg)
[12] Baranov V T, Bellucci S, Biryukov V M, Britvich G I, Chepegin V N *et al* 2006 *Nucl. Instrum. Methods* **252** 32–35
[13] Backe H, Krambrich D, Lauth W, Buonomo B, Dabagov S B, Mazzitelli G, Quintieri L, Hansen J. Lundsgaard Uggerhøj U K I, Azadegan B, Dizdar A and Wagner W 2011 *Nuovo Cimento* C **34** 175–180
[14] Backe H, Kunz P, Lauth W and Rueda A 2008 *Nucl. Instrum. Methods* B **266** 3835–3851
[15] Backe H, Krambrich D, Lauth W, Hansen J Lundsgaard and Uggerhøj U K I 2011 *Nuovo Cimento* C **34** 157-165
[16] Backe H, Lauth W, Kunz P, Rueda A, Esberg A *et al* 2010 in *Charged and Neutral Particles Channeling Phenomena* (Eds: S B Dabagov, L Palumbo, A Zichichi) (Singapore, World Scientific) pp 281–290
[17] Mikkelsen U, Uggerhøj E 2000 *Nucl. Instrum. Methods* B **160** 435–439
[18] Krause W, Korol A V, Solov'yov A V, Greiner Walter 2002 *Nucl. Instrum. Methods* A **483** 455–460
[19] Uggerhøj U 2012 Private communication
[20] Kostyuk A, Korol A V, Solov'yov A V, Greiner Walter 2011 *J. Phys. B: At. Mol. Opt. Phys.* **44** 075208
[21] Backe H and Lauth W 2013 Private communication
[22] Tsai Yung-Su 1974 *Rev. Mod. Phys.* **46** 815–851
[23] Korol A V, Solov'yov A V, Greiner Walter 2001 *J. Phys. G: Nucl. Part. Phys.* **27** 95-125
[24] Baier V N, Katkov V M, Strakhovenko V M 1998 *Electromagnetic Processes at High Energies in Oriented Single Crystals* (World Scientific, Singapore)
[25] Bogdanov O V and Dabagov S N 2012 *J. Phys.: Conf. Ser.* **357** 012029

International Conference on Dynamics of Systems on the Nanoscale (DySoN 2012) IOP Publishing
Journal of Physics: Conference Series **438** (2013) 012019 doi:10.1088/1742-6596/438/1/012019

Simulations of electron channeling in bent silicon crystal

G B Sushko[1], V G Bezchastnov[1,4], A V Korol[1,2], Walter Greiner[1], A V Solov'yov[1,5], R G Polozkov[3], V K Ivanov[3]

[1] Frankfurt Institute for Advanced Studies, Ruth-Moufang-Str. 1, 60438 Frankfurt am Main, Germany

[2] St. Petersburg State Maritime University, Leninsky ave. 101, 198262 St. Petersburg, Russia

[3] St. Petersburg State Polytechnic University, Politekhnicheskaya str. 29, 195251 St. Petersburg, Russia

E-mail: korol@fias.uni-frankfurt.de

Abstract. We report on the results of theoretical simulations of the electron channeling in a bent silicon crystal. The dynamics of ultra-relativistic electrons in the crystal is computed using the newly developed part [1] of the MBN Explorer package [2,3], which simulates classical trajectories of in a crystalline medium by integrating the relativistic equations of motion with account for the interaction between the projectile and crystal atoms. A Monte Carlo approach is employed to sample the incoming electrons and to account for thermal vibrations of the crystal atoms. The electron channeling along Si(110) crystallographic planes are studied for the projectile energies 195–855 MeV and different curvatures of the bent crystal.

1. Introduction

Channeling is a phenomenon which is known to arise when a charged particle enter a crystal at angles small enough with respect to crystallographic planes or axes [4]. The particles become confined and forced to move through the crystal preferably along the corresponding crystallographic direction experiencing the collective action of electrostatic field of the lattice ions. The field is repulsive for positively charged particles and, therefore, they are steered into the inter-atomic region, while negatively charged projectiles move in the close vicinity of ion strings or planes.

Under certain conditions [5] the channeling effect persists even if a crystal is bent. In this case, the particle deviates from its initial direction of motion due to extremely strong electrostatic field in the crystal. The field strength is typically of the order of 10^{10} V/cm which is equivalent to the magnetic field of approximately 3000 T. Therefore, bent crystal can steer particles much more effectively than any existing magnet. In particular, in experimental high energy physics the crystals with bent crystallographic planes are used to steer high-energy charged particle beams replacing huge dipole magnets (see, e.g. [10]).

[4] On leave from Department of Theoretical Astrophysics, Ioffe Physical-Technical Institute, Politechnicheskaya Str. 26, 194021 St. Petersburg, Russia and from St. Petersburg State Polytechnical Universty, Politechnicheskaya 29, 195251 St. Petersburg, Russia

[5] On leave from A.F. Ioffe Physical-Technical Institute, St. Petersburg, Russia

A comprehensive review of theoretical and experimental achievements in the investigation of the channeling effect in straight and bent crystals one can find in [4, 6–10].

Recently, the crystalline undulator (CU) concept was formulated for producing undulator-like electromagnetic radiation in the sub hundreds of keV up to the MeV photon energy range [11,12]. In a CU the radiation is emitted by a beam of ultra-relativistic charged particles which undergo channeling in a periodically bent crystal. As a result, in addition to a well-known channeling radiation, there appears the radiation due to the undulating motion of the particles which follow the periodic bending of crystallographic planes. The intensity and characteristic frequencies of the CU radiation can be varied by changing the type of channeling particles, the beam energy, the crystal type and the parameters of periodic bending. We refer to Ref. [13] which represents the underlying fundamental physical ideas as well as the theoretical, experimental and technological advances made during the fifteen years in exploring various features of CUs and its radiation.

Several experimental attempts were made [14] or planned to be made [15] to detect the radiation from a positron-based CU. So far, the attempts have not been successful due to various reasons (see discussion in Refs. [13, 15]). However, quite recently the first signatures showing evidence for the CU radiation were experimentally observed for 195–855 MeV electrons at the Mainz Microtron (Germany) facility [16, 17]. Another set of experiments is scheduled for the year 2013 at the SLAC facility (USA) with 10 – 20 GeV electron beam [18].

Theoretical support of ongoing and future experiments as well as accumulation of numerical data on channeling and radiative processes of ultra-relativistic projectiles in straight, bent and periodically bent crystals must be based on an accurate procedure which allows one to simulate the trajectories corresponding to the channeling and non-channeling regimes. The procedure must include a rigorous description of the particle motion and an efficient algorithm of its numerical realization. It is strongly desirable to make the procedure as much model-independent as possible. This means that the only allowed parameters are those which describe *pairwise* interactions of the projectile with constituent atoms.

Most of the existing codes, capable to simulate channeling process, can be categorized as following. Some of them (see, for example, Refs. [19–21]) are based on the concept of the continuous potential [4]. This approximation, being adequate in describing the channeling motion, becomes less accurate and more model-dependent when accounting for uncorrelated scattering events. The accurate description of the latter is essential for a quantitative analysis of the dechanneling and rechanneling processes. Other group of the codes [22–24] utilizes the scheme of binary collisions which assumes that the motion of a projectile at all times is influenced by the force due to the nearest atom. Computer facilities available at present allow one to go beyond this limitation and to account for the interaction with larger number of the crystal atoms. Such an extension of the binary collisions algorithm was implemented in the recent code for electron channeling described in Refs. [27]. The code, however, was based on the specific model for electron–atom scattering which results in a noticeable overestimation of the mean scattering angle. In more detail this topic was addressed in Ref. [1] (see also comments in Section 3).

To simulate propagation of ultra-relativistic particles through media, the channeling process in particular, one can approaches and algorithms used in modern molecular dynamics codes (see Ref. [2] for the comparative review of the codes). The latter allow one to model the dynamics of various molecular system by efficient numerical integration of classical equations of motion for all atoms in the system. The interaction between atoms is implemented in terms of interatomic potentials, the types and parameters of which can be chosen from a broad range to ensure the most adequate quantitative description of the simulated molecular system.

Recently, the molecular dynamics concept was applied to describe the motion of a single projectile in the static field of atoms which constitute a scattering medium [1]. For this purpose we have used the MBN Explorer software package [2,3] and supplied it with additional module which treats classical relativistic equations of motion and generates the crystalline environment dynamically in the course of particle propagation. This module, combined with the variety of interatomic potentials implemented in MBN EXPLORER, makes the program a unique tool for studying relativistic phenomena in various environments, such as crystals, amorphous bodies, biological medium. Verification of the code against available experimental data was carried out for 6.7 GeV projectile electrons and positrons in channeling in straight Si(110) channels [1].

In the present paper we present the results obtained with the new code [1] on the Monte Carlo simulation of the channeling of electrons along the (110) crystallographic plane in straight and bent single silicon crystal. The calculations were performed for $\varepsilon = 195, 345, 600$ and 855 MeV electrons which correspond to the experimental conditions at the Mainz Microtron (Germany) [30]. We have restricted the current analysis to the bent crystals with constant curvature, although these results will help one to estimate the reasonable parameters of periodically bent crystal channels and, therefore, will facilitate future simulations and experimental study of the electron-based crystalline undulator.

2. Physical model for channeling

Upon entering a crystal, a particle with the charge q experiences an electrostatic field $U(\mathbf{r})$ of the crystalline environment. To compute the motion of projectiles in the latter potential, we employ the relativistic classical dynamics and solve the following pair of equations

$$\mathrm{d}\mathbf{r}/\mathrm{d}t = \mathbf{v}, \quad \mathrm{d}\mathbf{p}/\mathrm{d}t = -q\,\partial U/\partial \mathbf{r} \tag{1}$$

for the particle coordinate $\mathbf{r}(t)$ and velocity $\mathbf{v}(t)$ as the functions of propagation time t. The momentum $\mathbf{p}$ relates to the velocity $\mathbf{v}$ according to the relativistic expression $\mathbf{p} = m\gamma\mathbf{v}$, where m is the mass of the particle, $\gamma = [1-(v/c)^2]^{-1/2} = \varepsilon/(mc^2)$ is the Lorenz factor, ε is the projectile energy and c is the speed of light.

To incorporate the many-body interaction of the projectile with the crystalline constituents, the potential U is represented by the sum of atomic potentials $U_{\rm at}$,

$$U(\mathbf{r}) = \sum_j U_{\rm at}\left(\mathbf{r} - \mathbf{R}_j\right), \tag{2}$$

where $\mathbf{R}_j$ is the position vector of the j-th atom in the crystal. The atomic potentials are evaluated according to the Molière approximation [25]. The computations with MBN EXPLORER can also be performed, if demanded, employing a more elaborated Pacios approximation for the atomic potentials [26].

Formally, the sum in Eq. (2) is carried out over all atoms of the sample. However, the atomic potentials decay with increasing distance from the atoms, and the decay becomes exponential when the distances exceed the atomic sizes. This allows one to restrict the number of atoms interacting with the projectile by accounting only for those separated from the projectile by distances shorter than some cut-off distance. Selection of these atoms for a given location of the projectile is facilitated by a linked cells algorithm implemented in MBN EXPLORER [2,3].

The module for channeling simulations, implemented in MBN EXPLORER, has been designed to describe the crystalline structure acting on a moving projectile "on the fly" in the course of integrating the dynamical equations (1). This is achieved by introducing a "dynamic simulation box" which shifts following the propagation of the particle (see Ref. [1] for more details). Within the simulation box, the equilibrium positions of the crystal nodes are determined according to the lattice structure under study. Then, the thermal vibrations of the crystal are accounted for by displacing the atoms with respect to the equilibrium positions. For each atom, a displacement Δ_i along a given Cartesian axis $i = x, y, z$ is selected randomly according to the normal distribution

$$w(\Delta_i) = \left(2\pi u_T^2\right)^{-1/2} \exp\left[-\Delta_i^2/2u_T^2\right], \tag{3}$$

where u_T is the amplitude of the thermal vibrations. For a Silicon crystal at the room temperature $u_T = 0.075$ Å [6].

To study the channeling motion along a particular crystallographic plane one needs to specify the orientation of the crystal with respect to the incoming particles. To do this, we introduce the laboratory reference frame with the z-axis along the beam. In the case of a straight crystal we orient the crystallographic plane parallel to the (xz)-plane. A care is taken to avoid accidental orientations of crystallographic axes along the beam, as this is also done in the most common experiments in order to exclude the axial channeling. To simulate a bent crystal, the coordinates x', y', z' of each lattice node are obtained from the coordinates x, y, z of the same node in the straight crystal according to the relations $x' = x$, $y' = y + \delta(z)$, $z' = z$, where

$$\delta(z) = R - \sqrt{R^2 - z^2} \tag{4}$$

is the shape of the bent crystallographic plane and R is the bending radius.

A convenient parameter showing impact of the curvature R^{-1} on the channeling for the particles with the energy ε is the ratio C of two opposite forces acting on the projectile: the centrifugal force ε/R and a maximal interplanar force $F_{\max}$ stabilizing the motion of the particle transversely to the crystallographic planes. The value $C = 0$ corresponds to the straight crystal and $C = 1$ is the critical value at which the potential barrier between the channels disappear. The value $F_{\max}$ has been estimated withing the continuous approximation for the planar potentials and found to be 6.67 GeV/cm for the electron channeling along the 110 plane in Silicon crystal. We use the latter value as the reference to introduce the bending parameter

$$C = 1.57 \times 10^{-4}(\varepsilon_{\mathrm{MeV}}/R_{\mathrm{cm}}), \tag{5}$$

where $\varepsilon_{\mathrm{MeV}}$ is the electron energy in MeV and R_{cm} is the bending radius in cm.

Having specified the orientation and bending of the crystal with respect to the beam direction, we use a Monte Carlo approach to sample the beam electrons. The particles are assumed to enter the crystal at $t = 0$ with $z = 0$ and the aperture (x, y) coordinates being randomly selected from the intervals $-1.5\,d \le x \le 1.5\,d$ and $-0.5\,d \le y \le 0.5\,d$, where d is the interplanar distance (e.g., $d = 1.92$ Å for the Si (110) planes). The corresponding initial velocities are set to be $v_x = v_y = 0$ which correspond to the ideal case of a zero-emittance beam. Thus, at the entrance the particles are directed along the crystalline planes, and the particular sampling of the initial y-coordinates covered uniformly the interplanar extension for the channel which can guide the particles into the crystal. With the above initial conditions, the dynamical equations (1) are integrated numerically employing the forth-order Runge-Kutta algorithm. The integration proceeds until z exceeds the length L of the crystalline sample.

3. Numerical results

Using the approach described in the preceding section we have simulated propagation of the electrons and positrons with energies $\varepsilon = 195$, 345, 600 and 855 MeV along the (110) planes in straight and bent single crystal of the length L from 25 up to 300 micron. Most of the simulations have been performed for $L = 75$ μm.

The Monte Carlo approach to sampling the incoming particles and including the thermal fluctuations of the crystalline lattice yields a manifold of different trajectories. By analyzing the dependencies $y(z)$ for each trajectory one concludes whether the channeling occurs and, if so, what parts of this trajectory are in the channeling mode. The analysis is similar to that used for the straight crystal [1], whereas now the shape of bent channels is described by Eq. (4).

Although the particles are initially directed along the crystalline planes, not all of the trajectories show the channeling. Thus, the important parameter to estimate is the acceptance

$$A = N_{\mathrm{acc}}/N_0, \tag{6}$$

which is the ratio of the number N_{acc}, of particles captures into the channeling mode at the crystal entrance (the accepted particles) to the total number N_0 of the simulated trajectories (the incident particles). In general, the acceptance depends on the angular distribution of the particles at the entrance, in particular it decreases drastically for the incident angles exceeding the Lindhard's critical angle. In our simulations the incident beam has zero emittance, i.e. the projectiles are collimated "optimally" to get into the inside of the crystal. Therefore, the acceptance we calculate represents an upper bound estimate for a possible experimental outcome.

Fig. 1 shows the acceptance of the electrons of different energies directed into the Si(110) channels. It is seen that for the same values of the bending parameter C the acceptance does not vary noticeably with the electron energy. Notice that the bending parameter is energy dependent itself, and the same values of this parameter correspond to larger bending radius with increasing energy (see Eq. (5). For each electron energy studied in Fig. 1, the acceptance is maximal for the straight channel ($C = 0$) and decreases with C due to increasing centrifugal force acting on the projectile at the entrance.

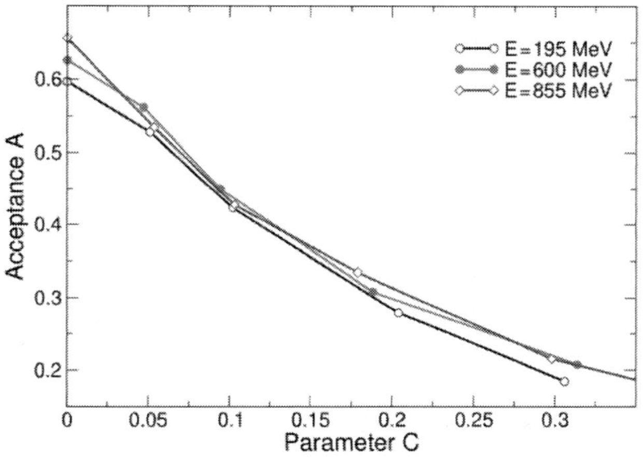

Figure 1. Acceptance A of the planar Si (110) channel as a function of the bending parameter C for several energies of electrons as indicated.

To describe the channeling properties quantitatively, we have computed the numbers $N_{ch0}(z)$ and $N_{ch}(z)$ of the electrons that are in the channeling mode at distance z from the crystal entrance. The former quantity stands for the number of electrons that propagate to z channeling in the channel where they were accepted at the entrance. The quantity $N_{ch}(z)$ is the number of electrons that are in the channeling mode inside the crystal, irrespective of the channel which guides the electronic motion at given z. With increasing z the number $N_{ch0}(z)$ decreases as the accepted electrons leave the entrance channel, i.e. a *dechanneling* takes place. In the contrast, the number $N_{ch}(z)$ can increase when the electrons, including those not accepted at the entrance, can become channeling in some other channel as a result of *rechanneling*. The latter process results from random collisions with the crystal constituents experienced by the electrons that have left the channeling regime. The collisions can occasionally scatter the projectiles into the crystalline channels. If the scattering is accompanied by a reduction in the transverse energy, the projectile can become re-channeled.

By normalizing $N_{ch0}(z)$ with respect to the number N_0 of all simulated simulated and to the number N_{acc} of accepted electrons we determine the channeling fractions "1" and "2" which characterize the dechanneling process. We also introduce the channeling fraction "3" as the ratio of $N_{ch}(z)$ to the number of accepted electrons. The latter fraction is determined by both dechanneling and rechanneling that happen to propagating electrons.

Fig. 2 shows the fractions defined above for the electrons with energy 855 MeV channeling in Si(110). Four separate plots in the figure correspond to different values of the bending parameter. Fraction "1" differs from fraction "2" by a constant factor - the acceptance A, and is included in the plots with an illustrative purpose only. For the same propagation distance z, all three fractions become progressively smaller with increasing the bending parameter. More interesting is to compare fractions "2" and "3". Here, a large enhancement of fraction "3" with respect to "2" is seen for the straight ($C = 0$) channel as well for bent channels corresponding to small values of the bending parameter, $C \lesssim 0.05$. In particular, for the straight channel at the distance $z = 10$ microns the number of electrons channeling in *any channel* exceeds the number of electrons staying in the *entrance channel* by the factor of two, as a result of the dechanneling process. With C increasing, the events of dechanneling become more rare, and the difference between the fractions diminishes (see the plot for $C = 0.2$).

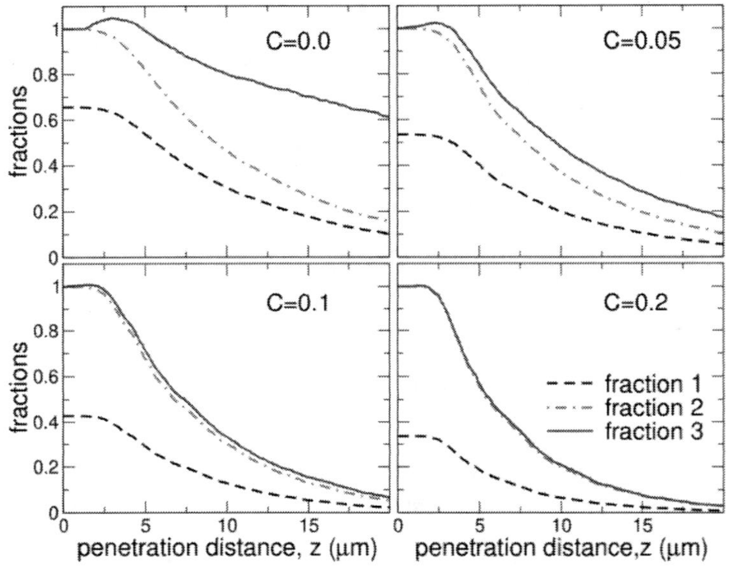

Figure 2. Fractions of channeling electrons with energy 855 MeV for different values of the bending parameter C as indicated. Fractions "1", "2" and "3" designate the ratios $N_{ch0}(z)/N_0$, $N_{ch0}(z)/N_{acc}$ and $N_{ch}(z)/N_{acc}$, respectively.

Fig. 3 shows the fractions $N_{ch}(z)/N_{acc}$ of channeling electrons with the energies $\varepsilon = 600$ and 855 MeV for different values of the bending parameter C. Comparing the fractions for same values of C one notices , that that the portion of channeling electrons is higher for higher ε. With C increasing the channeling fractions gradually decrease and change the character of the decay at large distances: from the inverse power-law decay at small C to the exponential one at the larger values.

By analyzing the channeling segments for each simulated trajectory we have calculate two characteristic penetration lengths introduced in Ref. [1]. The first one, L_{p1}, is a mean value of the primary channeling segments which start at the crystal entrance. The other length, L_{p2}, is defined as a mean path with respect to all channeling segments, including those which result due to the rechanneling process. Generally speaking, L_{p1} depends on the angular distribution of

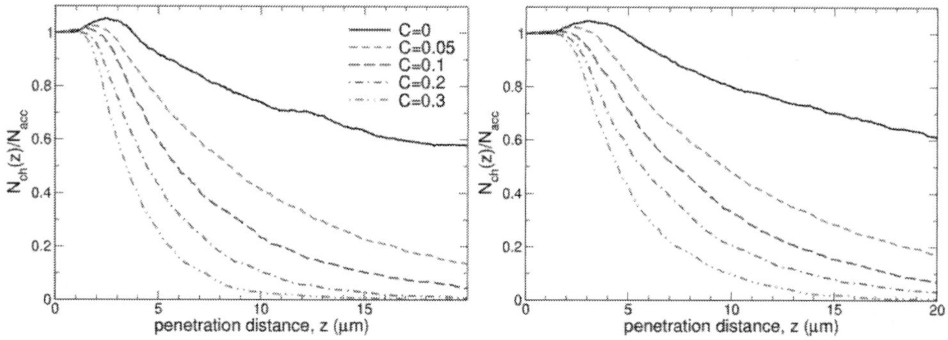

Figure 3. The channeling fraction $N_{\rm ch}(z)/N_{\rm acc}$ for $\varepsilon = 600$ MeV (left plot) and $\varepsilon = 800$ MeV (right plot) electrons versus the penetration distance. Different curves correspond to different values of the bending parameter as indicated.

Table 1. Acceptance A and penetration lengths $L_{\rm p1}$ and $L_{\rm p2}$ for different energies ε of electrons channeling in bent Si(110). The results refer to the crystalline sample of the length $L = 75$ μm.

ε(MeV)	R (cm)	C	N_0	A	$L_{\rm p1}$ (μm)	$L_{\rm p2}$ (μm)
195	∞	0	1708	0.583	4.08 ± 0.29	4.08 ± 0.11
	3	0.010	1570	0.575	3.90 ± 0.30	4.08 ± 0.14
	0.6	0.051	1710	0.525	3.91 ± 0.29	3.91 ± 0.23
	0.3	0.102	1550	0.423	3.19 ± 0.25	3.21 ± 0.24
	0.1	0.306	1728	0.196	1.99 ± 0.22	1.99 ± 0.21
600	∞	0	1420	0.609	9.38 ± 0.75	8.70 ± 0.33
	2	0.047	1502	0.560	8.81 ± 0.72	8.25 ± 0.62
	1	0.095	1461	0.446	7.05 ± 0.63	7.12 ± 0.60
	0.5	0.188	1432	0.330	5.28 ± 0.52	5.28 ± 0.52
855	∞	0	1128	0.656	11.90 ± 1.11	11.13 ± 0.53
	2.5	0.054	1462	0.553	9.79 ± 0.81	9.87 ± 0.72
	1.3	0.103	1419	0.423	8.55 ± 0.81	8.50 ± 0.78
	0.7	0.192	1460	0.303	6.59 ± 0.68	6.61 ± 0.68
	0.5	0.268	1504	0.241	5.38 ± 0.55	5.43 ± 0.56
	0.3	0.447	1376	0.134	4.50 ± 0.61	4.48 ± 0.61

the particles at the entrance. Our simulations, performed for of the zero emittance beams, yield an upper bound estimate for this length. However, the rechanneling electrons are statistically distributed over the incident angles that do not exceed the Lindhard critical angle. Therefore, the length $L_{\rm p2}$ can be thought for quantifying the penetration depth for a beam with non-zero emittance.

Either one from $L_{\rm p1}$ and $L_{\rm p2}$ provides an estimate of the *dechanneling length*. Our results on these quantities for different electron energies and bending parameters are presented in Table 1. Within the statistical uncertainties, the values for both lengths yield the same estimates for the dechanneling length. Also included in this table are the numbers N_0 of the simulated trajectories along with the corresponding acceptances A.

It is worth noting that the values of dechanneling lengths reported in the table noticeably exceed those estimated earlier for $\varepsilon = 855$ MeV electrons channeled in straight [27, 28] and bent [29] Si (110) channels. The calculations performed in the cited papers were based on the peculiar model of the elastic scattering of an ultra-relativistic projectile from the crystal constituents. The model assumes that due to the high speed of the projectile, its interaction interval with a crystal atom is short enough to substitute the atom with its "snapshot" image: instead of the continuously distributed electron charge the atomic electrons are treated as point-like charges placed at fixed positions around the nucleus. Next, the model implies that the interaction of an ultra-relativistic projectile (e.g., an electron) with each atomic constituent can be reduced to the classical Rutherford scattering. In Ref. [1] it was demonstrated, that such a "snapshot" model noticeably overestimates the mean scattering angle in the process of elastic scattering in a single electron-atom collision. Qualitatively, it is clear that substituting a "soft" electron cloud with a set of point-like static electrons must lead to the increase of the scattering angle simply because each electron acts as a charged scatterer of an infinite mass. As a result, the projectile experiences, on average, harder collisions with electrons as compared to the case when they are continuously distributed in space. The mean square angle for a single scattering is a very important quantity in the multiple-scattering region, where there is a large succession of small-angle deflections symmetrically distributed about the incident direction. In particular, the mean square angle due to soft collisions defines the diffusion coefficient which, in turn, is proportional to the dechanneling length (see, for example, Refs. [8,30]). It was noted in Ref. [1] that the "snapshot" approximation underestimates the dechanneling length of 855 MeV electrons in straight Si (110) by approximately 30 per cent. Comparing the $L_{\rm p1}$ and $L_{\rm p2}$ values presented in Table 1 for $C > 0$ with the data from Ref. [29] one concludes that the discrepancy becomes more pronounced for larger values of the bending parameter.

Let us introduce another characteristic length, the total channeling length $L_{\rm ch}$, which can be used to quantify the channeling process. This quantity is defined as an average length of all channeling segments per trajectory. Due to the rechanneling process, the total channeling length $L_{\rm ch}$ is an increasing function of the crystal length L. Under the assumption that the probability of rechanneling does not depend on the penetration length one can expect that $L_{\rm ch} \propto \sqrt{L}$ for $L \gg L_{\rm p1}, L_{\rm p2}$. In this limit the number of rechanneling events is large enough to justify the diffusion approach in estimating $L_{\rm ch}$. The calculated values of $L_{\rm ch}$ (open circles) versus L for 855 MeV electrons in straight Si(110) along with the fitting curves are presented in Figure 4. The solid line stands for the dependence $L_{\rm ch} = 3.68\sqrt{L}$ which seems to be adequate for large values of L but overestimates $L_{\rm ch}$ in the range $L < 100$ micron. The dashed line stands for the best fit which is described by the formula $L_{\rm ch} = 2L^{0.61}$.

4. Concluding remarks

Using the newly developed code [1], which was implemented as a module in the MBN Explorer package [2], we have performed the Monte Carlo simulations of trajectories of ultra-relativistic electrons in oriented (110) straight and bent single Si crystal. The description of the particle motion is given in classical terms by solving the relativistic equations of motion which account for the interaction between the projectile and the crystal atoms. The probabilistic element is due a random choice of transverse coordinates and velocities of the projectile at the crystal entrance as well as by accounting for the random positions of the crystal atoms due to thermal vibrations.

The calculations were carried out for several values of the bending radius R and crystal length L. The chosen energies 195, 600 and 855 MeV of electrons are of interest in connection with the ongoing experiments with electron beams at Mainz Microtron [16,17].

The simulated trajectories are used as the input data for numerical analysis of channeling conditions and properties. The calculated data include the channel acceptance as a function of the bending parameter C and of electron energy ε; the number of primarily channeled particles $N_{\rm ch0}$ and the number of particles in the channeling mode $N_{\rm ch}$ as functions of the penetration distance z. By comparing the dependencies $N_{\rm ch0}(z)$ and $N_{\rm ch0}(z)$ we have demonstrated the

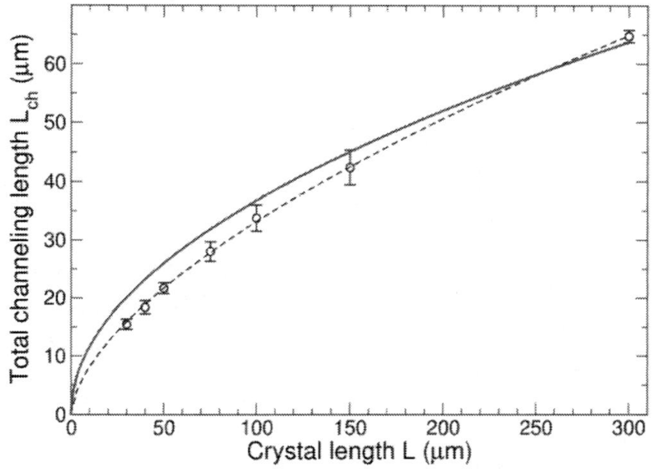

Figure 4. Total channeling length $L_{\rm ch}$ for 855 MeV electrons in straight Si(110) channel versus crystal length. Open circles stand for the calculated values, solid and dashed curves present the dependencies $L_{\rm ch} \propto L^{0.5}$ and $L_{\rm ch} \propto L^{0.61}$, respectively. See also explanations in the text.

important role of rechanneling for projectile electrons.

By analyzing the simulated trajectories we estimated of the electron dechanneling lengths for various ε and C. The obtained results, which are summarized in Table 1, are more accurate than those calculated recently [27, 29] since the latter data were based on the erroneous model of ultra-relativistic electron–atom elastic scattering. We have demonstrated that the model overestimates the mean scattering angle,

Finally, we have performed quantitative analysis of the total length $L_{\rm ch}$ of all channeling segments per a simulated trajectory. This quantity can be measured experimentally by analyzing the radiation emitted within a large cone, $\theta \gg \gamma^{-1}$, as a function of the crystal length. As a prime further step in application of the simulated trajectories we plan to perform the calculations of the emission spectra in a broad range of the photon energies. This will allow us to carry out comparison with the available experimental data and to analyze the influence of the crystal bending on the spectra. Similar calculations are to be performed for the case of positron channeling.

Another important set of calculations concerns simulation and analysis of the channeling process and computation of the emission spectra from sub-GeV electrons and positrons in crystalline undulators. The results of this work, which is currently in progress, will be published elsewhere.

Acknowledgments

We are grateful to Hartmut Backe and Werner Lauth for fruitful and stimulating discussions. The work was supported by the European Commission CUTE (the CUTE-IRSES project, grant GA-2010-269131). The possibility to perform complex computer simulations at the Frankfurt Center for Scientific Computing is gratefully acknowledged.

References

[1] Sushko G B, Bezchastnov V G, Solov'yov I A, Korol A V, Greiner W and Solov'yov A V 2013 *J. Comp. Phys.* (in print)

[2] Solov'yov I A, Yakubovich A V, Nikolaev P V, Volkovets I and Solov'yov A V 2012 *J. Comp. Chem.* **33** 2412–2439
[3] http://www.mbnexplorer.com/
[4] Lindhard J 1965 *K. Dan. Vidensk. Selsk. Mat. Fys. Medd.* **34** 1-64
[5] Tsyganov E N 1976 Fermilab Preprint TM-682, TM-884 (Fermilab, Batavia)
[6] Gemmell D S 1974 *Rev. Mod. Phys.* **46** 129–227
[7] Sørensen A H 1996 *Nucl. Instum. Methods B,* **119** 2-29
[8] Biryukov V M, Chesnokov Yu A and Kotov V I 1996 *Crystal Channeling and its Application at High-Energy Accelerators* (Springer-Verlag, Berlin, Heidelberg)
[9] Uggerhøj U I 2005 *Rev. Mod. Phys.* **77** 1131–1171
[10] Uggerhøj U I 2001 Doctoral Dissertation. (http://www.phys.au.dk/ ulrik/Doct_dis_UIU.pdf)
[11] Korol A V, Solov'yov A V and Greiner W 1998 *J. Phys. G: Nucl. Part. Phys.* **24**, L45–L53
[12] Korol A V, Solov'yov A V and Greiner W 1999 *Int. J. Mod. Phys.* E **8** 49-100
[13] Korol A V, Solov'yov A V and Greiner W 2013 *Channeling and Radiation in Periodically Bent Crystals,* (Springer-Verlag Berlin Heidelberg)
[14] Baranov V T, Bellucci S, Biryukov V M, Britvich G I, Chepegin V N *et al* 2006 *Nucl. Instrum. Methods* B **252** 32–35
[15] Backe H, Krambrich D, Lauth W, Buonomo B, Dabagov S B, Mazzitelli G, Quintieri L, Hansen J. Lundsgaard Uggerhøj U K I, Azadegan B, Dizdar A and Wagner W 2011 *Nuovo Cimento* C **34** 175–180
[16] Backe H, Krambrich D, Lauth W, Hansen J Lundsgaard and Uggerhøj U K I 2011 *Nuovo Cimento* C **34** 157-165
[17] Backe H and Lauth W 2012 Private communication
[18] Uggerhøj U 2012 Private communication
[19] Artru X 1990 *Nucl. Instrum. Methods* B **48** 278–282
[20] Biryukov V M 1995 *Phys. Rev.* E **51** 3522–3528
[21] Maisheev V A 1996 *Nucl. Instrum. Methods* B **119** 42–47
[22] Bak J, Jensen P R, Madsøll H, Møller S P, Schiøtt H E *et al* 1984 *Nucl. Phys.* B **242** 1–30
[23] Smulders P J M and Boerma D O 1987 *Nucl. Instrum. Methods* B **29** 471-489
[24] Shul'ga N F and Syshchenko V V 2005 *Nucl. Instrum. Methods* B **227** 125–131
[25] Molière G 1947 *Z. f. Naturforsch* A **2** 133–145
[26] Pacios L. Fernandes 1993 *J. Comp. Chem.* **14** 410 – 421
[27] Kostyuk A, Korol A V, Solov'yov A V and Greiner W 2011 *J. Phys. B: At. Mol. Opt. Phys.* **44** 075208
[28] Kostyuk A, Korol A V, Solov'yov A V and Greiner W 2001 *Nuov. Cim.* C **34** 167–174
[29] Kostyuk A, Korol A V, Solov'yov A V and Greiner W 2011 *Monte-Carlo Simulations of Electron Channelling a Bent (110) Channel in Silicon* arXiv.org: arXiv:1104.3890 (http://arxiv.org/abs/1104.3890)
[30] Backe H, Kunz P, Lauth W and Rueda A 2008 *Nucl. Instrum. Methods* B **266** 3835–3851

Dynamics of martensitic structure at TiNb-based quenched alloys under heating and loading

Mikhail Petrzhik

National University of Science and Technology “MISIS”,

4 Leninskiy prospekt, Moscow 119049, Russian Federation

E-mail: petrzhik@shs.misis.ru

Abstract. Phase competition at TiNb-based Ni-free shape memory alloys caused by changes of alloying, temperature, external stresses is considered by analysis of experimental results on elastic modulus determination, X-ray diffraction, thermal analysis, mechanical tensile and indentation testing. Martensitic $\beta \leftrightarrow \alpha''$ transformation undergoes due to a reversible displacements of atoms at (002) plane of orthorhombic unit cell. The martensitic structure has high stability against thermal and mechanical cycling.

1. Introduction

It is well known that there are only two stable phases in pure Ti and solid solutions on its base: high temperature cubic (bcc) β-phase and low temperature hexagonal α-phase. However, some other metastable phases could be formed dependent on alloying, heat treatment and pressure processing, as well as due to contamination by impurities [1].

Ternary and quaternary Ti–Nb-Zr, Ti–Nb-Ta, Ti–Nb-Ta-Zr alloys have been extensively investigated during the last years and the results confirm the possibility to obtain Ni-free, Ti-based solid solution alloys with shape memory effects (SME) connected with reversible $\beta \leftrightarrow \alpha''$ martensitic transformation [2,3].

The tailoring of metastable structures via reasonable processing yields inelastic phenomena as shape memory, pseudo-elasticity, high damping capacity, negative temperature dependence of electric resistivity etc. The shape memory alloys are attractive for medical applications as metallic implant materials combining superelasticity with biocompatibility comparable to that of pure Ti [4]. It is due to the both reasons: all the metals and their oxides are friendly for human body and the quenched alloys manifested excellent functional properties, including low elastic moduli closed to that of the human trabecular bone.

Other interesting feature of the Ti-Nb based alloys is a high thermal stability of the orthorhombic α″’ martensite. This is good for some technical applications based on high temperature SME as well as to study model alloys having reversible martensitic transformation in wide temperature range.

In this work phase transitions and products of martensitic transformation of biocompatible quenched Ti-Nb-Ta(Zr) alloys, especially metastable orthorhombic α″ and cubic β phases have been studied by elastic E and G moduli determination, thermal analysis (DTA), X-ray diffraction, tensile tests and nanoindentation to monitor the their structure dynamics and reversibility.

2. Factors of martensitic transformation

2.1. Alloying of solid solution by substitution elements

Pure titanium (α-Ti) has hexagonal lattice cell ($P6_3/mmc$) and undergoes at 882 °C at heating a phase transformation to body centered cubic, or β Ti (bcc, Im3m). Alloying of Ti by bcc isomorphous elements (Nb, Ta, Mo etc.) retards the diffusion and degreases the temperature of transformation as one can observe at Fig.1a, monitoring the temperature range of endothermal peak.

Fast cooling of the high temperature bcc β-Ti yields metastable α"-martensitic phase due to a shear-like formation of the orthorhombic unit cell (Fig.1b) according to Bagariatskii schematic [5]. It implies that martensitic β↔α" transformation undergoes due to a reversible subnano scale displacements of atoms at (002) plane of orthorhombic unit cell. So, alloying of Ti by bcc elements (Nb, Ta, Mo etc.) changes structure type of quenched solid solution on the way of rearrangement from hexagonal (alfa) phase till (cubic) beta phase.

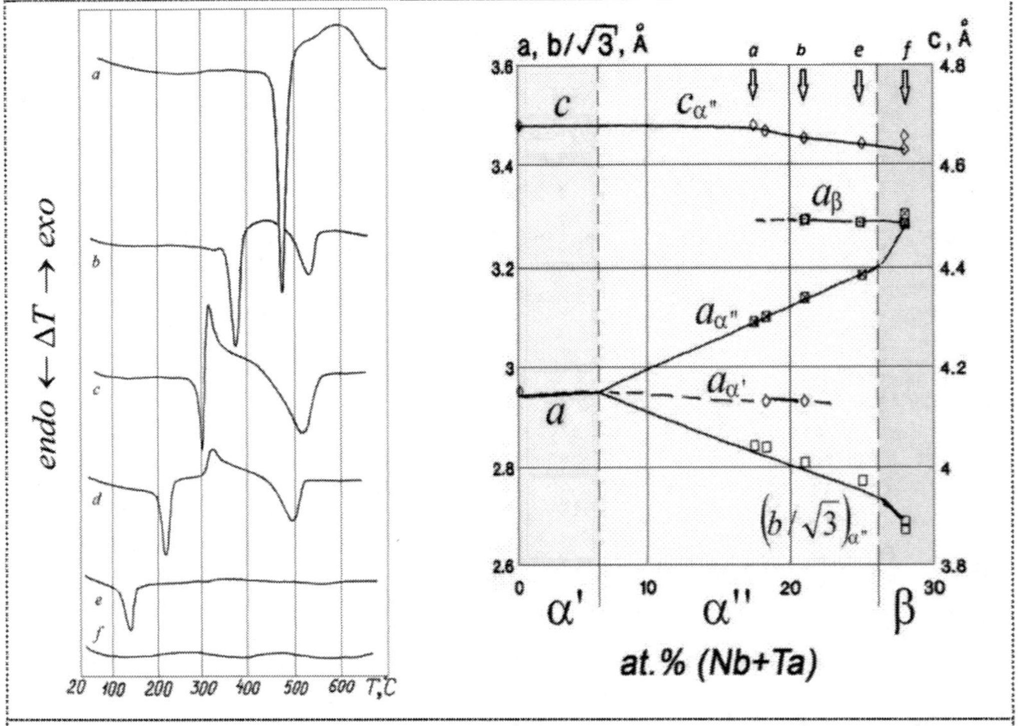

Figure 1. The effect of alloying on endothermal effect and temperature range of martensitic transformation studied by differential thermal analysis, DTA (a) and distortion of orthorhombic lattice cell (b) on data of X-ray diffraction. Thermal curves marked ***a, b, e, f*** correspond to four typical alloys shown at right plot and at Table 1.

As shown at Fig.1b and at Table 1 and Table 2 α"-martensite is found at wide range of concentrations intermediated between stable α-Ti and β-Ti phases. Each composition can be characterized by unique set of parameters of lattice cell as well as a degree of its rhombic distortion ***R***.

Table 1 Experimental parameters of unit cell (x10 nm) in quenched Ti-Nb-Ta alloys

Phase		α'	α"	β	β'
Space Group		p6₃/mmc	Cmcm	Im3m	Im3m
Conventional Presentation	a	2.93	3.051-3.198	3.285	3.303
	b	---	4.949-4.785	---	---
	c	4.77	4.673-4.650	---	---
Orthorhombic Presentation	a	2.93	3.051-3.198	3.285	3.303
	b	5.07	4.949-4.785	4.645	4.657
	c	4.77	4.673-4.650	4.645	4.657
Degree of Rhombic Distortion $R=\sqrt{3}a/b$		1	1-1.16	1.225	1.225

Table 2. Chemical composition of the quenched Ti-Nb-Ta alloys

Alloy composition	Symbol	Rhombic distortion, *R*
$Ti_{82.5}Nb_{4.4}Ta_{13.1}$	*a*	1.08
$Ti_{82.5}Nb_{7.5}Ta_{7.5}$	*b*	1.08
$Ti_{82.5}Nb_{17.5}$	*c*	1.10
$Ti_{79}Nb_{18.9}Ta_{2.1}$	*d*	1.12
$Ti_{75.0}Nb_{8.75}Ta_{6.25}$	*e*	1.15
$Ti_{75}Nb_{25}$	*f*	1.225

2.2. Dynamics of martensitic structure at heating

Distortion ***R*** of orthorhombic α"-martensite is also sensitive to temperature changes with the same way as to alloying. As one can see at Fig.2 ***R*** increases with increasing of temperature and degreases at cooling back. Heating is also causes the softening of elastic E and G modulus, they go down to the minimal values (E<60 GPa) within the A_s-A_f range of reversible martensitic transformation β↔α". However, above A_f temperature β-phase became the stable one and elastic modulus increases.

It was found [6] two composition ranges of the formation orthorhombic martensite which is stable to heating to secure the reversible martensitic transformation at heating, namely at range 80-240 °C and 450-550 °C. Between the ranges (250-450 °C) a concurrent metastable ω-phase forms at heating.

Formation of hexagonal ω-phase above 300 °C could be monitored at Fig. 1a (curves ***c*** and ***d***) and at Fig. 2 as pronounced exothermal effect on ΔQ-curve which is correlated with sharp increase of E-modulus.

2.3. Mechanical behaviour of quenched alloys at loading

The mechanical behavior of the alloys depends on the chemical, phase and structure composition. Quenched alloys with β-phase structure (like ***f***-alloy) manifest superelastic behavior due to stress-induced β↔α" transformation started at low (ca. 100 MPa) external stresses as shown at Figure 3. The alloys with α"+β martensitic structure (like ***e***, ***d*** alloys), are able to accumulate deformation via rearrangements of martensite structure, and then undergo shear transition induced by external stresses. They recover the accumulated deformation by shape memory effect at subsequent heating.

Crystallographic reserve of recoverable deformation at the β↔α" transformation in the alloys was estimated up to 3.5% returnable deformation [7]. The results were analyzed taking into account XRD data on phase composition and degree of distortion R=√3a/b of orthorhombic alfa''-martensite (Cmcm space group) according to Yu.A. Bagariatskii's model of β→α"→α' transition [5].

It was found that experimental realization of superelasticity (up to 2,5 % of returnable deformation) was close to its theoretical limit [7] and it was observed at alloys having extremely low Young's modulus (about 60 GPa and less) and the largest rhombic distortion (about R=1.16) of unit cell.

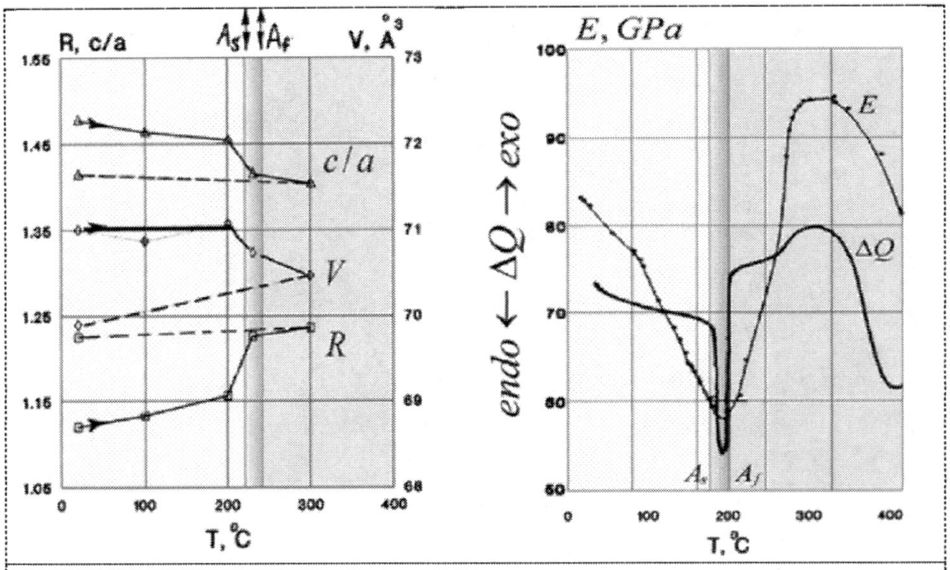

Figure 2. Thermal effects and changes of Young modulus at heating of quenched Ti-Nd-Ta alloys (left) and progressive rhombic distortion of unit cell (right) at heating through A_s-A_f range.

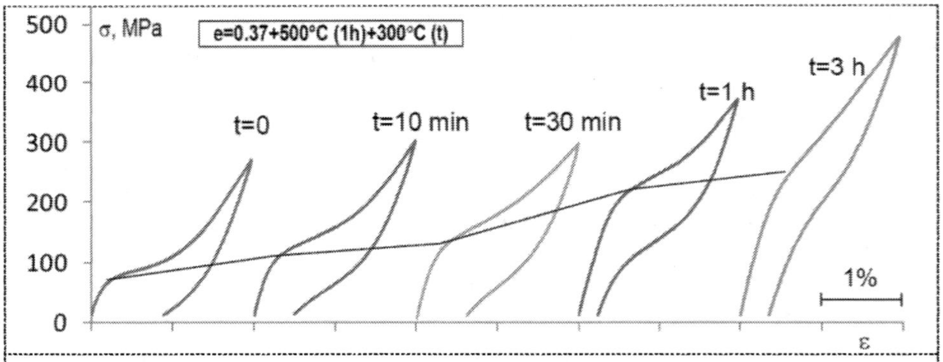

Figure 3. Examples of superelastic behaviour of tempered $Ti_{74.5}Nb_{19.7}Ta_{5.8}$, the starting pseudo yield stress is shown for each curve.

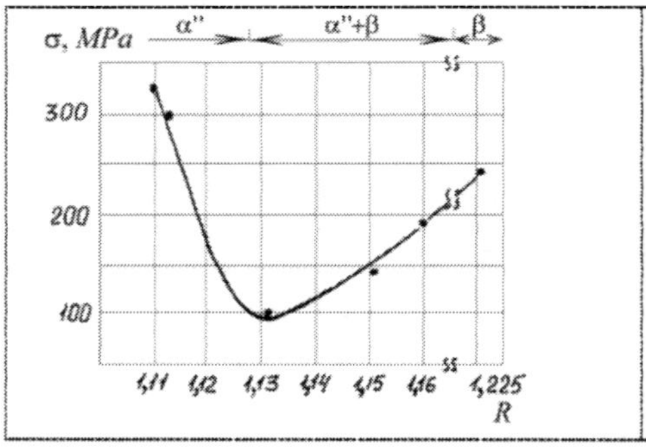

Figure 4. The effect of distortion of orthorhombic unit cell at quenched Ti-Nb-Ta alloys on starting pseudo yield stress.

3. Conclusions

Reversible martensitic β↔α" transition is observed at the Ti-Nb-Ta quenched alloys at thermal and loading cycling. There are two sequential mechanisms of reversible accumulation of deformation: lattice distortion of orthorhombic martensite and β↔α" phase transition. Reversible phase transition is exhibited by the definite alloys characterized by high distortion of unit cell, extremely low elastic moduli and excluding a formation of concurrent metastable ω-phase.

References

[1] Collings E, "The Physical Metallurgy of Titanium Alloys", *ASM Handbook*, 3 (1984).

[2] Petrzhik M and Fedotov S 1995 *Proc. XVI Conf. on Applied Crystallography* (Cieszyin, Poland 22-26 August 1994) (Singapore:World Scientific) p. 273

[3] Miyazaki S, Kim H Y and Hosoda H 2006 *Mater. Sci. Eng.* A **438–440** 18

[4] Brunette D, Tengvall P, Textor M, Thomsen P, *Titanium in Medicine, Springer*, Berlin, 2001.

[5] Bagariatskii Yu, Nosova G and Tagunova T, "Factors in the Formation of Metastable Phases in Titanium-Base Alloys", *Sov. Phys. Dokl. [transl. of Dokl. Akad. Nauk. SSSR, Vol122, 1958, p. 593-596]*, **3**, 1014-1018 (1959)

[6] Petrzhik M, Fedotov S, Kovneristyj Yu and Zhebyneva N 1992 *Metall Science and Heat Treatment* **34** 190

[7] Zhukova Yu, Petrzhik M and Prokoshkin S 2010 *Russian Metallurgy (Metally)* **11** 1056

Molecular dynamics simulation of nanoindentation of nickel-titanium crystal

G B Sushko[1], **A V Verkhovtsev**[1,2], **A V Yakubovich**[1,3] **and A V Solov'yov**[1,3]

[1] Frankfurt Institute for Advanced Studies, Ruth-Moufang-Str. 1, 60438 Frankfurt am Main, Germany

[2] St. Petersburg State Polytechnic University, Politekhnicheskaya ul. 29, 195251 St. Petersburg, Russia

E-mail: verkhovtsev@fias.uni-frankfurt.de

Abstract. We present the results of molecular dynamics simulations of nanoindentation of a bimetallic nickel-titanium crystal in the austenitic (cubic) B2 phase. By considering three different types of indenters, namely of square, conical and spherical shapes, we observe the dependence of deformations of the crystalline structure on the type of the indenter. Various load-displacement curves are observed for different indenter types. We perform the molecular dynamics simulations of a full indentation cycle, which includes the loading and unloading stages. On the basis of such simulations we evaluate mechanical properties of the material, namely we calculate hardness and reduced Young's modulus. We observe variation of the calculated parameters depending on the indenter type and discuss the origin of occurring discrepancies.

1. Introduction

Various metallic materials (stainless steels, Co-based alloys, titanium and its alloys, Ni-Ti alloys) are widely used for medical implants in trauma surgery, orthopaedic and dental medicine [1, 2]. Successful incorporation of these materials into clinical practice requires that they meet several principal criteria. One of them is the ability of the material to provide sufficient mechanical strength, especially under the cyclic loading conditions, to ensure the durability of medical devices made thereof [2]. Finally, the material should be machinable and formable, thereby, enabling fabrication of devices at an affordable cost.

The deposition of multifunctional bioactive films is an effective way to promote the formation of bone-like layers on the implant surface, prevent toxic ion release, and improve mechanical and tribological characteristics of implants [3]. In order to determine the optimal directions of development and design of nanostructured titanium and titanium based alloys for medical applications, fundamental understanding of the deformation mechanisms and reliable computational models for the virtual, numerical testing of these materials are necessary.

In the present paper we investigate the nanoindentation process of the crystalline nickel-titanium alloy by large-scale molecular dynamics (MD) simulations. To describe interatomic interactions, we utilize the classical Finnis-Sinclair potential. We study the austenitic (B2) face-centered cubic (fcc) phase of the Ni-Ti alloy with equal concentration of nickel and titanium

[3] On leave from A.F. Ioffe Physical-Technical Institute, St. Petersburg, Russia

atoms. Three different types of indenters, namely of a square, conical and spherical shapes, are considered in the present study. For each indenter type, we perform MD simulations of a full indentation cycle, which includes the loading and unloading stages. On the basis of such simulations we evaluate mechanical properties of the material, namely we calculate hardness and reduced Young's modulus. We observe the dependence of the obtained values of the parameters on the type of the indenter and discuss the origin of this dependence.

The study presented is devoted to the modeling of nanoindentation and mechanisms of localized deformation of nanostructured biomaterials, Ni-Ti alloys in particular. The main aim of this study is to investigate the strength and mechanical properties of the films and the substrate/coating systems to be used in implants. Understanding the fracture mechanisms during indentation, scratching and impact testing is critical for understanding of how the film/substrate system will perform in service.

The prime advantage of the nanoindentation procedure is the possibility to test very small material volumes and its applicability to thin film characterization. Generally, the indenter is made of diamond in order to avoid spurious effects of indenter tip deformation. Only single crystal diamond, free of impurities or inclusions is used. The advantages of diamond for nanoindentation come from its exceptional properties of hardness, thermal conductivity and chemical inertness, which surpass any other known material. There are numerous geometries commercially available for the indenter shape such as three-sided pyramids, four-sided pyramids, wedges, cones, cylinders or spheres. The tip end of the indenter can be made sharp, flat, or rounded to a cylindrical or spherical shape. A typical size of the tip is of the order of a few tens of nanometers for the sharpest tips and of a few hundreds of nanometers for the rounded tips. The residual imprint on the sample surface after nanoindentation is on the order of a micrometer diameter even with high force levels and a blunt indenter type with a conical geometry (with a nominal radius of curvature of a few μm) leaving a rather large imprint.

In contrast to traditional hardness testers, the nanoindentation system allows the application of a specified force on the indenter (of the order of μN) while simultaneously monitoring the indenter displacement (of the order of nm) to obtain the applied force-indenter displacement curve as the output (see, e.g., [4, 5]). The experiment follows a predefined loading sequence where the applied force on the indenter displacement as a function of time is specified. Generally, three parts of the loading sequence are distinguished: the loading part where the applied force is increased until a peak value, the holding part where this peak load is maintained for a prescribed amount of time, and finally the unloading part where the applied force is decreased gradually to zero. A nanoindentation experiment lasts generally only for some tens of seconds in order to remain in the high precision domain of the measuring equipment. In experiments the indenter velocity varies between tens of nm/s to μm/s.

2. Numerical simulations of the nanoindentation process

Depending on the purpose of the simulation, different aspects of the nanoindentation procedure are addressed in different numerical models. Every numerical model is adapted to specific requirements, therefore it is hardly possible to compare directly the simulation results issued from different models. In Ref. [6], different numerical models and techniques applied at different length scales of the nanoindentation process are described.

When the numerical frame has to be coupled directly to experiments, most frequently the continuum models are preferred because of their computational efficiency. These models are well adapted to reproduce the overall average response of the material in nanoindentation and to conduct parametric studies concerning parameters which are experimentally difficult to access (e.g. surface roughness, contact geometry etc. [7]). It is worth mentioning the widely used finite element method (FEM) for the simulation of the nanoindentation procedure in the large and moderate indentation depths domain [8]. This method allows one to identify the key parameters of the nanoindentation process and various material parameters using dimensional analysis together with the continuum numerical models. In order to describe the more complex physics of shallow nanoindentations, where the size-dependent response of the material has to be taken into account, the continuum methods are specially adapted to microscale simulations, using higher order theories in the finite element method [9, 10].

However, when the purpose of investigation is a deeper understanding of the plastic deformations one needs to deal with the models on a smaller scale. The descent to the scales where continuum mechanics is not applicable anymore implies the use of more computationally expensive simulations as they require more information to be included in the model itself. One of the relatively computationally cheap numerical methods uses the discrete dislocation plasticity model which focuses on the plastic deformation of the crystalline material considering only the slip planes without details on the positions of the atoms [11, 12].

When the details of the onset of the plastic deformation and the dislocation activity are in the focus of interest, the numerical models have to descend to the lowest scale: atomistic simulations are conducted. Generally, atomistic models are used for system sizes not larger than 100 nm (around the upper limit of the range of thin film nanoindentations) from which scale on other numerical methods are more adequate to describe the behavior of the material [6]. The atomistic models have the indisputable advantage of identifying the trends concerning the main physical variables of a problem and give important qualitative information for the understanding of experimentally observed complex phenomena, such as the size effects, that cannot be reproduced by coarser models.

The main drawback of purely atomistic models for mechanical applications is that they are computationally expensive. Therefore, generally, they handle length (in the order of tens of nm) and time scales of some ns, i.e. several orders of magnitude smaller than in the experiments. Due to computing power limitations, the speeds of the indenter tip reported in the literature are on an order of $1 - 100$ m/s for most MD simulations of nanoindentation [13, 14]. Even though atomistic calculations are computationally expensive, in the past years the dynamic increase in computational power has allowed the adaptation of atomic level models to interesting mechanical applications [15]. Different approaches based on molecular dynamics addressing the problem of nanoscratching [16] or nanoscale machining [17] have been used for the identification of different wear regimes [18], adhesion [19] and the explication of the tribological phenomenon. Recently, very large systems (up to several million atoms [13, 20]) have become accessible for modeling using parallel computational techniques, but the computational effort, directly related to the enormous quantity of information of atomic level models becomes very large.

3. Modeling the nickel-titanium crystal

In the present study, atomistic simulations of nanoindentation of the Ni-Ti crystal were conducted using molecular dynamics (MD) methods. The geometrical structure of the system of interest was constructed on the basis of experimental crystallographic structures. For the theoretical investigation of the nanoindentation process on the nanoscale, we created a three-dimensional nickel-titanium sample of $260\ \mathring{A} \times 260\ \mathring{A} \times 260\ \mathring{A}$ consisting of about 1 million atoms.

The interaction between the nickel and titanium atoms was modeled using a Finnis-Sinclair-type potential [21], an empirical many-body potential which allows one to describe interatomic interactions in transition-metal materials. The total potential energy U of a system of nickel and titanium atoms, located at positions $\mathbf{r}_i$, can be written as:

$$U = \sum_{i=1}^{N} \left(\sum_{\substack{j=1 \\ (i \neq j)}}^{N} A_{\alpha\beta} \, e^{-p_{\alpha\beta}\left(\frac{r_{ij}}{d_{\alpha\beta}}-1\right)} - \sqrt{\sum_{\substack{j=1 \\ (i \neq j)}}^{N} \xi_{\alpha\beta}^{2} \, e^{-2q_{\alpha\beta}\left(\frac{r_{ij}}{d_{\alpha\beta}}-1\right)}} \right), \qquad (1)$$

where $(\alpha, \beta = \mathrm{Ni}, \mathrm{Ti})$, N stands for the total number of considered atoms in the system, r_{ij} is the distance between the ith and j^{th} atoms, and $d_{\alpha\beta}$, $p_{\alpha\beta}$, $q_{\alpha\beta}$, $A_{\alpha\beta}$ and $\xi_{\alpha\beta}i$ are adjustable parameters of the potential. The parameter $d_{\alpha\beta}$ is the first-neighbor distance, $\xi_{\alpha\beta}$ results from an effective hopping integral and $q_{\alpha\beta}$ describes the distance dependence of the hopping integral. In equation (1), the first term on the right-hand side describes a repulsive part of the potential which results from the repulsion between core electron of neighboring atoms, and the second term stands for the attractive density-dependent many-body term. The parameters of the potential were fitted to reproduce the properties (the cohesive energy, lattice parameters, elastic constants and the unrelaxed vacancy formation energy) of the B2 NiTi phase at 0 K. The MD simulations were performed using the parametrization proposed by Liu *et al.* [22]. Values of the parameters, used in the calculations are as follows: $d = 2.607$ Å, $A = 0.3$ eV, $p = 7.9$, $\xi = 2.48$ eV and $q = 3.0023$ [22]. In order to reduce the computational time, the linked cell method for the short range potential was used (see [23] for details). To use this method, the cutoff distance for the interatomic potential was chosen to be equal to 5.5 Å. In the range between 4.2 and 5.5 Å we used the spline interpolation for attractive and repulsive parts of the potential (see [23] for details). Coefficients of the splines were chosen to correspond to the value and the first derivative of the potential at 4.2 Å and equal to zero at 5.5 Å.

4. Modeling the indenter

We investigated the indentation process using three different shapes of the indenter, namely a square indenter with a cross section of 18 $\mathring{A} \times 18$ $\mathring{A}$, a conical indenter with the cone angle of 90°, and a spherical indenter of the radius of 70 Å. The indenters were modeled as absolutely rigid bodies constructed from carbon atoms in the fcc lattice with a unit cell size of 1 Å. Such a tight packing of atoms of the indenters was chosen in order to avoid possible penetration of nickel and titanium atoms into the indenter. The inner atoms from the indenters were removed in order to decrease the computational time. Although the tip radius of standard nanoindenters is on the order of tens-hundreds of nanometers, in the present simulations we used the indenters of smaller size because of computational limitations. The deformability of the indenters and the thermal oscillations of their atoms were neglected. The hardness of a diamond is much higher than that of NiTi alloys, therefore the rigid body model for the indenter is fully applicable in the case of a nickel-titanium sample.

The interaction of the indenter with the titanium sample was described using the following repulsive potential:

$$U = \varepsilon \left(\frac{r_m}{r} \right)^{12} \qquad (2)$$

with the following values of the parameters: $\varepsilon = 0.1$ eV and $r_m = 1.5$ Å. The cutoff radius for the Ni-C and Ti-C potential (the linked cell algorithm was used in the calculations) was set to 3 Å.

5. Results and discussion

The molecular dynamic simulations were performed using the `MBN Explorer` computer package [23], a universal program for structure optimization, simulation of dynamics, and growth processes in various nanosystems [24–28].

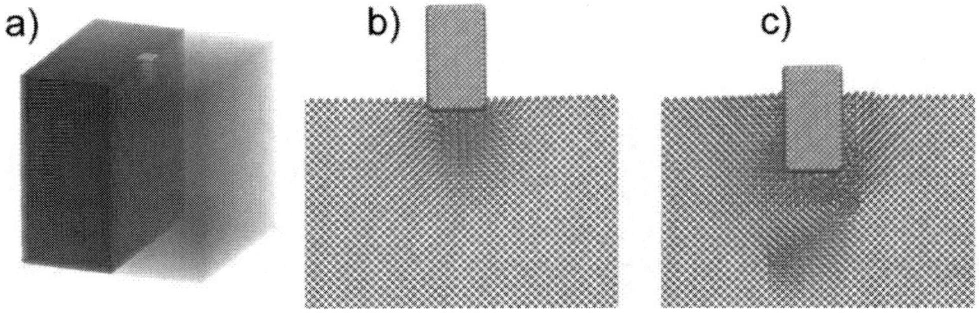

Figure 1. (a) Nanoindentation of the sample with a square indenter. Panels (b) and (c) show deformation of the crystalline structure at intermediate, ~ 0.6 nm, and the largest, 3 nm, penetration depths, respectively. Titanium and nickel atoms are shown by red and blue colors, respectively.

MD simulations of nanoindentation comprise the equilibration and indentation stages. In the equilibration stage, the indenter tip and the crystal surface were positioned ~ 0.3 nm apart from each other, and the system was simulated with a Langevin thermostat at 30 K for ~ 60 ps. After equilibration, the atomic positions of the two bottom layers of the crystal were "frozen" in order to fix the system in space. In the indentation stage of the simulations, the process is modeled by moving the indenter approximately 3 nm downward at a constant speed of 40 m/s, and then retracting the tip upwards at a constant speed of 10 m/s. These values correspond to the typical range of velocities which are used nowadays in the MD simulations of nanoindentation.

To describe the loading stage we performed two independent sets of simulations for each type of the indenter with various initial random seeds for Maxwell velocity distributions. The time integration of motion is performed by the velocity-Verlet algorithm [29]. After initial annealing of the crystalline structure at 30 K, the penetration of the indenter into the crystal was simulated in the canonical ensemble using the time integration step of 2 fs and the thermostat damping coefficient of 0.5 ps. The total vertical force acting on the indenter and coordinates of all the atoms in the system were recorded each 20 and 200 fs of the simulation, respectively. Each of the indenters was moved to the maximum indentation depth of approximately 3 nm into the crystal.

Firstly, we consider the simulations performed with the square indenter. The crystalline structure and the structural deformations occurring at various indentation depths are presented in figure 1. Panel (a) shows the penetration of the square indenter into the constructed crystalline sample. Panels (b) and (c) show the deformation of the structure at the indentation depth of 0.6 nm and 3 nm, respectively. The latter case corresponds to the maximal penetration of the indenter into the crystal.

The dependence of the force, exerted on the square indenter by the crystal, on the indentation depth is shown in figure 2. Solid black and blue lines represent two independent sets of simulations of the loading stage. The unloading stage was simulated in the first set of simulations (see black line). At the initial stage of indentation the load-displacement curve has a linear dependence, which means that at this stage the indentation is performed in the linear elastic regime. The dashed red line represents the analytic fit of the loading curve by a linear function.

After the initial stage of the loading, the load-displacement curve reaches the maximum and then fluctuates around a constant value of ~ 0.08 μN/nm during the further displacement of the indenter. Such a behavior of the curve can be explained by the geometry of the indenter. Due to the constant cross section of the square indenter, its further penetration into the crystal

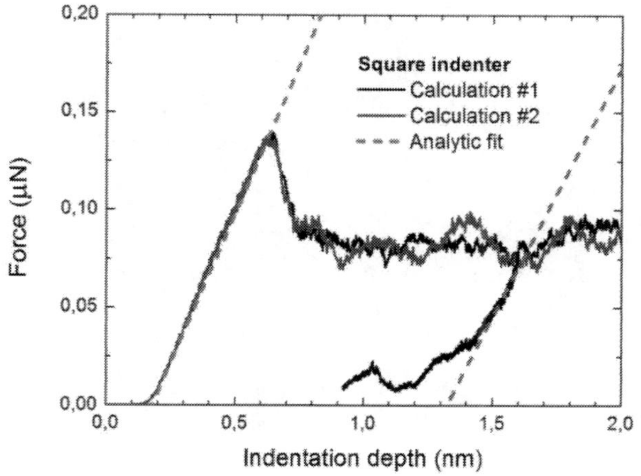

Figure 2. Dependence of the force acting on the square indenter on the indentation depth. Solid black and blue lines represent two independent sets of simulations of the loading stage. The unloading stage, studied within the first set of simulations, is shown also by the black line. Dashed red line represents the analytic fit to the pressure created by the indenter by a linear function. Green dashed line shows the fit to the initial part of the unloading curve.

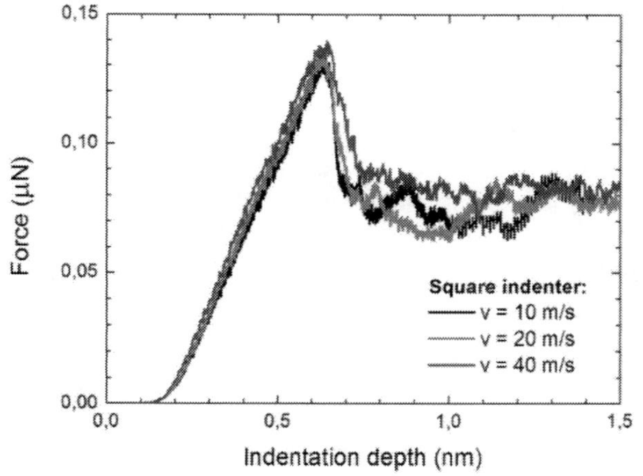

Figure 3. Load-displacement curves obtained for the case of the square indenter moving with the constant speed of 10 (black line), 20 (red line) and 40 m/s (blue line).

does not affect significantly the force acting on the indenter. The fluctuations are caused by the relaxation of stresses in the system, such as migration of dislocations from the deformed sites.

To evaluate the dependence of the obtained load-displacement curves on the indenter speed, we performed several independent simulations of the process with the square indenter. In the course of simulations the indenter was moving with the constant speed of 10, 20 and 40 m/s for each case study. The results of the comparison are presented in figure 3. Load-displacement curves obtained for the indenter speed of 10, 20 and 40 m/s are shown by the black, red and blue lines, respectively. It is seen that the indenter speed does not affect significantly the shape of the load-displacement curve.

Figure 4 represents the indentation process with a conical indenter. Panel (b) shows the structural deformations which occur in the system at the maximal penetration depth of 3 nm. The dependence of the force acting on the indenter is presented in figure 5. It indicates that the force acting on the indenter gradually increases with the increase of the penetration depth.

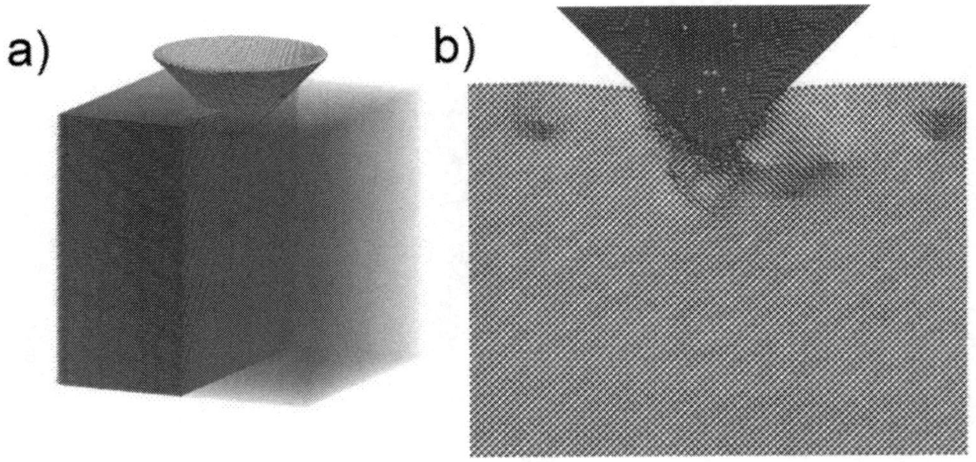

Figure 4. (a) Nanoindentation of the sample with a conical indenter. (b) Deformation of the crystalline structure of titanium at the penetration depth of 3 nm. Titanium and nickel atoms are shown by red and blue colors, respectively.

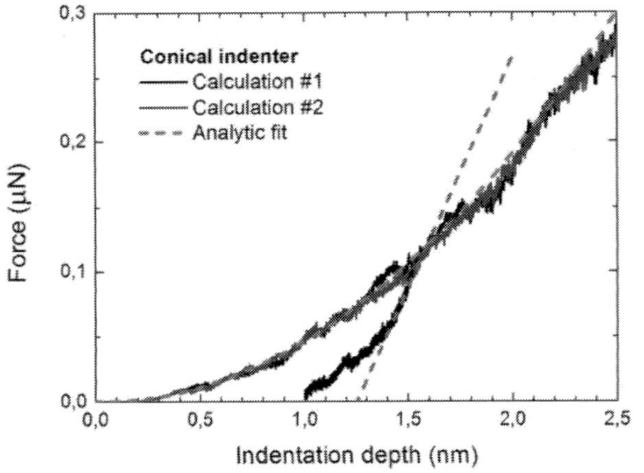

Figure 5. Dependence of the force acting on the conical indenter on the indentation depth. Solid black and blue lines represent two independent sets of simulations of the loading stage. The unloading stage, studied within the first set of simulations, is shown also by the black line. Dashed red line represents the analytic fit to the pressure created by the indenter by a quadratic function. Green dashed line shows the fit to the initial part of the unloading curve.

In the case of the conical indenter, the obtained load-displacement curves can be fitted with a quadratic function. In this case, there is no elastic regime, and the indentation process enters the plastic regime just after the initial contact of the indenter tip with the sample.

Figure 6 shows a general view of the crystalline structure and the occurring structural deformation during indentation with a conical indenter. The dependence of the force acting on the spherical indenter on the penetration depth is shown in figure 7. In case of the spherical indenter, the elastic regime is not observed since the spherical tip induces plastic deformations even at very small penetration depths. Nevertheless, for the small indentation depths, less than ~ 1 nm, the obtained load-displacement curves can be fitted quite well by the expression [30]

$$F = \alpha h^{3/2} , \tag{3}$$

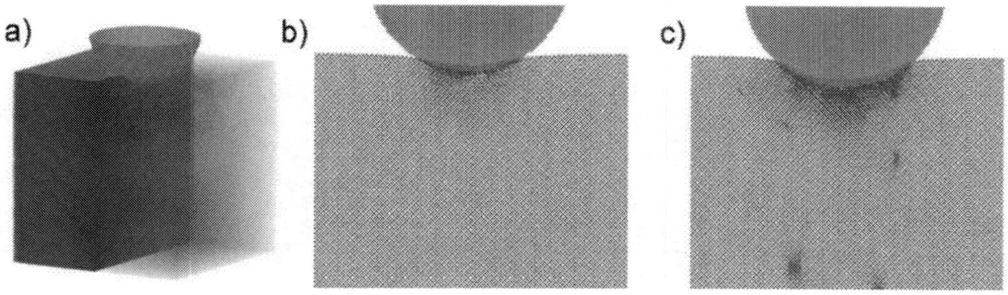

Figure 6. (a) Nanoindentation of the sample with a spherical indenter. Panels (b) and (c) show deformation of the crystalline structure at intermediate, $\sim$ 1.2 nm, and the largest, 2.5 nm, penetration depths, respectively. Titanium and nickel atoms are shown by red and blue colors, respectively.

where F is the indenter load, h is the displacement of the indenter, and α is a proportionality factor. The analytic fit of the load-displacement curve by such a power function is shown by the dashed red line in figure 7. Although equation (3) is valid for the elastic indentation regime, we observed that the power function $F \propto h^{3/2}$ is valid also for extended indentation depths. Movement of the spherical indenter at distances exceeding $\sim$ 1.5 nm leads to the deviation from the $h^{3/2}$ dependence. This deviation can be explained by the fact that the indenter radius becomes comparable with the indentation depth, as well as the finite size of the whole sample. Indeed, at high indentation depths dislocations move to the boundaries of the crystal and the system as a whole expands (see the right panel of figure 6). For the spherical indenter we observed the largest value of the force acting on the indenter. This can be explained by the largest deformations imposed on the system by the spherical indenter as compared to the square and conical ones.

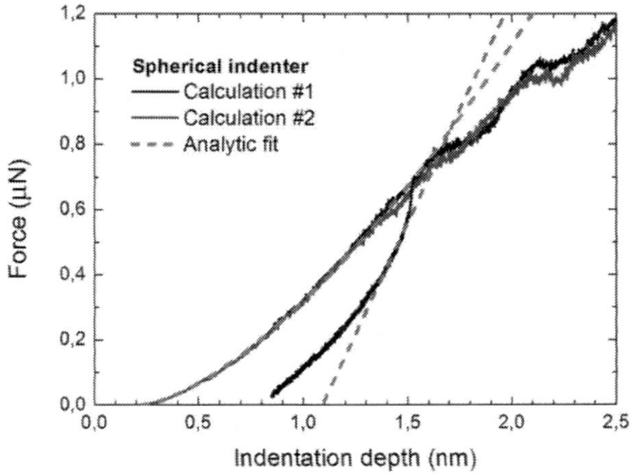

Figure 7. Dependence of the force acting on the spherical indenter on the indentation depth. Solid black and blue lines represent two independent sets of simulations of the loading stage. The unloading stage, studied within the first set of simulations, is shown also by the black line. Dashed red line represents the analytic fit to the pressure created by the indenter by a power function. Green dashed line shows the fit to the initial part of the unloading curve.

In the present study, we used the results of the performed MD simulations to calculate hardness of the material, which is defined as the ratio of the force acting on the indenter divided by the contact area of the indenter. Hardness was calculated at $\sim$ 1.3 nm of the indenter path from the beginning of the simulations for each of the indenters. The corresponding values of hardness are presented in table 1.

Table 1. Hardness of the NiTi alloy calculated for the square, conical and spherical indenters.

Indenter	Square	Conical	Spherical
Hardness (GPa)	39	24	15

As it is seen from table 1, the calculated value of hardness depends on the form of the indenter. This can be explained by the fact that for small indenters the channel of stress release from the deformation site associated with dislocations migration is suppressed, while for a large spherical indenter much higher dislocation activity is observed. Indeed, for small indenters it requires substantial relative energy (as compared to the total work done by indenter to deform the material) to activate the dislocation migration, therefore material has extended elastic deformation region, which results in higher values of hardness.

In the present study, we calculated also the reduced Young's modulus of the material, which is defined as:

$$\frac{1}{E_r} = \frac{1-\nu^2}{E} + \frac{1-\nu_i^2}{E_i} \,, \tag{4}$$

where E and ν are Young's modulus and Poisson's ratio for the sample material, and E_i and ν_i are the same parameters for the indenter, respectively. In the present calculations the second term in equation (4) can be omitted since we treat the indenter as an infinitely stiff body.

The reduced Young's modulus can be calculated from the initial part of the unloading curve using the following formula proposed by Oliver and Pharr [30]:

$$\frac{dF}{dh} = \frac{2}{\sqrt{\pi}} E_r \sqrt{A} \,, \tag{5}$$

where dF/dh is the initial slope of the unloading curve and A is the contact area at the initial moment of unloading. One can also apply the aforementioned equation to the initial part of the loading curve of the square indenter, when the indentation proceeds in the elastic regime. The results of calculation of reduced Young's modulus for the considered nickel-titanium sample are presented in table 2.

Table 2. Reduced Young's modulus of the bulk NiTi sample, E_r, calculated for the square, conical and spherical indenters at the beginning of the unloading stage, as well as the for the square indenter at the initial part of the loading, when the indentation proceeds in the linear elastic regime. See the text for details.

Indenter	Square (load)	Square	Conical	Spherical
E_r (GPa)	164	125	137	169

As it is seen from table 2, the value calculated for different indenters are about 130 – 170 GPa, which is somewhat larger than reported in experimental measurements for the NiTi B19' phase. In Ref. [31], the measured values of reduced modulus are about 114 – 140 GPa (with Poisson ratio 0.35). The discrepancies between the results of the present simulations and experimental measurements can be attributed to different phases of the NiTi alloy (B2 in MD simulations and B19' in experiments) as well as insufficient accuracy of derivation of the initial slope of the unloading curve, absence of constant force segment in the simulations before unloading, low ($\sim$ 30 K) temperature of the thermostat in the simulations and higher indentation velocity, as compared to that one used in experiments.

6. Conclusion

In this paper we presented the results of MD simulations of the nanoindentation process of a bimetallic nickel-titanium crystal. We considered the system consisting of approximately 1,000,000 atoms packed in the fcc lattice and studied its interaction with an infinitely stiff indenter of three different shapes, namely the square, conical and spherical ones. In the performed simulations, the indenter was moving with the constant speed of 40 m/s during the loading stage and 10 m/s during the unloading. The maximal penetration depth into the crystal was approximately 3 nm.

We observed the appearance of different deformation regimes depending on the geometry of the indenter. We found that the initial stage of indentation with a square indenter is performed in the linear elastic regime. Later on, the load-displacement curve reaches the maximum and fluctuates around a constant value during the further displacement of the indenter, which is caused by the constant cross section of the square indenter. In case of the conical and spherical indenters we did not observe the elastic regime and the indentation process enters the plastic regime just after the initial contact of the indenter tip with the sample. In both cases, the force, acting on the indenter, gradually increases with the increase of the penetration depth. For the spherical indenter we observed the largest value of the force acting on the indenter. This can be explained by the largest deformations imposed on the system by the spherical indenter as compared to the square and conical ones.

For each type of the indenter we calculated hardness and reduced Young's modulus. We found that the calculated value of hardness depends on the form of the indenter. This can be explained by the suppression of the channel of stress release from the deformation site associated with dislocations migration for small square and conical indenters, and a much higher dislocation activity for a large spherical indenter. The calculated reduced Young's modulus also depends slightly on the type of the indenter and corresponds to experimentally measured values for the nickel-titanium crystal.

In our further work, we will investigate the influence of grain boundaries present in the sample on the nanoindentation load-displacement curve. A detailed and comprehensive analysis of the results of MD simulations of nanoindentation will be performed. In particular, we will focus on the processes of phase transitions in the materials under the applied pressure and the associated shape-memory effects.

Acknowledgments

This work is supported by the FP7 European Program "Theoretical analysis, design and virtual testing of biocompatibility and mechanical properties of titanium-based nanomaterials (VINAT)". In part, the work was supported by the European Commission within the CUTE-IRSES project (grant GA-2010-269131). A.V.Y. acknowledges Stiftung Polytechnische Gesellschaft Frankfurt am Main. The authors acknowledge the Frankfurt Center for Scientific Computing for the opportunity of carrying out complex computer simulations.

References

[1] Davis J R 2003 *Handbook of Materials for Medical Devices* (Materials Park: ASM International)
[2] Kramer K-H 2005 *Implants for surgery – a survey on metallic materials* (Materials for Medical Engineering vol 2) eds H Stallforth and P Revell (Weinheim: Wiley-VCH Verlag)

[3] Shtansky D V, Glushankova N A, Sheveiko A N, Kharitonova M A, Moizhess T G, Levashov E A and Rossi F 2005 *Biomaterials* **26** 2909
[4] Tam E, Petrzhik M, Shtansky D and Delplancke-Ogletree M-P 2009 *J. Mater. Sci. Technol.* **25** 63
[5] Berke P, Delplancke-Ogletree M-P, Lyalin A, Semenikhina V V and Solov'yov A V 2004 *Latest Advances in Atomic Cluster Collisions* eds A V Solov'yov and J-P Connerade (London: Imperial College Press) pp 205-224
[6] Gouldstone A, Chollacoop N, Dao M, Li J, Minor A M and Shen Y-L 2007 *Acta Mater.* **55** 4015
[7] Jeong S-M and Lee H-L 2005 *Thin Solid Films* **492** 173
[8] Cheng Y-T and Cheng C-M 2004 *Mat. Sci. Eng.* R **44** 91
[9] Fleck N A and Hutchinson J W 2001 *J. Mech. Phys. Solids* **49** 2245
[10] Psakhie S G *et al* 2001 *Theor. Appl. Fract. Mec.* **37** 311
[11] Shilkrot L E, Miller R E and Curtin W A 2004 *J. Mech. Phys. Solids* **52** 755
[12] Kreuzer H G M and Pippan R 2004 *Mat. Sci. Eng.* R **387** 254
[13] Lee Y, Park J Y, Kim S Y and Jun S 2005 *Mech. Mater.* **37** 1035
[14] Jian S-R, Fang T-H, Chuu D-S and Ji L-W 2006 *Appl. Surf. Sci.* **253** 833
[15] Rafii-Tabar H 2000 *Phys. Rep.* **325** 239
[16] Komanduri R, Chandrasekaran N and Raff L M 2000 *Wear* **240** 113
[17] Lin Z-C, Huang J-C and Jeng Y-R 2007 *J. Mat. Process. Tech.* **192** 27
[18] Zhang L and Tanaka H 1997 *Wear* **211** 44
[19] Song J and Srolovitz D J 2006 *Acta Mater.* **54** 5305
[20] Vashishta P, Kalia R K and Nakano A 2006 *J. Phys. Chem.* B **110** 3727
[21] Finnis M W and Sinclair J E 1984 *Philosophical Magazine* A **50** 45
[22] Li J H, Dai X D, Wang T L and Liu B X 2007 *J. Phys.: Condens. Matter* **19** 086228
[23] MBN Explorer package: `http://www.mbnexplorer.com/`, see also Solov'yov I A, Yakubovich A V, Nikolaev P V, Volkovets I and Solov'yov A V 2012 *J. Comput. Chem.* **33** 2412
[24] Solov'yov I A, Solov'yov A V, Greiner W, Koshelev A and Shutovich A 2003 *Phys. Rev. Lett.* **90** 053401
[25] Solov'yov I A, Solov'yov A V and Greiner W 2003 *Int. J. Mod. Phys.* E **13** 697
[26] Solov'yov I A, Mathew M, Solov'yov A V and Greiner W 2008 *Phys. Rev.* E **78** 051601
[27] Geng J, Solov'yov I A, Reid D G, Skelton P, Wheatley A E H, Solov'yov A V and Johnson B F G 2010 *Phys. Rev.* B **81** 214114
[28] Dick V V, Solov'yov I A and Solov'yov A V 2011 *Phys. Rev.* B **84** 115408
[29] Frenkel D and Smit B 2002 *Understanding Molecular Simulation*, 2nd ed.
[30] Oliver W C and Pharr G M 1992 *J. Mater. Res.* **7** 1564
[31] Rajagopalan S and Little A L 2005 *Appl. Phys. Lett.* **86** 081901

Institute of Physics
Dirac House, Temple Back
Bristol BS1 6BE UK

CURRAN ASSOCIATES INC
proceedings.com